Mängelexemplar

ex libris

Springer Series in Computational Neuroscience

Volume 10

Series Editors

Alain Destexhe
CNRS
Gif-sur-Yvette
France

Romain Brette
École Normale Supérieure
Paris
France

For further volumes:
www.springer.com/series/8164

Boris Gutkin · Serge H. Ahmed
Editors

Computational Neuroscience of Drug Addiction

 Springer

Editors
Boris Gutkin
Group for Neural Theory
LNC, DEC, ENS
75005 Paris
France
boris.gutkin@ens.fr

Serge H. Ahmed
CNRS UMR 5293 IMN
Université Victor Segalen
33076 Bordeaux
France
sahmed@lnpb.u-bordeaux2.fr

ISBN 978-1-4614-0750-8   e-ISBN 978-1-4614-0751-5
DOI 10.1007/978-1-4614-0751-5
Springer New York Dordrecht Heidelberg London

Library of Congress Control Number: 2011940849

© Springer Science+Business Media, LLC 2012
All rights reserved. This work may not be translated or copied in whole or in part without the written permission of the publisher (Springer Science+Business Media, LLC, 233 Spring Street, New York, NY 10013, USA), except for brief excerpts in connection with reviews or scholarly analysis. Use in connection with any form of information storage and retrieval, electronic adaptation, computer software, or by similar or dissimilar methodology now known or hereafter developed is forbidden.
The use in this publication of trade names, trademarks, service marks, and similar terms, even if they are not identified as such, is not to be taken as an expression of opinion as to whether or not they are subject to proprietary rights.

Organization of Computational Neuroscience: www.cnsorg.org

Printed on acid-free paper

Springer is part of Springer Science+Business Media (www.springer.com)

Foreword

One only has to read the newspapers to understand the extent to which addictions are among the scourges of the day. They lead to wastage and wasting of innumerable individual lives, and a huge cost to the body politic, with gargantuan sums of illicit money supporting edifices of corruption. The blame for other modern solecisms, such as burgeoning obesity, is increasingly being laid at the same door.

From the perspective of neuroscience, addictions present a critical challenge. Substances with at least initially relatively immediate effects on more or less well-defined sets of receptors, have, in some individuals, a panoply of physiological and psychological consequences that unravel over the course of years. Understanding each domain of inquiry by itself, and the links between them, is critical for understanding the course of addictions, and in the longer run, conceiving more effective options for palliation or even cure. Although there is a near overwhelming volume of data, the complexities of the subject mean that there are also many apparent inconsistencies and contradictions.

The understanding of addiction rests on analyses over multiple scales. For instance, we not only have to understand the progressive effects of long-term drug use on receptor characteristics and density, we must also grasp the changes this leads to in the neurons concerned, and then in the dynamical operation and information processing of the circuits and systems those neurons comprise. Equally, we have to understand how the effects on decision-making play out in terms of the economic choices made by complex, human, decision-makers.

The need to tie together phenomena at these multiple scales is a critical force leading to the current book's focus on theoretical ideas. Indeed research in addiction is a paradigmatic example of modern systems biology. The task of providing a formal scaffolding for understanding the links across levels of inquiry, is the topic of one of three wings of theoretical neuroscience. In the top-down direction, this is a case of a formal scientific reduction, explaining phenomena observed at one scale by mechanistic models built from components that live at finer scales. These components are characterized either by descriptive models, or are themselves explained by models at yet finer scales. Building and proving such multi-scale models is a perfectly normal role for mathematical and (increasingly) simulation-based modeling in a natural science. One might only cavil that, compared with some of its cousins, experimental neuroscience has sometimes seemed a little tardy in playing ball.

Chapter 4 is perhaps the purest example of this sort of analysis—providing, for instance, a formal way of resolving the apparent conflict between *in vivo* and *in vitro* data as to whether nicotine's main action on the activity of dopaminergic cells (believed to be key to the drug's addictive potential) is direct or indirect. However, many of the other chapters also contain elements of this *modus operandi* too, applied at different levels. For instance, the sophisticated agent modeling of Chap. 11 reminds us about the complexity of interactions between addicts and the environment which facilitates and hinders their addiction. Rich patterns of positive and negative feedback emerge. Thus, it is possible to examine and predict the effects of making manipulations at single points in the nexus of interactions—woe would, for instance, betide the policy-maker who attempts to intervene too simplistically in a system that is sufficiently non-linear as to be chaotic.

However, the chapters of the book also attest to the power of a second wing of theoretical neuroscience. This is the concept that brains must solve computational, information-processing and control-theoretic problems associated with surviving in a sometimes nasty and brutish world. Decision-making, in its fullest sense, is perhaps *the* critical competence for survival. The idea is to start from an understanding of the various ways that systems of any arbitrary sort can (learn to) make good decisions in the face of rewards and punishments. This topic lies at the intersection of economics, statistics, control theory, operations research and computer science. The resulting computational and algorithmic insights provide a foundation for, and constraints on, how humans and other animals actually make choices. Further, the sub-components of these models provide a parameterization of failure—points at which addictive substances can exert their intricately malign effects.

Somewhat alternative versions of these normative, and approximately normative, notions are apparent in many of the remaining chapters, differing according to the degree of abstraction, the intensity of focus on the computational level versus aspects of the algorithms and implementations of those computations, and also the extent to which the complexity of the neural substrate is included. There are also many different ways to formalize control theoretic problems, in terms of (a) the nature of the world (and the possible internal model thereof); (b) the goals of control, for instance the homeostatic intent of keeping variables within appropriate bounds as opposed to finding optimal solutions in the light of costs and returns; along with (c) assumptions about the possible solutions.

The relatively purer control theoretic approaches discussed in Chaps. 1, 2, 3 lie nearer one end of the spectrum. They pay particular attention to one of the central ideas in control theory, namely feedback as a way of keeping systems in order. Although one can imagine computational-level renditions, this is mostly an algorithmic idea, being rather divorced from possible computational underpinnings in terms of such things as being an optimal way of preventing divergence from a suitable operating point according to a justifiable cost function. Drugs can knock systems out of kilter, and so inspire immediate or predictive feedback to correct the state; the negative effects of withdrawal can also lead to a corrective policy of self-administration.

In these chapters, we can see something of the power of quantitatively characterizing the dynamical effects of drugs in systems with substantial adaptation.

The behavioral economic modeling of Chap. 10 has a somewhat similar quality. Here, the aliquots of computational analysis concern the relationship between price and demand and the effect of temporal delay on value. These then play out through psychological data and algorithmic realizations of these data.

The chapters based on modern reinforcement learning (RL) ideas (parts or all of Chaps. 5, 6, 7, 8) span the other end of the spectrum. As will become apparent in reviewing these chapters, reinforcement learning, which comprises forms of optimal, adaptive control, has become a dominant theoretical paradigm for modeling human and animal value-based decision-making. In some ways, it has taken over this role from cybernetics, its close control-theoretic cousin, which historically exerted a strong influence over systems thinking in areas in which value has played a lesser part. Aspects of the activity of dopaminergic cells offered one of the early strong ties between theory and experiment; it is only surprisingly recently that addiction, with its manifold involvement of dopamine, has become a target for this sort of effort.

Different RL approaches, such as model-based methods (which make choices by building and searching models of the decision-making domain) and model-free methods (which attempt to make the same optimal choices, but by learning how to favor actions from experience, without building models) are suggested as being realized in structurally and functionally-segregated parts of the brain, and have each been provided with normative rationales. The various chapters also speak to a fecund collaboration between theory and experiment, for instance winkling out potentially suboptimal interactions between the systems, and roles for evolutionarily prespecified control in the form of Pavlovian influences.

These chapters take rather different perspectives on the overall problem of addiction, even if generally adopting a rather common language. Indeed, although this language is powerful, uniting as it does information-processing notions with psychological and anatomical, physiological and pharmacological data, it is fair to say that it reveals rather than reduces the complexity of the individual systems, a complexity which is then hugely magnified in the way the systems interact. However, the multiple influences of drugs acting over the systems and their interactions all at diverse timescales, can at least be laid bare, and reasoned about, given this formal vocabulary.

Chapter 9 comes from another, subtly different, tradition of normative modeling, associated originally more with unsupervised learning (or probabilistic modeling of the statistical structure of the environment with no consideration of valence) than RL, and indeed comes to suggest a rather different (and, currently rather restricted) way that drugs can act. Nevertheless, there are ties both the RL models and indeed the control theoretic ones—for instance, Chap. 3's discussion of the limits of homeostasis is nicely recapitulated in Chap. 9's notion of itinerant policies, for which fixed points are an anathema.

Finally, note a critical lesson from the diversity of the chapters. Chapter 4's whole contribution concerns important (but still doubtless not comprehensive) details about the complex effects over time of one particular drug of addiction on one key circuit. Consider how these effects would be rendered in the abstract, impressionistic, terms of pretty much all the other chapters. These latter models simply lack the sophistication to capture the details—but, sadly, without being able to provide a guarantee that in the complex systems they do model, omitting the details is benign.

If one regards the diversity as a mark of adolescent effervescence in this field of computational addiction; the fact that such a fascinating book is possible is a mark of impending maturity. Theoretical ideas and mathematical and computational models are rapidly becoming deeply embedded in the field as a whole, and, most critically, are providing new, and more powerful, ways to conceive of the multiple, interacting, problems wrought by addictive substances. Providing the guarantees mentioned above, and indeed a more systematic tying together of all the different levels and types of investigation across different forms and causes of addiction, is the challenge for early adulthood.

Acknowledgements

I am very grateful to the Gatsby Charitable Foundation for funding.

London, UK   Peter Dayan

Acknowledgements

The editors would like to acknowledge the support from the French Research Council (CNRS) (BSG and SHA), the Université Victor Segalen-Bordeaux (SHA), the Conseil Régional d'Aquitaine (SHA), Ecole Normale Superieure and INSERM (BSG), the French National Research Agency (ANR) (BSG and SHA), NERF (BSG) and Fondation Pierre-Gilles de Genes. Boris Gutkin would like to thank Prof. J.P. Changeux for launching into the field of addiction research and his continuing support.

Contents

Part I Pharmacological-Based Models of Addiction

1 Simple Deterministic Mathematical Model of Maintained Drug Self-administration Behavior and Its Pharmacological Applications . . 3
Vladimir L. Tsibulsky and Andrew B. Norman

2 Intermittent Adaptation: A Mathematical Model of Drug Tolerance, Dependence and Addiction 19
Abraham Peper

3 Control Theory and Addictive Behavior 57
David B. Newlin, Phillip A. Regalia, Thomas I. Seidman, and Georgiy Bobashev

Part II Neurocomputational Models of Addiction

4 Modelling Local Circuit Mechanisms for Nicotine Control of Dopamine Activity 111
Michael Graupner and Boris Gutkin

5 Dual-System Learning Models and Drugs of Abuse 145
Dylan A. Simon and Nathaniel D. Daw

6 Modeling Decision-Making Systems in Addiction 163
Zeb Kurth-Nelson and A. David Redish

7 Computational Models of Incentive-Sensitization in Addiction: Dynamic Limbic Transformation of Learning into Motivation 189
Jun Zhang, Kent C. Berridge, and J. Wayne Aldridge

8 Understanding Addiction as a Pathological State of Multiple Decision Making Processes: A Neurocomputational Perspective . . . 205
Mehdi Keramati, Amir Dezfouli, and Payam Piray

Part III Economic-Based Models of Addiction

9 Policies and Priors 237
Karl Friston

10 Toward a Computationally Unified Behavioral-Economic Model of Addiction 285
E. Terry Mueller, Laurence P. Carter, and Warren K. Bickel

11 Simulating Patterns of Heroin Addiction Within the Social Context of a Local Heroin Market 313
Lee Hoffer, Georgiy Bobashev, and Robert J. Morris

Index 333

Contributors

J. Wayne Aldridge Department of Psychology, University of Michigan, Ann Arbor, MI, USA

Kent C. Berridge Department of Psychology, University of Michigan, Ann Arbor, MI, USA, berridge@umich.edu

Warren K. Bickel Advanced Recovery Research Center, Virginia Tech Carilion Research Institute and Department of Psychology, Virginia Tech, 2 Riverside Circle, Roanoke, VA 24016, USA, wkbickel@vtc.vt.edu

Georgiy Bobashev RTI International, Research Triangle Park, NC, 27709, USA, bobashev@rti.org

Laurence P. Carter Center for Addiction Research, Psychiatric Research Institute, University of Arkansas for Medical Sciences, Little Rock, AR 72205, USA

Nathaniel D. Daw Center for Neural Science and Department of Psychology, New York University, New York, NY, USA, nathaniel.daw@nyu.edu

Amir Dezfouli Neurocognitive Laboratory, Iranian National Center for Addiction Studies, Tehran University of Medical Sciences, Tehran, Iran, a.dezfouli@ut.ac.ir

Karl Friston The Wellcome Trust Centre for Neuroimaging, University College London, Queen Square, London WC1N 3BG, UK, k.friston@ucl.ac.uk

Michael Graupner Group for Neural Theory, Laboratoire de Neurosciences Cognitives, INSERM Unité 969, Départment d'Etudes Cognitives, École Normale Supérieure, 29, rue d'Ulm, 75005 Paris, France, michael.graupner@ens.fr

Boris Gutkin Group for Neural Theory, Laboratoire de Neurosciences Cognitives, INSERM Unité 969, Départment d'Etudes Cognitives, École Normale Supérieure, 29, rue d'Ulm, 75005 Paris, France

Lee Hoffer Department of Anthropology, Case Western Reserve University, 11220 Bellflower Rd., Cleveland, OH, 44106, USA

Mehdi Keramati Group for Neural Theory, Ecole Normale Superieure, Paris, France

Zeb Kurth-Nelson Department of Neuroscience, University of Minnesota, 6-145 Jackson Hall, 321 Church St. SE, Minneapolis, MN 55455, USA, kurt0073@umn.edu

Robert J. Morris RTI International, Research Triangle Park, NC, 27709, USA

E. Terry Mueller Advanced Recovery Research Center, Virginia Tech Carilion Research Institute, Virginia Tech, 2 Riverside Circle, Roanoke, VA 24016, USA

David B. Newlin RTI International, Baltimore, MD, USA, dnewlin@rti.org

Andrew B. Norman Department of Psychiatry, College of Medicine, University of Cincinnati, Cincinnati, OH 45267-0583, USA

Abraham Peper Department of Medical Physics, Academic Medical Centre, University of Amsterdam, Amsterdam, The Netherlands, a.peper@planet.nl

Payam Piray Neurocognitive Laboratory, Iranian National Center for Addiction Studies, Tehran University of Medical Sciences, Tehran, Iran

A. David Redish Department of Neuroscience, University of Minnesota, 6-145 Jackson Hall, 321 Church St. SE, Minneapolis, MN 55455, USA, redish@umn.edu

Phillip A. Regalia Catholic University of America, Washington, DC, USA

Thomas I. Seidman University of Maryland Baltimore County, Baltimore, MD, USA

Dylan A. Simon Department of Psychology, New York University, New York, NY, USA

Vladimir L. Tsibulsky Department of Psychiatry, College of Medicine, University of Cincinnati, Cincinnati, OH 45267-0583, USA, tsibulvr@ucmail.uc.edu

Jun Zhang Department of Psychology, University of Michigan, Ann Arbor, MI, USA

Part I
Pharmacological-Based Models of Addiction

Chapter 1
Simple Deterministic Mathematical Model of Maintained Drug Self-administration Behavior and Its Pharmacological Applications

Vladimir L. Tsibulsky and Andrew B. Norman

Abstract Currently dominating psychological theories of drug self-administration explain some experimentally observed facts, however, at the same time they lead to logical contradictions to other accepted facts. A quantitative pharmacological theory of self-administration behavior states that cocaine-induced lever pressing behavior occurs only when cocaine concentrations are within a certain range, termed the compulsion zone. These concentrations can be calculated using standard pharmacokinetic mathematical models. The lower and upper limits of this range of concentrations are the priming and satiety thresholds, respectively. This pharmacological theory explains all phases of the self-administration session but is particularly useful at explaining the intervals between drug injections during maintained self-administration in terms of a mathematical model that contains only three parameters: the drug unit dose, the satiety threshold and the drug elimination rate constant. The satiety threshold represents the equiactive agonist concentration. Because competitive antagonists increase equiactive agonist concentrations and the agonist concentration ratio increases linearly with antagonist concentration, the satiety threshold model allows the measurement of the in vivo potency of dopamine receptor antagonists using the classical mathematical method of Schild. In addition, applying pharmacokinetic models to the time course of the magnitude of the antagonist-induced increase in the satiety threshold allows the pharmacokinetics of antagonists to be calculated. Pharmacokinetic/pharmacodynamic models explain self-administration behavior and make this paradigm a rapid and high-content bioassay system useful for screening agonists and antagonists that interact with important neurotransmitter systems in the brain.

1.1 Introduction

Drug abuse/addiction research is part of the field of behavioral pharmacology. The first animal model of drug-taking behavior that featured feedback for intravenous

V.L. Tsibulsky (✉) · A.B. Norman
Department of Psychiatry, College of Medicine, University of Cincinnati, Cincinnati, OH 45267-0583, USA
e-mail: tsibulvr@ucmail.uc.edu

self-injections of morphine in freely moving rats was introduced in the pivotal paper by Weeks (1962). This method could be easily used for any other species or drug of abuse and it became a favorite model for psychologists. Despite the critically important role of drugs in this paradigm, the method has been largely ignored by pharmacologists. There are two principal approaches to the study of any behavior—physiological and psychological. Psychological pharmacology (psychopharmacology) has been the dominant approach. The numerous paradoxes that continue to confuse the interpretation of self-administration behavior (Norman and Tsibulsky 2001) are mainly due to the combination of incompatible psychological and physiological principles. An alternative approach to study this behavior is to restrict the methodology and theory to the principles of physiological pharmacology. We have found that this physiological approach resolves all of the existing paradoxes. Physiological pharmacology has the inherent advantage of expressing measured parameters in well defined terms and units. This chapter will illustrate the usefulness of the physiological approach by presenting a mathematical model of maintained cocaine and apomorphine self-administration behavior. The ability of the model to obtain quantitative data on the pharmacodynamic and pharmacokinetic potencies of agonists and antagonists of dopamine receptors will then be demonstrated.

The model presented herein was developed using intravenous cocaine self-administration in Sprague-Dawley rats. Procedural details can be found in our previous publications (Tsibulsky and Norman 1999, 2005; Norman and Tsibulsky 2006). Briefly, silicone catheters were implanted into jugular or femoral veins and rats were trained to press a lever on a fixed ratio 1 (FR1) schedule of drug injections. The unit dose of drug was changed randomly between sessions within the range of 0.3–12.0 µmol/kg for cocaine and 0.075–0.6 µmol/kg for apomorphine by changing the duration of activation of a syringe pump. The time of every press was registered and the inter-injection intervals were analyzed. The cocaine and apomorphine levels in the body were calculated every second during the sessions.

1.1.1 Definitions of Terms

There are two basic characteristics of cocaine self-administration during the maintenance phase of the session: (1) self-injections occur regularly, (2) the rate of self-injections is inversely related to the unit dose of cocaine (Pickens and Thompson 1968). Building a mathematical model of the relationship between the dose and the rate of self-administration of cocaine will help to understand the mechanisms regulating drug taking behavior.

In any pharmacological paradigm, the independent variable is the drug concentration at the site of action. As this cannot be readily measured in vivo, in this model, the drug unit dose can be readily controlled and is convenient to use as the independent variable. Although, when a lever-press occurs immediately after the end of an injection it will initiate the next injection and the resulting dose will be equal to two unit doses. Such an event creates some difficulty for the application of the model.

Therefore, the actual dose rather than the unit dose is the appropriate independent term in the model.

In order to define the dependent variable in the paradigm, it is necessary to identify the response. There are two different aspects in any drug-induced response: a dynamic component, that is, the magnitude of the response, and a kinetic component, that is, the duration of the response. In self-administration studies, the response is defined by the manipulandum selected (typically a lever-press or a photo-beam interruption). Therefore, the response used in the self-administration paradigm is quantal. Consequently, there is no dynamic term for response in our mathematical model. In this model, the dependent variable has only a kinetic component as the inter-response intervals are controlled exclusively by the animal. The inter-injection intervals are partially controlled by the experimenter by means of a schedule of drug delivery. Using complex schedules greatly complicates any pharmacological model of self-administration behavior. For the sake of simplicity, we use the FR1 schedule where the pump injects a unit dose of cocaine in response to one lever-press. Under this schedule, inter-response and inter-injection intervals should be indistinguishable. Infrequent lever-presses occurring when the pump is already activated are registered but do not have any consequence. Therefore, the inter-injection interval is the true dependent variable in this model.

In our physiological pharmacology approach, the response is quantal and the dose-inter-injection interval relationship is treated as a dose–duration function. In contrast, in the traditional psychological approach the rate of lever-pressing is assumed to reflect the magnitude of a graded response. Therefore, the dose–rate relationship is treated as a dose–response function.

1.1.2 Principles

In behavioral pharmacology, it is assumed that the drug exerts its effects according to the principles of pharmacology. There are two basic pharmacological principles. Pharmacodynamics relates the magnitude of drug effect to the occupancy of a receptor population. The fractional occupancy of a receptor population depends on the affinity of the drug for this receptor and on the concentration of the drug. Pharmacokinetics studies the drug concentration over time which is determined by the absorption, distribution, metabolism and excretion of the drug. Although these basic pharmacological principles should be common for both psychological and physiological pharmacology, they are largely ignored by psychopharmacology.

The assumption intrinsic only to the psychological approach is that lever-pressing behavior is emitted in order to receive a reinforcer. In the case of self-administration, the animal is working to obtain the next dose of drug, which is still in the syringe. In the physiological pharmacology approach, the drug in the syringe cannot exert actions in the body and, therefore, the psychological and pharmacological approaches are incompatible. Strictly speaking, according to operant theory the lever-press is not a response to any stimulus but an operant. The magnitude of an operant is measured by the rate. In psychopharmacology, the lever-pressing behavior is a graded dependent variable and the rate of presses should be proportional to the magnitude of reinforcer, in this case, to the dose of the drug. This approach leads to a dose–rate paradox (see below).

The assumption intrinsic to the physiological approach is that lever-pressing behavior is a drug-induced response which depends on the cumulative drug concentration at the site of action. According to occupancy theory, the inter-injection interval should be proportional to the dose (or concentration) of the drug which is in the body. It should be emphasized that approaching the paradigm from the point of view that animals are seeking the drug is intrinsically teleological where the effect precedes the cause. This is incompatible with the view that responses are induced by the drug that is already in the body where the cause precedes the effect. Therefore, these approaches must be clearly differentiated and cannot be combined or used interchangeably.

It is well established that the rate of lever-pressing decreases as a function of cocaine dose (Pickens and Thompson 1968; Mello and Negus 1996; Zernig et al. 2007). These results are not consistent with the prediction from operant theory that assumes the direct proportionality between the rate of the reinforced behavior and the dose. This is the major paradox that arises from the application of operant theory to the drug self-administration paradigm. It should be noted though, that in some studies in rats and mice and in many studies in monkeys researchers report the so-called ascending limb of the dose–rate curve: within a very narrow range of small cocaine doses the rate can be proportional to the dose. However, closer scrutiny of these data reveals that the reported ascending limb contains a maximum of only three data points: the rate of pressing at zero cocaine dose (saline injections), the maximal rate and occasionally one intermediate point (Wilson et al. 1971; Mello and Negus 1996; Flory and Woods 2003). To the best of our knowledge, there is no report showing that activity at this intermediate dose is regular and can be stably maintained for a reasonable length of time (more than half an hour in the case of cocaine). Some researchers suggest that the ascending limb represents an artifact that results from the incorrect averaging of periods with maximal activity and periods with no activity (Wise 1987; Sizemore and Martin 2000; Norman and Tsibulsky 2001; Zittel-Lazarini et al. 2007).

1.1.3 Critical Review of Existing Models

There is one mathematical model describing both the ascending limb of the dose–rate curve that is predicted by psychologically oriented theories and the descending limb (Zernig et al. 2004). According to this theory, the inverted U-shaped dose–rate curve results from superposition of two opposite effects: one leads to an increase in the rate of lever-pressing behavior (variously termed a reinforcing effect, reward, euphoria, pleasure, craving, wanting or liking) and the second one leads to a suppression of the rate (direct effect, depression, satiety, aversion or disliking). Because the former effect starts at relatively lower doses than the latter, the overall dose–rate curve has the inverted U-shape. According to this mathematical model, both effects are proportional to the fractional occupancy of different populations of receptors. This attempt to explain the dose–rate function in pharmacological terms reasonably applied the classical Hill equation.

The first problem with this model is that the lever-press rate, i.e., the kinetic component of the response, is used instead of the response amplitude, i.e., the dynamic dependent variable which is required in the Hill equation. The second important mistake typical for the psychological approach is that the cocaine unit dose is used as the independent variable instead of required drug concentration. Therefore, the application of the Hill equation is inappropriate. Not surprisingly given the large magnitude of effect over a narrow range of doses the Hill equation generated slopes (Hill coefficient) in the range of 7–8 for the ascending limb (Zernig et al. 2007). This indicates that the dose–rate function does not follow the law of mass action for a response mediated by a single population of receptors. Although Hill slopes in the range of 2–3 are not uncommon for agonist dose–response curves, slopes in the range of 7–8 suggest that the rate on the ascending limb is all-or-nothing and not a graded response.

Sizemore and Martin (2000) also attempted to develop a mathematical model in which the ability of a drug to maintain responding was described in terms of receptor occupancy theory. The rate of lever-pressing was treated as a response and, respectively, the Hill equation was used to describe the dose–response relationship in terms of classical occupancy theory. Obviously, all criticism of the previous model is relevant in this case. In addition, the "pharmacological reinforcement function" was used to incorporate a description of how an organism's behavior affects the amount of drug in the animal over time. Although the final equations were not derived, the main predictions of the theory were that: (1) there should be no ascending limb, (2) the running rate of responding does not depend on the unit dose and, (3) pause duration should be an exponential function of the dose. The first two predictions are consistent with most experimental data. However, the third prediction is not, as will be shown in the next section of this chapter.

Ahmed and Koob (2005) published the most comprehensive mathematical model of maintained cocaine self-administration combining a two-compartment pharmacokinetic model of distribution and elimination of cocaine with a pharmacodynamic E_{max} model of cocaine effect on reward threshold. However, the authors made two important assumptions in the course of developing their model: (1) The equation could not be derived unless the two-compartment model was partially reduced to a one-compartment model by the assumption $e^{-\alpha t} \approx 1$ and (2) the drug-specific pharmacodynamic parameter T_{50} became redundant after the assumption $T_{max} = T_0$ (as $T_0 = 2T_{50}$). Unfortunately, these simplifications undermined the intent to provide a comprehensive pharmacokinetic/pharmacodynamic model of acquired self-administration behavior. Interestingly, by substituting $\beta \approx k_{el}$, $K \approx 1/V_d$ and $T_{50} \cdot (T_0 - T_S)/T_S = C_S$ into their final equation it can be demonstrated that it is invariant with the satiety threshold model proposed by Tsibulsky and Norman (1999) using a one-compartment model. Thus, the satiety threshold model remains the simplest (see the model development in Sect. 1.2) and the most useful model developed over the last decade (see the application in Sect. 1.3).

1.2 Mathematical Model of Maintenance

As maintained, cocaine self-administration is characterized by regular inter-dosing intervals, a quasi-steady-state should be eventually established. By definition, at steady-state the drug concentration is constant in spite of ongoing pharmacokinetic processes that strive to change it. In a *quasi-steady-state*, drug concentration changes cyclically. This state is typically characterized by three constant concentrations: the mean, the maximal, and the minimal concentration.

1.2.1 Assumptions

For the sake of simplicity, let's assume that:

1. The whole body of the animal can be represented by one compartment.
2. The process of drug elimination is first order $dC/dt = -k \cdot C$ where C = drug concentration, t = time and k = first-order elimination rate constant.

    During quasi-steady-state, the three constant concentrations depend on the drug unit dose and the dosing interval according to the following equations:

$$C_{\text{mean}} = D/(V_d \cdot k \cdot T) \tag{1.1a}$$

$$C_{\max} = C_{\max} \cdot e^{-k \cdot T} + D/V_d \tag{1.1b}$$

$$C_{\min} = (C_{\min} + D/V_d) \cdot e^{-k \cdot T} \tag{1.1c}$$

where C_{mean}, $C_{\max}$ and $C_{\min}$ are the mean, maximal, and minimal concentrations of the drug maintained at the unit dose (D) administered regularly every T seconds and V_d = the volume of distribution. Rearranging these equations to fit the dose–duration curve $T = f(D)$ yields:

$$T = D/(V_d \cdot k \cdot C_{\text{mean}}) \tag{1.2a}$$

$$T = \ln[C_{\max}/(C_{\max} - D/V_d)]/k \tag{1.2b}$$

$$T = \ln[(C_{\min} + D/V_d)/C_{\min}]/k \tag{1.2c}$$

3. The final assumption of the model is that one of these three concentrations is independent of the cocaine dose. Nonlinear regression analysis applied to the experimentally derived dose–duration curves using Eqs. (1.2a), (1.2b), or (1.2c) shows that at any cocaine dose the response is induced when cocaine concentration declines to the same minimal concentration ($C_{\min}$) maintained during the quasi-steady-state (Fig. 1). The $C_{\max}$ and C_{mean} models failed to adequately account for the observed dose–duration relationship and, therefore, these quasi-steady-state parameters are not constant during self-administration at different unit doses.

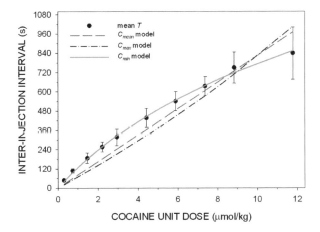

Fig. 1 Dose–duration function for cocaine during maintained self-administration in rats. Self-administration was under an FR1 schedule with no time out. The *filled circles* represent mean inter-injection intervals in the group of 6 rats. The *bars* represent the standard deviation. The *lines* represent the best approximation to the data points according to three mathematical models for the mean, maximal and minimal concentrations (see Eqs. (1.2a), (1.2b) and (1.2c)) with correlation coefficients: 0.966, 0.934 and 0.999, respectively. The satiety threshold $C_{min} = 5.1$ μmol/kg, the elimination half-life $t_{1/2} = 494$ s

1.2.2 Dose–Duration Curve

During the self-administration of higher doses of cocaine, it takes more time for the higher concentrations to decline back to C_{min} when the next dose is injected. Furthermore, cocaine's elimination is first order, where the absolute rate of drug elimination is linearly proportional to the absolute drug concentration. Consequently, the drug will be eliminated faster at higher drug concentrations. This accounts for the relatively shorter intervals observed at increasing unit doses. Therefore, the curvature of the dose–duration function, that is, progressive deviation from a straight line, is due to first order drug elimination.

The shape of the dose–duration function depends on both parameters: C_{min} (the pharmacodynamic parameter) and on the elimination constant k (the pharmacokinetic parameter). This makes it possible to estimate both independent parameters by nonlinear regression analysis using Eq. (1.2c) (Tsibulsky and Norman 1999).

It is interesting that if the drug elimination rate does not depend on the absolute drug concentration (zero order kinetics: $dC/dt = -k_0$ where $k_0 =$ zero-order elimination rate constant) then the dose–duration function would be strictly linear and independent of the value of C_{min}:

$$T = D/(V_d \cdot k_0) \qquad (1.3)$$

1.2.3 Satiety Threshold

During the maintenance phase of a cocaine self-administration session, lever-pressing behavior occurs only when the cocaine level is at C_{min}, which we termed the satiety threshold. The probability of a lever-press is close to zero when the cocaine level is above the satiety threshold. Therefore, the satiety threshold represents the lower limit of the range of cocaine concentrations when the animal is in the state of apparent satiety. This upper range represents the satiety zone. The adjacent lower range of concentrations below the satiety threshold, but above the priming threshold (see Norman et al. 1999), was called the compulsion zone because cocaine within this range of concentrations induces the highest rate of compulsive lever-pressing activity (Norman and Tsibulsky 2006). Therefore, the satiety threshold represents the upper limit of the compulsion zone. Because the switch between the state of lever-pressing activity and the satiety state occurs over a narrow range of agonist concentrations implies that this event represents a quantal response.

The satiety threshold is apparent not only at the first lever-press occurring after long inter-injection pauses during the maintenance phase, but also at the last lever-press of a series of short-interval loading injections that precede the first long pause. In addition, the model assumes that the magnitude of the satiety threshold is independent of whether cocaine concentrations are rising or falling.

The C_{min} cannot be readily measured in the brain. But the product of $C_{min} \cdot V_d$ can be calculated from the dose–duration function during the maintenance phase (Eq. (1.2c)). The proposed mathematical model of self-administration suggests that the inter-injection interval represents the apparent dependent measure which allows for the calculation of the satiety threshold. This latter parameter represents the fundamental dependent measure in the self-administration paradigm. Because the satiety threshold is independent of the unit dose, the only reason to measure inter-injection intervals at different unit doses is to calculate the satiety threshold if the drug first-order elimination rate constant (k) is unknown. If k is known, then the satiety threshold (D_{ST}) can be calculated using the inter-injection interval measured at any single unit dose according to the equation:

$$D_{ST} = C_{min} \cdot V_d = D/(1 - e^{-k \cdot T}) \tag{1.4}$$

1.3 Application of the Model

It has been demonstrated that the concentration of dopamine in the brain is linearly proportional to the concentration of cocaine (Nicolaysen et al. 1988). Fluctuations of dopamine levels are parallel to the fluctuations of cocaine levels during maintained self-administration in rats (Pettit and Justice 1989; Wise et al. 1995). Based on receptor occupancy theory, we can assume that the satiety threshold corresponds to a certain fractional occupancy of a fixed population of dopamine transporters by cocaine and, correspondingly, to a certain occupancy of receptors by dopamine.

Competitive receptor antagonists increase the equiactive concentration of an agonist. The satiety threshold represents an equiactive cocaine concentration and is increased in the presence of competitive dopamine receptor antagonists (Norman et al. 2011b).

1.3.1 Measurement of Antagonist K_{dose}

The potency of a series of antipsychotic drugs, measured as the dose that increased the rate of cocaine self-administration by 25%, correlated with antipsychotic potency (Roberts and Vickers 1984). This promising approach to using self-administration behavior as a pharmacological assay system has remained underdeveloped. The magnitude of competitive antagonist-induced increases in the equiactive agonist concentration, measured as a concentration ratio, should be directly proportional to the antagonist concentration (Schild 1949; Colquhoun 2007). Theoretically, the dose of antagonist required to increase the satiety threshold by a factor of two would represent 50% fractional occupancy of the receptor population mediating the agonist-induced satiety effect. This represents the antagonist K_{dose} and is a measure of the absolute potency of the antagonist in vivo. This value can also be expressed as the antagonist apparent pA_2, the negative logarithm of the K_{dose}. The following simple equations describe the relationship between this in vivo and the conventional in vitro terms: $K_{dose} = V_d \cdot C_{min}$ and apparent $pA_2 = -\log K_{dose}$. The potency of antagonists measured by their pA_2 values in vitro has provided a basis for receptor classification (Rang 2006).

Occasional reports have provided apparent pA_2 values in vivo that have been used to identify the receptor subtypes mediating the self-administration of alfentanil (Bertalmio and Woods 1989) and heroin (Rowlett et al. 1998) which are direct receptor agonists. The theory of competitive antagonism assumes that the competitive antagonist and the agonist bind reversibly to the same site. Therefore, indirect receptor agonists in general may not be suitable for this type of investigation. However, because the concentration of dopamine is directly proportional to the concentration of cocaine in the nucleus accumbens (Nicolaysen et al. 1988) the occupancy of dopamine transporters by cocaine should be directly proportional to fractional occupancy of dopamine receptors in the basal ganglia. Therefore, it should be possible to use the Schild method to measure the potencies of antagonists on the cocaine satiety threshold during self-administration. Additionally, the direct dopamine receptor agonist apomorphine is also self-administered by rats (Baxter et al. 1974) making this agonist an appropriate model system that can be compared with cocaine (Norman et al. 2011a).

1.3.2 Calculation the Level of Cocaine in the Body

The cumulative cocaine (and in separate experiments apomorphine) level in the body was calculated every second by summation of the amount of agonist that was administered to the animals minus the amount of the drug eliminated per unit time according to the simplified linear equations for the zero-order input and first-order elimination kinetics for a one or a two-compartment model (Tsibulsky and Norman 2005). The agonist's volume of distribution was assumed to be constant. The kinetic constants of distribution and elimination were taken from the literature.

1.3.3 Agonist Concentration Ratios

The magnitude of a pharmacological response should be proportional to the fractional occupancy (f) of a receptor population by an agonist (A). A competitive antagonist (B) decreases f according to the equation:

$$f = (X_A/K_A)/((X_A/K_A) + (X_B/K_B) + 1) \tag{1.5}$$

where X_A and X_B are the concentrations of agonist and antagonist, respectively and K_A and K_B are the equilibrium dissociation constants of the agonist and antagonist, respectively (for overviews see Rang 2006; Colquhoun 2007). A useful method of determining antagonist potency is to measure the antagonist-induced increase in the equiactive agonist concentration according to the method devised by Schild (1957). In the case of our model, the lever-press occurs at the same agonist fractional occupancy. Therefore, in the presence of a competitive antagonist at the time of lever-press the Schild equation holds:

$$r = (X_B/K_B) + 1 \tag{1.6}$$

where $r = $ the ratio by which X_A is increased in order to produce the same agonist fractional occupancy produced in the absence of the antagonist. The concentration of antagonist that is required to increase the equiactive agonist concentration by 2-fold (agonist concentration ratio, $r = 2$) would occupy 50% of the receptors mediating the agonist-induced responses if it is administered alone. This concentration of antagonist has been termed the constant of dissociation K_d (or K_B in the original notations used for Eq. (1.5)) and the antagonist $pA_2 = -\log K_d$ (Arunlakshana and Schild 1959).

The mean of the values for the level of apomorphine or cocaine at the satiety threshold during the maintenance phase and prior to the injection of antagonist represented the baseline. The level of agonist at the time of each lever-press after the injection of antagonist was divided by the baseline value for that session and the resulting value represented the agonist concentration ratio. These agonist concentration ratios minus one were plotted as a function of time after the injection of antagonist. The subtraction of 1 makes the initial value 0, corresponding to the absence of antagonist, making it convenient to apply standard pharmacokinetic models.

1.3.4 Measurement of Antagonist Pharmacokinetics

As stated above, the magnitude of a competitive antagonist-induced increase in the equiactive agonist concentration, in our model C_{min}, should be directly proportional to the concentration of antagonist at its site of action. If so, then the time course of the change in the magnitude of C_{min} should reflect the change in the concentration of the antagonist in vivo, that is, its pharmacokinetics.

1.3.5 Pharmacokinetic and Pharmacodynamic Models

Because the maximum agonist concentration ratio was found to be approximately linearly proportional to antagonist dose, the concentration ratio minus one was assumed to be linearly proportional to the antagonist concentration at any time after dynamic equilibrium was approached. This allowed the application of pharmacokinetic models to the antagonist-induced changes in the agonist concentration ratio.

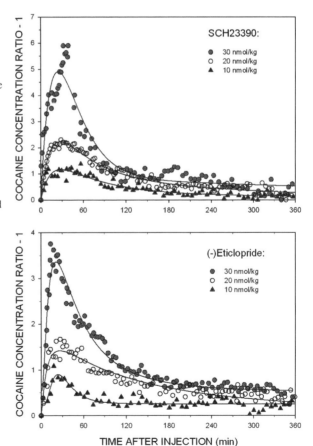

Fig. 2 The dose-dependent antagonist-induced increases of the cocaine satiety threshold. *Each point* represents the cocaine concentration ratio at the time of each self-administration after the injection of antagonist. The ratios represent the cocaine concentration at the time of each lever-press after the injection of antagonists divided by the mean baseline values of the satiety threshold in each session. The *curve* through each data set is the best fit generated by a pharmacokinetic model with first order absorption into a single compartment

Fig. 3 The absolute pharmacodynamic potencies of SCH23390 and (−)-eticlopride on apomorphine and cocaine satiety thresholds. The concentration ratio values represent the mean maximum concentration ratio at each dose of antagonist calculated from sessions conducted as described in Fig. 2. The K_{dose} values for the antagonists were calculated according to the method of Schild. For SCH23390 the slopes of the linear regressions were 1.3 and 1.0 on apomorphine and cocaine, respectively. For eticlopride slopes of the linear regressions were 1.2 and 1.1 on apomorphine and cocaine, respectively

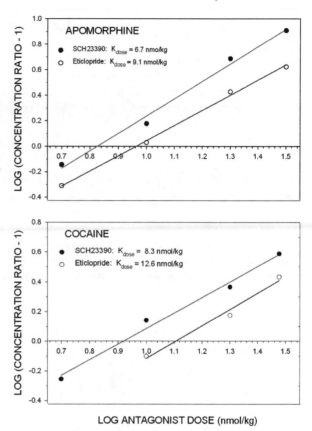

The bimodal SCH23390 and (−)-eticlopride-induced changes in agonist concentration ratio as a function of time were analyzed using a series of single and multiple-compartment pharmacokinetic models. These models provided reasonable fits to the data (for illustration purposes, one of these models is shown in Fig. 2).

Linear regression analysis using Eq. (1.6) was applied to the data on the Schild plots (Fig. 3) and the slope and intercept on the abscissa of the linear regressions were assessed.

1.4 Discussion

The mathematical model of the maintained cocaine self-administration is developed on the basis of only three assumptions. The possibility that animals "adjust their response rates to maintain an optimum blood level" was discussed 40 year ago (Wilson et al. 1971). This pharmacological approach to explain the dose–duration function was further developed in the seminal work of Yokel and Pickens (1974). They were the first to demonstrate that the amphetamine level at the time of lever-press is independent of the drug unit dose. However, this promising pharmacological approach was sacrificed in favor of the psychological approach which now dominates the field.

Application of the Schild method to the satiety threshold model of maintained drug self-administration shows that the Schild equation is obeyed exactly by a competitive antagonist in vivo. Surprisingly, the Schild equation holds even for the indirect agonist cocaine and if the true equilibrium is not reached. The Schild method appears to be even better than Colquhoun realized (Colquhoun 2007).

1.4.1 Indirect Agonist

Cocaine is an antagonist of the dopamine transporter and because the dopamine in the synaptic cleft continues to act and to induce a number of behaviors, cocaine works as an indirect agonist. Strict linear proportionality between the exogenous indirect agonist (cocaine) and endogenous direct agonist (dopamine) (Nicolaysen et al. 1988) makes the condition of competitive antagonism at the same site required by the Schild method applicable to dopamine antagonists and cocaine. This is because the ratio of cocaine concentrations in the presence and in the absence of the antagonist will be equal to the ratio of corresponding dopamine concentrations.

1.4.2 Quasi-Equilibrium

An important assumption of the Schild equation is that the measurements of antagonist-induced shifts in the agonist concentration ratio are made at equilibrium (Kenakin 1997; Colquhoun 2007). Because both agonist and antagonist concentrations are constantly changing in vivo it is not in equilibrium at most times. However, the reasonable approximation to the data of a slope of unity on the Schild plot may indicate that the major assumptions of the Schild method are satisfied. An important caveat is that non-equilibrium situations can produce Schild plots with slopes of unity but provide erroneous measures of pA_2 (Kenakin 1997). Whether this caveat will prove to be a significant limitation of this method will only become apparent from additional studies (Norman et al. 2001a).

Drug concentrations in the body are influenced by several simultaneous processes. Some of them are reversible, for example, distribution and redistribution, association and dissociation. Some are irreversible, for example, administration and elimination. Quasi-steady state of the agonist during maintained self-administration is achieved because of the quasi-equilibrium between administration and elimination. According to the satiety threshold model, regardless of the unit dose, the rate of dosing equals the mean rate of cocaine elimination. Although, the antagonist concentration is never in quasi-steady-state because only a single injection is administered, the antagonist concentration at the site of action is in *quasi-equilibrium* exactly at the time when the fractional occupancy of the receptors is at maximum.

This is because the rates of association and dissociation are equal at this moment. As the maximum occupancy produces the maximum effect, we can use the highest satiety threshold to compare effects of different doses of antagonist under conditions satisfying the Schild assumptions.

1.4.3 Volume of Distribution

If the maximum concentration ratio is measured close to equilibrium binding conditions, the linear increase in the peak agonist concentration ratio as a function of antagonist dose provides a measure of the pharmacodynamic potency of eticlopride and SCH23390. When the cocaine or apomorphine concentration ratio is two this should correspond to the dose of the antagonist required to occupy 50% of the receptors that mediate the satiety response, i.e. the K_{dose} of the antagonist. The antagonist K_{dose} value should be the product of its affinity constant measured in vitro and its apparent volume of distribution (V_d). As the apparent V_d of the antagonist can usually be assumed to be constant across a range of doses, the K_{dose} of antagonists should be constant for all responses mediated by the same receptors in vivo. If so, then knowing the K_d from radiolabel binding assay and the K_{dose} from self-administration assay allows the apparent V_d for an antagonist to be estimated. In the case of SCH23390, assuming a K_d of 0.3–0.5 nmol/L and with a K_{dose} of 7–8 nmol/kg the apparent V_d would be approximately 14–27 L/kg. For eticlopride assuming a K_d of 0.2–0.5 nmol/L and with a K_{dose} of 9–12 nmol/kg the apparent V_d would be approximately 18–40 L/kg. It is possible that the rank order of antagonist potencies for a specific receptor measured in vitro and in vivo may be different if there are significant differences in their apparent V_d. However, because the apparent V_d of an antagonist should be constant, the rank order of in vivo potencies for a series of antagonists should also be constant for different responses mediated by the same receptor population.

1.4.4 Summary

The self-administration paradigm represents a sensitive and rapid assay system to measure the pharmacokinetic and pharmacodynamic potencies of self-administered agonists and antagonists of the receptor populations in the brain mediating this agonist-induced behavior. This method may be useful for validating pharmacokinetic/pharmacodynamic models of agonist and antagonist effects in the brain. The ability of this system to measure drug pharmacokinetics has not been investigated previously and may offer a useful adjunct to standard analytical pharmacokinetic methods that rely on measuring drug concentrations in timed plasma samples.

1.5 Conclusions

- Psychological and physiological pharmacology approaches to study animal behavior are not compatible and mixing the two approaches creates paradoxes. Restricting investigations of self-administration behavior to the physiological pharmacology approach provides explanations of the major phenomena while avoiding all paradoxes.
- The experimental data accumulated during the first decade of intravenous drug self-administration studies in rats (1962–1974) were consistent with the principles of physiological pharmacology. This approach was abandoned apparently because of the assumption that this behavior must follow psychological principles.
- The satiety threshold model represents a return to the neglected application of pharmacological principles and provides a rational theoretical basis for studying the mechanisms underlying drug self-administration and addictive behavior. A mathematical model of this phenomenon may be used to calculate the pharmacokinetic and pharmacodynamic potencies of agonists.
- The satiety threshold is an equiactive concentration of the self-administered agonist at which a quantal response is induced. This equiactive agonist concentration is increased in the presence of competitive receptor antagonists and the degree of this increase is directly proportional to the antagonist concentration. A mathematical model of this phenomenon may be used to measure the pharmacokinetic and pharmacodynamic potencies of antagonists.
- Self-administration behavior represents a sensitive and rapid bioassay system useful for high-content screening of agonists and antagonists that interact with several clinically relevant neurotransmitter systems.

References

Ahmed SH, Koob GF (2005) Transition to drug addiction: A negative reinforcement model based on an allostatic decrease in reward function. Neuropsychopharmacology 180:473–490

Arunlakshana O, Schild HO (1959) Some quantitative use of drug antagonists. Br J Pharmacol Chemother 14:48–58

Baxter BL, Gluckman MI, Stein L, Scerni RA (1974) Self-injection of apomorphine in the rat: Positive reinforcement by a dopamine receptor stimulant. Pharmacol Biochem Behav 2:387–391

Bertalmio AJ, Woods JH (1989) Reinforcing effect of alfentanil is mediated by mu opioid receptors: Apparent pA2 analysis. J Pharmacol Exp Ther 251:455–460

Colquhoun D (2007) Why the Schild method is better than Schild realised. Trends Pharmacol Sci 28:608–614

Flory GS, Woods JH (2003) The ascending limb of the cocaine dose–response curve for reinforcing effect in rhesus monkeys. Neuropsychopharmacology 166:91–94

Kenakin TP (1997) Pharmacological analysis of drug-receptor interaction, 3rd edn. Lippincott-Raven, Philadelphia

Mello NK, Negus SS (1996) Preclinical evaluation of pharmacotherapies for treatment of cocaine and opioid abuse using drug self-administration procedures. Neuropsychopharmacology 14:375–424

Nicolaysen LC, Pan HT, Justice JB Jr (1988) Extracellular cocaine and dopamine concentrations are linearly related in rat striatum. Brain Res 456:317–323

Norman AB, Norman MK, Hall JF, Tsibulsky VL (1999) Priming threshold: a novel quantitative measure of the reinstatement of cocaine self-administration. Brain Res 831:165–174

Norman AB, Tsibulsky VL (2001) Satiety threshold regulates maintained self-administration: comment on Lynch and Carroll (2001). Exp Clin Psychopharmacol 9:151–154

Norman AB, Tsibulsky VL (2006) The compulsion zone: A pharmacological theory of acquired cocaine self-administration. Brain Res 1116:143–152

Norman AB, Tabet MR, Norman MK, Tsibulsky VL (2011a) Using the self-administration of apomorphine and cocaine to measure the pharmacodynamic potencies and pharmacokinetics of competitive dopamine receptor antagonists. J Neurosci Methods 194:152–258

Norman AB, Norman MK, Tabet MR, Tsibulsky VL, Pesce AJ (2011b) Competitive dopamine receptor antagonists increase the equiactive cocaine concentration during self-administration. Synapse 65:404–411

Pettit HO, Justice JB Jr (1989) Dopamine in the nucleus accumbens during cocaine self-administration as studied by in vivo microdialysis. Pharmacol Biochem Behav 34:899–904

Pickens R, Thompson T (1968) Cocaine-reinforced behavior in rats: Effects of reinforcement magnitude and fixed-ratio size. J Pharmacol Exp Ther 161:122–129

Rang HP (2006) The receptor concept: Pharmacology's big idea. Br J Pharmacol 147(Suppl 1):S9–S16

Roberts DCS, Vickers G (1984) Atypical neuroleptics increase self-administration of cocaine: An evaluation of a behavioural screen for antipsychotic activity. Neuropsychopharmacology 82:135–139

Rowlett JK, Wilcox KM, Woolverton WL (1998) Self-administration of cocaine-heroin combinations by rhesus monkeys: Antagonism by naltrexone. J Pharmacol Exp Ther 286:61–69

Schild HO (1949) pAx and competitive drug antagonism. Br J Pharmacol Chemother 4:277–280

Schild HO (1957) Drug antagonism and pAx. Pharmacol Rev 9:242–246

Sizemore GM, Martin TJ (2000) Toward a mathematical description of dose-effect functions for self-administered drugs in laboratory animal models. Neuropsychopharmacology 153:57–66

Tsibulsky VL, Norman AB (1999) Satiety threshold: A quantitative model of maintained cocaine self-administration. Brain Res 839:85–93

Tsibulsky VL, Norman AB (2005) Real time computation of in vivo drug levels during drug self-administration experiments. Brain Res Brain Res Protoc 15:38–45

Weeks JR (1962) Experimental morphine addiction: Method for automatic intravenous injections in unrestrained rats. Science 138:143–144

Wilson MC, Hitomi M, Schuster CR (1971) Psychomotor stimulant self administration as a function of dosage per injection in the rhesus monkey. Neuropsychopharmacology 22:271–281

Wise RA (1987) Intravenous drug self-administration: A special case of positive reinforcement. In: Bozarth MA (ed) Methods of assessing the reinforcing properties of abused drugs. Springer, New York, pp 117–141

Wise RA, Newton P, Leeb K, Burnette B, Pocock D, Justice JB Jr (1995) Fluctuations in nucleus accumbens dopamine concentration during intravenous cocaine self-administration in rats. Neuropsychopharmacology 120:10–20

Yokel RA, Pickens R (1974) Drug level of d- and l-amphetamine during intravenous self-administration. Neuropsychopharmacology 34:255–264

Zernig G, Wakonigg G, Madlung E, Haring C, Saria A (2004) Do vertical shifts in dose–response rate-relationships in operant conditioning procedures indicate "sensitization" to "drug wanting"? Neuropsychopharmacology 171:349–351

Zernig G, Ahmed SH, Cardinal RN, Morgan D, Acquas E, Foltin RW, Vezina P, Negus SS, Crespo JA, Stockl P, Grubinger P, Madlung E, Haring C, Kurz M, Saria A (2007) Explaining the escalation of drug use in substance dependence: Models and appropriate animal laboratory tests. Pharmacology 80:65–119

Zittel-Lazarini A, Cador M, Ahmed SH (2007) A critical transition in cocaine self-administration: Behavioral and neurobiological implications. Neuropsychopharmacology 192:337–346

Chapter 2
Intermittent Adaptation: A Mathematical Model of Drug Tolerance, Dependence and Addiction

Abraham Peper

Abstract A model of drug tolerance, dependence and addiction is presented. The model is essentially much more complex than the commonly used model of homeostasis, which is demonstrated to fail in describing tolerance development to repeated drug administrations. The model assumes the development of tolerance to a repeatedly administered drug to be the result of a process of intermittently developing adaptation. The oral detection and analysis of endogenous substances is proposed to be the primary stimulus triggering the adaptation process. Anticipation and environmental cues are considered secondary stimuli, becoming primary only in dependence and addiction or when the drug administration bypasses the natural—oral—route, as is the case when drugs are administered intravenously. The model considers adaptation to the effect of a drug and adaptation to the interval between drug taking to be autonomously functioning adaptation processes. Simulations with the mathematical model demonstrate the model's behaviour to be consistent with important characteristics of the development of tolerance to repeatedly administered drugs: the gradual decrease in drug effect when tolerance develops, the high sensitivity to small changes in drug dose, the rebound phenomenon and the large reactions following withdrawal in dependence.

2.1 Introduction

If a drug is administered repeatedly, the effect it has on the organism decreases when the organism develops tolerance to the drug. Many models of drug tolerance have been developed in the past. Most of these models are qualitative only and do not illuminate the mechanism underlying the effect very much. Those models that do attempt to describe the process mathematically are often too simple and only consider the effect of a single drug administration. A proper model describing how drug tolerance develops should account for a gradual decrease in the drug effect when a drug is administered repeatedly and should include a triggered response

---

A. Peper (✉)
Department of Medical Physics, Academic Medical Centre, University of Amsterdam,
Amsterdam, The Netherlands
e-mail: a.peper@planet.nl

to the drug administrations. The slow build-up of tolerance during successive drug administrations and the triggered response necessarily imply the presence of long term memory for the properties and the effects of the drug.

A variety of theories and models have been proposed to explain the mechanism relating the various aspects of drug taking. Very important has been the concept of homeostasis. In 1878, Bernard wrote: *"It is the fixity of the 'milieu interieur' which is the condition of free and independent life. All the vital mechanisms however varied they may be, have only one object, that of preserving constant the conditions of life in the internal environment"* (Bernard 1878, cited by Cannon 1929). Cannon translated Bernard's observation into the model of homeostasis (Cannon 1929). Fundamental in Cannon's theory is the presumption that physiological processes are regulated and that their functioning is in a "steady state": their conditions are stable and held constant through feedback. Homeostasis has been the basis of important theories like Bertalanffy's Systems Theory and Norbert Wiener's Cybernetics, which propose that physiological processes can be simulated by electronic feedback models (Wiener 1948; von Bertalanffy 1949, 1950). In the mathematical models of drug tolerance developed on the basis of these theories, the effects produced by drugs are assumed to be counteracted by a feedback mechanism which keeps the processes involved functioning at a preset level, thus causing tolerance to develop (Goldstein and Goldstein 1968; Jaffe and Sharpless 1968; Martin 1968; Kalant et al. 1971; Snyder 1977; Poulos and Cappell 1991; Dworkin 1993; Siegel 1996; Siegel and Allan 1998).

Besides the theories of drug tolerance based on homeostasis, there are theories which do not consider tolerance development the result of a regulated process. An influential theory was developed by Solomon and Corbit, the Opponent-Process theory (Solomon and Corbit 1973, 1974; Solomon 1977, 1980). In this theory, the drug is thought to trigger a response known as the A-process. The A-process induces a reaction called the B-process which opposes the A-process and increases in magnitude by repeated elicitation of the A-process. The A-process is fast, while the B-process is delayed and slow. As the difference between the A-process and the (negative) B-process is the ultimate effect of the drug, the drug effect will slowly decrease.

Several theories are based on a model of habituation developed by Rescorla and Wagner, which attributes tolerance to a learned diminution of the response (Rescorla and Wagner 1972; Wagner 1978, 1981; Tiffany and Baker 1981; Baker and Tiffany 1985; Tiffany and Maude-Griffin 1988). Dworkin incorporated this theory in a feedback model of drug tolerance (Dworkin 1993).

Another influential theory was proposed by Siegel (Siegel 1975, 1996, 1999; Siegel and Allan 1998; Siegel et al. 1982). In Siegel's theory, drug tolerance is assumed to be caused by Pavlovian conditioning: the compensatory response of the organism to the administration of a drug is triggered by environmental cues paired to the drug taking. Poulos and Cappell augmented Siegel's theory of drug tolerance by incorporating homeostasis, which was adopted by Siegel (Poulos and Cappell 1991; Siegel 1996; Siegel and Allan 1998).

In what follows, a model of drug tolerance, dependence and addiction will be presented which is different from the theories outlined above. The model is based on the assumption that most processes in a living organism are regulated, which is in accordance with homeostasis. It will be argued that the slow build-up of tolerance during repeated drug administrations, combined with a triggered response to those administrations, requires a complex adaptive regulation mechanism which, although incorporating feedback, is essentially different from homeostasis. The model presented is a general model of drug tolerance and drug dependence where "general" indicates that the model is based on principles which are thought to be more or less applicable to all processes of tolerance development. The model assumes the development of tolerance to a drug to be a process of intermittent adaptation to the disturbing effects of the drug: *during* the disturbances the body gradually learns to counteract these effects (Peper et al. 1987, 1988; Peper and Grimbergen 1999; Peper 2004a, 2004b, 2009a, 2009b). It also assumes that when processes in living organisms are disturbed, they adapt in a way that is fundamentally the same for all processes. Knowledge about adaptation in one process, therefore, teaches us about adaptation in other processes. The latter hypothesis is defended by many writers (Thorpe 1956; Kandel 1976; Koshland 1977; Poulos and Cappell 1991; Siegel and Allan 1998). It allows us to use our knowledge of the body's adaptation to changing environmental temperature equally well as, for instance, knowledge about adaptation to colour stimuli (Siegel and Allan 1998) to solve problems in modelling the organism's adaptation to drugs.

2.2 Properties of Adaptive Regulated Physiological Processes

2.2.1 *Homeostasis*

Homeostasis has made an invaluable contribution to our understanding of how physiological processes function by introducing the concept of the regulated physiological process: the presumption that most processes in a living organism are, one way or another, regulated. Regulation implies that the behaviour of a certain process in the organism is ultimately determined by an aim set by the organism itself, which in a highly simplified process is the process set point or process reference. In a simple regulated process, the output of the process—i.e. what is produced or obtained—is observed by a sensor and compared with a desired value, the process reference. When the output is not at the desired level, the process parameters are changed until the output is—within certain margins of accuracy—equal to the process reference. In this way the process is maintained at the desired level through feedback. There are many forms of feedback. In general, the feedback is negative. Negative feedback of a process in its most simple form means that the process output is subtracted from (negatively added to) the process input. The effect of negative feedback is that the regulation error—the deviation of the process output from the desired value—is reduced, the remaining error depending on the amplification of the feedback loop. When delay and stability problems can be managed, negative feedback can be very effective in counteracting the effects of disturbances to the process, making the process output less responsive to changing parameter values or changes in its environment.

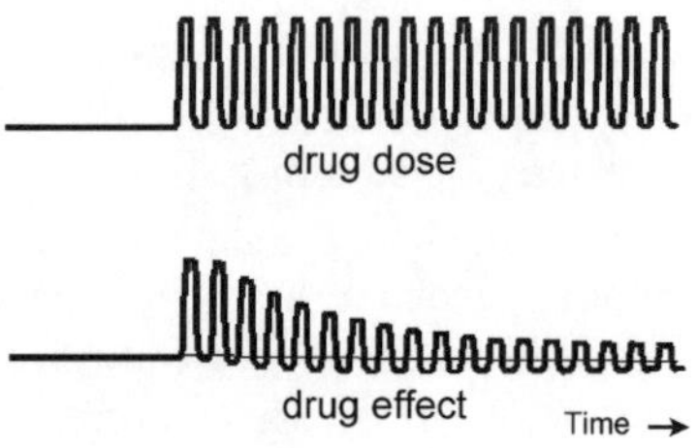

Fig. 2.1 Development of tolerance to the repeated administration of a drug

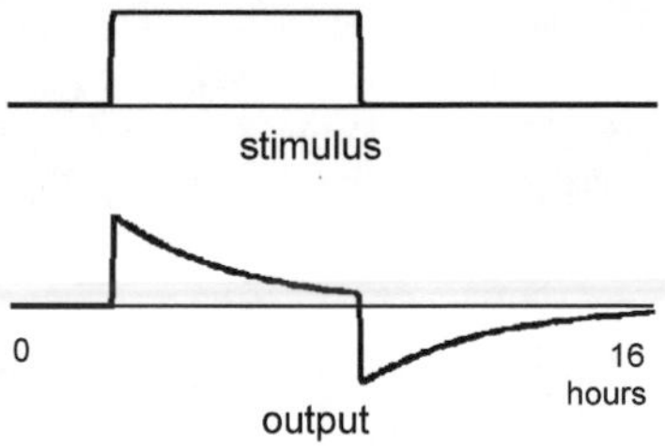

Fig. 2.2 Computer simulation of the effect of a single disturbance on the process output of a simple linear negative feedback circuit

Homeostasis has made clear that most physiological processes are regulated, and that regulation implies feedback. This has resulted in numerous models using negative feedback systems as a description of their behaviour. However, the incorporation of negative feedback in itself does not suffice to obtain a model describing the behaviour of adaptive physiological processes like the development of drug tolerance, as will be demonstrated with the response of negative feedback systems to regularly occurring disturbances.

Figure 2.1 illustrates the effect of tolerance development on the drug effect when a drug is administered repeatedly. The gradual build-up of tolerance is reflected in a gradual decrease in drug effect. It is accompanied by reactions during the interval between two drug administrations (the signal going below the base line), representing the rebound phenomenon: opposite symptoms after the drug effect has ended.

Figure 2.2 shows a computer simulation of the effect of a disturbance on the output of a simple (first order) linear negative feedback circuit. The length of the stimulus and the time constant τ of the circuit are set at 6 and 3 hours, respectively. The vertical axes are in arbitrary units. The initially large effect of the stimulus on the output decreases over time at a speed determined by τ. This decrease more or less resembles the development of acute tolerance: tolerance to the effect of a single drug administration. When the stimulus ends, there is an effect in the opposite direction, which could be regarded as representing the rebound mechanism.

If the same stimulus is applied repeatedly to this simple regulated system, the model's response does not resemble the development of tolerance shown in Fig. 2.1. This is demonstrated in the simulation depicted in Fig. 2.3, where the stimulus is applied twice a day. Every time the stimulus is applied, the effect of the stimulus on the output (Fig. 2.3b) appears to be the same as in Fig. 2.2. The stimuli are all suppressed to the same degree, which does not reflect the decrease in drug effect over time as the organism develops tolerance. If the time constant of the regulation is increased from 3 hours to 3 days, the sole effect of the regulation is that the average

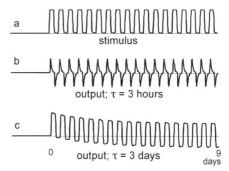

Fig. 2.3 Effect of a repeatedly applied stimulus on a simple feedback circuit

value of the signal drifts towards the base line (Fig. 2.3c). Although this example of a simple regulated process shows some qualities of tolerance development and might give an acceptable description of acute tolerance, it apparently lacks the capacity to adapt to recurring disturbances. The above simulation uses a simple, first order negative linear feedback circuit. When a mathematical model combines systems to form a complex, higher order feedback circuit, it will generate a response which differs from that of Fig. 2.2b. However, the effect of repeatedly applied stimuli will always give the pattern displayed in Fig. 2.3. Apparently, feedback does not suffice to describe the development of tolerance to repeatedly applied disturbances and, consequently, the model of homeostasis cannot describe drug tolerance.

An attempt to modify the model of homeostasis to account for its obvious shortcomings is the model of allostasis (Koob and Le Moal 2001; Ahmed et al. 2002; Schulkin 2003; Sterling 2004). Allostasis challenges the basis of homeostasis that processes are functioning at a steady state and proposes that the goal of regulation is not constancy, but rather, 'fitness under natural selection' (Sterling and Eyer 1988; Sterling 2004). Yet, in spite of its criticism of the homeostatic model, allostasis assumes that while the set points of process regulations are controlled by the organism to meet its overall goal these processes themselves are regulated in a homeostatic manner. Allostasis is predominantly a qualitative model (Ahmed and Koob (2005) set out a quantitative model which controls the intravenous administration of cocaine in rats) and there is no indication that it can describe the effects of repeated drug administrations.

2.2.2 The Properties of Adaptive Processes

When the development of drug tolerance cannot be described by homeostasis, or in general, by simple feedback systems, what then is the mechanism which does describe it? The model presented here posits that the development of drug tolerance is an expression of the general process of adaptation to environmental disturbances. Homeostasis and adaptive regulation are often assumed to be synonymous. In reality, these concepts are very different. The basis of homeostasis is that processes continue functioning at a preset level during changing environmental conditions, the "equilibrium" or "steady state" of Cannon. Adaptive processes, on the other hand, aim for optimal performance, which in a changed environment may imply functioning at a different level or even in a different way (Bell and Griffin 1969; Toates 1979). Moreover, as most processes in the organism interact with numerous other processes, environmental changes may affect the functioning of the entire organism.

Adaptation and habituation, too, are often used interchangeably even though they are essentially different concepts. Habituation is a multiplicative mechanism: the response to the stimulus is attenuated to reduce the effect of the stimulus. Adaptation, on the other hand, is an additive process: the disturbance is counteracted by a compensating mechanism. The applicability of additive and multiplicative mechanisms to the description of tolerance development has been analysed in an earlier publication (Peper et al. 1988).

Adaptation is often considered a relatively slow, continuous learning process. Drug tolerance, however, usually manifests itself as a relatively short lasting, but recurrent and triggered process and may therefore be seen as an intermittent learning process of the organism: during the disturbances it learns how to deal with recurrent changes in its environment to keep functioning optimally. If a drug is administered, the organism "remembers" the effect of the drug during previous administrations and takes measures to lessen its effect this time. When full tolerance is established, the organism has learned to deal with the disturbance as effectively as possible in the given circumstances. The organism's learning process during adaptation in response to the repeated administration of a drug inevitably presumes memory over an extended period of time: memory for the properties of the particular drug, memory for the effects exerted by the drug on previous occasions and memory for the measures it has to take to oppose the effect of the drug.

In the general process of adaptation, it is postulated that the organism remembers as separate facts changes in its functioning when these are caused by different changes in its environment. This seems obvious: different drugs elicit different adaptation processes. Yet the implications of such specificity are far-reaching as is demonstrated with a simplified example of how the body's thermogenesis reacts to temperature changes. When one leaves a warm room to stay in the cold outside for a few minutes, the warm room feels normal on returning. After a day in the cold outside, the warm room feels hot on entering. Apparently, after adaptation to the cold outside, adaptation to the warm room must develop again. This adaptation to the warm room could be interpreted as the transition phase back to the normal situation. However, when the length of the disturbance is increased, the concept of "normal situation" becomes ambiguous. For somebody who has lived rough on the street over a prolonged period, the cold outside has become the normal situation and entering a warm room a disturbance: there has been a shift in the normal situation from the high temperature in the room to the low temperature outside. This shift is only comprehensible when it is accepted that for an adaptive process there is no normal situation: every change in environmental condition results in a new situation to which the process adapts by seeking a new level of functioning.

When this analysis of how the organism adapts is translated to the administration of drugs, it implies that for the organism the beginning of the drug action and its

Fig. 2.4 General outline of the development of adaptation to a repeatedly occurring disturbance in an adaptive process

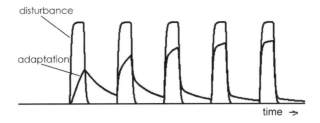

ending constitute different disturbances because they are the beginning of different (opposite) events, namely the drug effect and the interval between drug administrations. In existing models of drug tolerance, the interval between drug administrations is assumed to be the base line, the situation identical to the undisturbed situation before the first dose. In the model proposed, the organism's adaptation to the effect of a drug and its adaptation to the interval between drug administrations are considered autonomous processes.

Like homeostasis, the model adapts to a disturbance by opposing its effect. Figure 2.4 illustrates how this process of adaptation develops. The level of adaptation at any moment depends on the magnitude and length of the disturbance while it increases with every disturbance. Adaptation to the interval proceeds from the level acquired during the disturbance. In the above example of the body's thermoregulation, an increase in thermogenesis on entering the cold outside is the body's method of adapting to that disturbance. A return to the warm room will result in a decrease in heat production, accompanied by cooling if necessary through, for instance, sweat secretion. Figure 2.4 shows that after the body has learned to cope with this particular disturbance, the increase in thermogenesis on entering the cold and its decrease on return to the room will take place rapidly, while the level of adaptation has increased considerably.

2.2.3 The Detection of Exogenous Substances

The effects of drugs are for an important part determined by their disturbing effect on the information transfer within the organism's regulated processes. Consider a process which sends information about its level of functioning to the regulator of that process (this is detailed below in Fig. 2.5). The messenger used to transfer this information—a number of molecules of a certain substance—is detected by a sensor—receptors sensitive to that particular substance—which relays the information to the process regulator. If a drug interferes with the transport of this messenger, for instance by binding to the receptors, changing their affinity for the messenger, or simply by adding to the amount of the messenger substance, the information from the sensor will change and the effect will be a change in the output level of the process.

The disturbing effect of a drug on the regulation of a physiological process decreases when tolerance develops: the process regulator learns to counteract the effect of the drug on the information transfer. This antagonistic action of the regulator is operative mainly during the time the drug is present. This can be deduced from the fact that when a drug is given only occasionally, the effect during the intervals is very small, even though the organism may have developed a high level of tolerance to the drug (this subject is treated extensively in Peper et al. 1987, 1988). If tolerance to a drug manifests itself mainly during the time the drug is present, an important conclusion can be drawn: when a process is disturbed by a drug, its regulator must at that moment "know" that the change in the output of the sensor is due to the presence of the drug and not to a normal fluctuation in the process it regulates. From the output signal of the sensor alone the regulator will not be able to determine whether the receptors are bound to an endogenous or an exogenous substance or whether a drug has changed the sensitivity of the sensor to the messenger substance. It can distinguish between the various ways in which a drug may interfere only by acquiring additional information about the situation. If, for instance, the exogenous substance differs from substances usually found at the location of the sensor, the regulator might be able to acquire this information from the receptor site. If, however, the exogenous substance is of the same chemical composition as an endogenous messenger substance, this information cannot be acquired other than from the fact that the organism has detected the substance somewhere in the organism where it is normally not present or from oral or environmental information about the substance entering the body.

The organism has several ways to detect a drug. If administered orally, there are gustatory and olfactory mechanisms to record the presence of a drug and its chemical characteristics. At a later stage, when the drug is within the organism or if the drug is administered intravenously, there are other ways in which a process regulator may obtain information about its presence and characteristics: from chemical sensors which are sensitive to the drug, from information originating from processes in the organism which themselves are disturbed by the drug or from environmental cues which it has learned to associate with the presence of the drug. But regardless of how the information is acquired, to enable a process regulation to take measures to reduce the effect of an exogenous substance upon the process, information about the presence of the drug should reach the regulator at an early stage, before the drug actually reaches the receptor site. This implies that the regulator will attach greater value to oral information about the presence of the drug than to information from the surrounding tissue (Steffens 1976; Grill et al. 1984). Given, furthermore, that the natural route into the body is through the mouth, it can be assumed that the organism will regard the detection of exogenous substances in the mouth as the fundamental source of information about the presence of a drug.

2.2.4 Oral and Environmental Cues

In discussions about tolerance development, cues originating from environmental causes are usually considered more important than the administration of the drug itself. Although environmental cues can indeed dominate completely in certain situations, under closer scrutiny it becomes clear that the oral administration of a drug must be the primary and natural stimulus for the development of tolerance. One rational consideration is that for a living organism there is a relationship between oral drug-taking and the drug effect and that the organism will use such a relationship. After all, the natural route of an exogenous substance into the body is through the mouth. The mouth is - so to speak—made for that purpose. Together with the nose, it contains all the means needed to detect and analyse exogenous substances. The primary functions of the mouth and the nose—taste and smell—are there to allow the organism to recognise a substance when it enters the body, enabling it to anticipate its effect and to take appropriate measures in time.

An additional consideration indicating that oral administration is the fundamental stimulus in the tolerance process is that, when the organism is able to pair very different kinds of environmental cues with the drug effect as has been demonstrated in the literature, it will certainly relate the drug's presence to the drug effect. In fact, this relation must have been the first to develop in primitive organisms as it can also be observed at cell level where the mere presence of a drug can induce tolerance without the mediation of higher structures like the central nervous system. This has been demonstrated explicitly in isolated cell cultures, where repeated stimulation with toxic substances or changes in temperature induce tolerance (Peper et al. 1998; Wiegant et al. 1998).

There is ample evidence that the adaptive response—the compensatory action of the organism to the effect of a drug—is triggered by the oral administration of the drug. For instance, the oral administration of glucose almost immediately results in an increased release of insulin into the bloodstream (Deutsch 1974; Steffens 1976; Grill et al. 1984; Dworkin 1993; Loewy and Haxhiu 1993). In fact, the organism will make use of any cue it can find to anticipate disturbances of its functioning, and oral drug taking seems crucial in this mechanism.

These considerations do not mean that an oral stimulus is always the dominant stimulus for the tolerance process. Environmental cues become of prime importance when the natural—oral—route is bypassed through the injection of the drug directly into the bloodstream. Since much of the research into drug tolerance has been done with drugs administered intravenously, that is, without the fundamental—oral—cue being present, care should be taken in interpreting any results. When the oral drug cue is not present, the body will have to depend on environmental cues to trigger the tolerance mechanism, which may result in a different behaviour. In any research into the development of drug tolerance, it is therefore essential to understand the natural way in which the organism develops drug tolerance and the consequences of administering drugs directly into the bloodstream.

2.2.5 The Effect of Unknown Substances

When tolerance to a drug has developed, the organism apparently has enough information about the drug to reduce its disturbing effect. That information may include the chemical characteristics of the drug, the exact processes disturbed by the drug, the nature and the extent of the disturbance, the time taken by the drug to reach the receptor site, its effect on the sensor characteristics, and so on. When a drug enters the organism for the first time, the organism may be assumed not yet to have gathered this information and it must then establish the relationship between the taking of the unknown drug and subsequent disturbances in the organism.

As postulated above, the function of the mouth is to detect exogenous substances entering the body and to activate the processes which will be disturbed so they can generate a compensating response to the effect of the substance. Although no tolerance exists and no compensating response will be generated when a substance is unknown to the organism, for the organism to know that a drug is new implies that the substance first has to be analysed. It will, consequently, not make much difference for the organism whether a drug is new or whether there already exists a certain degree of tolerance to the drug: familiar or not, every drug entering the organism will be analysed. In case of an unknown substance, the changes in functioning of processes which follow will then be related to the composition of the substance and tolerance can develop.

In addition, it is quite conceivable that the organism has a built-in degree of tolerance to all (or most) substances in nature, in which case there are no "new" drugs and it is not a matter of analysis but of recognition. Every drug entering the organism is "recognised" and the organism "remembers" what the consequences for its functioning were on previous occasions when it detected that particular drug, where "previous" includes inheritance. The latter hypothesis is difficult to test, however, as in most cases it is not possible to determine the actual level of tolerance to a certain drug: the drug effect itself does not reveal information about the magnitude of the compensatory response or the level of tolerance.

2.2.6 The Magnitude of the Compensatory Response

The question now remains of why the organism requires so much time to develop tolerance to a drug when it apparently has all the information about the drug's chemical characteristics even when the drug enters the body for the first time. The answer to this question derives from the observation that, while a drug's chemical characteristics determine which processes are disturbed, it is the quantity of the drug which determines how much those processes are disturbed and hence the extent of the measures the organism must take to reduce the drug effect. This quantity, however, cannot be determined at an early stage. The organism is, for example, unable to determine the quantity of a medication before it is dissolved completely, or whether a cup of coffee is followed by a second or third. Such information becomes available only after a relatively long time, which is (or may be) too long for the processes involved to counteract the drug's disturbing effect in an effective way. The organism is thus confronted with a fundamental problem. It wants to counteract the drug effect but has no definite information about the magnitude of the measures it has to take. The approach the organism has adopted to solve this difficulty is to base the magnitude of the compensatory response on the drug dose it expects: the usual or habitual drug dose. In practise, this will be about the average dose of a number of previous drug administrations.

It then becomes clear that tolerance to a certain drug does not merely mean that the organism knows how to cope with the given drug, but that the organism knows how to cope with a certain *quantity* of the drug. A change in that quantity—a change in the habitual drug dose—will therefore result in a period of incomplete tolerance during which the effect of the drug on the organism differs substantially from the tolerant situation. The functioning of the organism will then remain disturbed until it has learned to cope with the new drug level and has become tolerant to the new drug dose.

It is difficult to find a rationale for the initial large drug effect and the long time it takes the organism to develop tolerance other than the assumption that the organism does not determine the quantity of a drug entering the body. Again, if the organism were able to determine the properties and the quantity of the drug at an early stage, it would have all the information needed to rapidly suppress any drug activity. The organism needs a relatively long period to make an approximation of the drug dose it can expect.

2.3 Modelling Tolerance Development in Physiological Processes

2.3.1 The Model

The initial effect of a disturbance upon a regulated physiological process will now be elucidated with a simplified model. Subsequently, the model will be expanded to describe the complex response of a regulated physiological process to repeated disturbances in its functioning. Figure 2.5 shows a simple model of a regulated physiological process and the way in which a drug may disturb its functioning. In the normal, undisturbed functioning of the process, an endogenous substance in the blood, e, which is a measure of the level of the substance in the bloodstream produced by the process, E, is detected by the sensor, receptors which have affinity

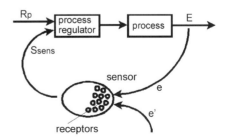

Fig. 2.5 Example of a simplified regulated process and the way in which a drug in the bloodstream may disturb its functioning

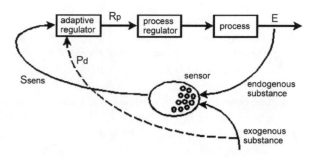

Fig. 2.6 Adaptive regulator added to the regulated process of Fig. 2.5

with the substance in question. The binding of this substance with the receptors results in a signal from the sensor to the process regulator, S_{sens}. The magnitude of S_{sens} is a measure of the number of bound receptors and thus of the amount of the substance in the bloodstream. The process regulator compares the level of S_{sens} with the level of the process reference, R_p, and regulates the process in such a way that S_{sens} and R_p are about equal. In this way the level of the substance in the bloodstream is kept at the desired level through negative feedback. If an exogenous substance, e', with which the receptors also show affinity (this may, but need not, be the same substance as the endogenous substance) is introduced into the bloodstream, the subsequent binding of this exogenous substance to the receptors will raise the level of S_{sens}. However, to keep S_{sens} at about the level of the reference, the negative feedback will reduce the process output, E—and consequently the level of the messenger substance, e—until the number of bound receptors is about the same as before the intervention.

In Sect. 2.2.1, it was argued that the development of drug tolerance cannot be described adequately in terms of simple feedback regulation. The mechanism responsible for tolerance development in the organism is fundamentally more complex and, hence, even a model which describes only the main characteristics of drug tolerance will be more complex. An adequate model of drug tolerance should possess the following characteristics:

- When a drug is administered repeatedly, the process should gradually learn how to readjust its functioning to oppose the effect of the drug.
- This adaptation process should be active mainly during the time the drug is present and should be activated upon the detection of the drug or associated cues.
- The drug's presence and the intervals between drug administrations should be considered different disturbances and should therefore initiate their own respective adaptation processes.

In Fig. 2.6, an "adaptive regulator" is added to the model of the regulated process in Fig. 2.5. This regulator is assumed to possess the qualities listed above. During successive drug administrations, it learns to change the process reference R_p during the presence of the drug in such a way that the effect of the disturbance on the level of the substance in the bloodstream, E, is reduced. To this end, it uses the output signal of the sensor, S_{sens}, and information about the drug administration, P_d. The dashed line indicates that P_d is information about the moment of drug administration only. In this model, the sensor output is assumed to be proportional to the sum

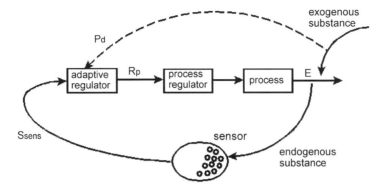

Fig. 2.7 Model of adaptive regulated process in which a drug increases the level of the produced substance

of the exogenous substance and the endogenous substance. The binding rates of the two substances with the receptors of the sensor are assumed to be equal.

2.3.2 Different Ways in Which Drugs Disturb the Body

A distinction has to be made between two fundamentally different ways in which drugs may disturb physiological processes:

Case 1: a drug changes the level of a regulated substance in the organism, increasing it when the drug and the substance are similar, or decreasing it, for instance by neutralisation.

Case 2: a drug disturbs the information transfer in the organism.

These two kinds of drug effects have essentially different consequences. If a drug increases the level of an endogenous substance of the same chemical composition, the long term effect will be a decrease in the production of that substance by the organism. When the low level of insulin in the blood of a diabetic is increased via the administration of exogenous insulin, the organism develops tolerance by gradually decreasing the insufficient insulin production of the pancreas even further, necessitating a gradual increase in the dose of the exogenous insulin (Heding and Munkgaard Rasmussen 1975; Mirel et al. 1980). If a drug interferes with the information transfer in a regulated process in the organism by affecting messenger-receptor interactions, or in general, the sensitivity of a sensor to an endogenous substance, the organism will learn to counteract the effect and, after a while, the process will more or less regain its normal functioning.

Figure 2.7 shows a model of an adaptive regulated process. The level of the substance produced by the process is increased by an exogenous substance of the same composition (case 1). The adaptive regulator gradually learns to suppress the effect of the drug during the period when the drug is in the bloodstream by lowering the process output. The adaptive regulator bases its action on information it receives from the sensor about the level of the regulated substance in the bloodstream, E,

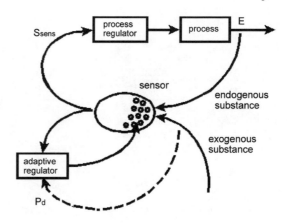

Fig. 2.8 Model of regulation in which a drug interferes with the information transfer in the regulation

and on information about the drug administration, P_d. In many models of drug tolerance, adaptation is assumed to be effected at the receptor site. However, if a drug changes the amount of a substance whose level is regulated, this information is crucial for the process regulator and should pass the sensor unaltered. It follows that the transfer function of the sensor (its input–output relation) must be kept constant. Consequently, when a drug changes the amount of a substance which is regulated at a preset level, the organism can be expected to counteract such a disturbance primarily by a readjustment of the process parameters.

When a drug interferes with the information transfer in the process regulation (case 2), it is not the level of the process which has to be corrected, but the change in input signal to the process regulator induced by the drug. As the feedback path in the regulation is affected here, the disturbance caused by the drug may be corrected via a change in the transfer function of the sensor, for instance by means of a change in the number of receptors sensitive to the drug. In this configuration, the adaptive regulator learns to change the transfer function of the sensor in a way that counteracts the effect of the drug on the sensor's sensitivity to the messenger.

Figure 2.8 shows a model of a regulated process in which the information transfer is disturbed by a drug. The adaptive regulator gradually learns to suppress the effect of the drug on the sensor signal by changing the sensitivity of the sensor. The adaptive regulator bases its action on information it receives from the sensor, S_{sens}, and on information about the drug administration, P_d.

The model in Fig. 2.7 describes the effect of a drug on the level of an endogenous substance which does not function as a messenger. The model in Fig. 2.8 describes the effect of a drug on messenger–receptor interactions and is therefore applicable to many of the effects associated with addictive drugs.

2.3.3 Fast and Slow Adaptation

The adaptive regulator treated above minimises the direct effect of a drug on the regulation. If it could suppress the drug effect completely, it would do all that is required. However, in general drug effects are only partially suppressed and in most cases substantial effects remain. An important additional function of an adequate regulator is minimising the effect of the remaining disturbance. The model achieves this by combining the fast regulator, which reduces the immediate effect of the disturbance, with a slow regulator, which minimises the magnitude of the remaining disturbance in the long run and which anticipates frequently occurring stimuli (see also Peper et al. 1987). After tolerance has been established, the slow adaptation is responsible for a shift in the output level to below normal in the interval between drug administrations. The magnitude of these negative reactions in the tolerant situation depends on the length of the interval. When a drug is taken infrequently, the organism is not much affected during the intervals; when the frequency of administration is high, the shift can become considerable. The fast regulator is a complex system and determines to a large extent how tolerance develops. The slow regulator has a small effect by comparison but is an essential component of the adaptive regulator. Slow regulation can manifest itself in different forms. For a human moving to a hot climate, it may imply a permanent increase in sweat secretion. The thermoregulation in animals moved to a colder climate may adapt through a slow increase in the growth of their fur. The time constant of the slow regulator may amount to weeks months or even years.

2.4 The Mathematical Model and Its Practical Significance

2.4.1 The Model

It is important to observe that the mathematical model supports the underlying theory. This contrasts with other published models of drug tolerance, which are generally qualitative only. The importance of conducting research into the behaviour of physiological systems using control theoretical principles cannot be overemphasised as the behaviour of regulated systems can only be understood from the behaviour of mathematical models describing them. Even the behaviour of the simplest regulated system cannot be described other than mathematically. The behaviour of more complex regulated systems can only be understood from simulations with computer programs using advanced, iterative methods to solve the differential equations involved. This implies that a model which is qualitative only may never include feedback systems as the behaviour of such systems cannot be predicted or understood qualitatively.

The mathematical implementation of the current model is discussed in the appendix, which addresses the complex structure of the components of the regulation loop and presents the equations describing them. The model is a nonlinear, learning, adaptive feedback system, fully satisfying the principles of control theory. It accepts any form of the stimulus—the drug intake—and describes how the physiological processes involved affect the distribution of the drug through the body.

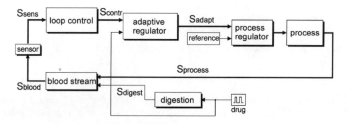

Fig. 2.9 Block diagram of the mathematical model of Fig. 2.7

A previous publication (Peper 2004b) derives the equations more fully and extensively discusses the control-theoretical basis of the regulation as well as the conditions for its stability.

The following model simulations are based on a number of simplifying assumptions:

- The parameters have been chosen to obtain a clear picture of the outcome of the simulations. Because in practise the stimulus—the drug intake—is extremely short in terms of repetition time, its duration has been extended for additional clarity.
- The mechanism of tolerance development will only function if it is triggered when the drug is administered. For the behaviour of the mathematical model, it is of no relevance whether it is triggered orally or by environmental cues. Hence, the simulations do not distinguish between different kinds of triggering.
- Whenever the paper discusses oral drug administration, the drug is assumed to be gustatorily detectable.
- As the model is a general model of tolerance development and does not describe a specific process, the vertical axes in the figures are in arbitrary units.

Figure 2.9 shows a block diagram of the mathematical implementation of the regulated adaptive process of Fig. 2.7. The process produces a hypothetical substance. Its regulation is disturbed by an exogenous substance of the same composition. The diagram comprises the digestive tract, the bloodstream, the process, the process regulator, a loop control function (see the Appendix) and the adaptive regulator. When the exogenous substance changes the level of the substance in the bloodstream, the adaptive regulator corrects for this disturbance by readjusting the output level of the process. The heavy arrows indicate the main route of the regulation loop. The thin arrows indicate the route of the disturbance: the transfer of the exogenous substance through the digestive tract to the bloodstream and the transfer of the information about the presence of the substance to the adaptive regulator. The block "reference" represents the reference level for the process regulator, which is set at a higher level in the hierarchical organisation of the organism. This subject will not be treated here.

When the exogenous substance enters the body, a series of activities readjusts the processes involved in order to reduce the disturbance. Figure 2.10 shows some signals from the block diagram which illustrate this mechanism (Peper 2004b). The

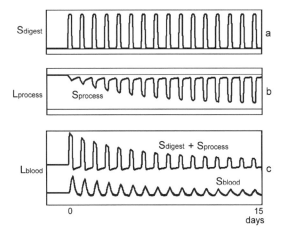

Fig. 2.10 Some signals from a process modelled with the mathematical model clarifying the functioning of the tolerance mechanism: (**a**) The exogenous substance when it enters the bloodstream, S_{digest}. (**b**) Process output during tolerance development, $S_{process}$. (**c**) $S_{process}$ and S_{digest} added in the blood stream and the resulting blood level, S_{blood}. The level of the process output and the resulting blood level before the drug is administered are $L_{process}$ and L_{blood}

endogenous substance is produced at a normally constant level, $L_{process}$. The resulting blood level is L_{blood}. When a similar substance is administered exogenously, the blood level will be disturbed. When the exogenous substance is administered repeatedly, the regulated process will develop tolerance to the disturbance. Trace (a) shows the exogenous substance, S_{digest}, when it enters the bloodstream. Trace (b) shows the process output: during the disturbances the output level will drop to counteract the induced rise in the level of the substance in the blood. The signal representing this change in process output level, $S_{process}$, represents the compensatory response of the process to the disturbance. In addition to these temporary changes in level, a permanent downward shift in the process output occurs. This shift of the curve to a level substantially lower than the baseline, $L_{process}$, represents a fundamental change in the functioning of the processes involved.[1] The two signals—S_{digest} and $S_{process}$—are added when the endogenous and exogenous substances mix in the bloodstream. The resulting signal is shown in trace (c) together with the resulting blood level, S_{blood}. The disturbance of the blood level gradually

---

[1]This downward shift in the functioning of the process represents the drug induced change in the functioning of processes involved in the drug effect. The shift depends mainly on the functioning of the slow regulator which can have a long time constant (see Sect. 2.4.2). As a result, the shift may remain a long time after a drug is withdrawn. This has important consequences as was first pointed out in a previous publication (Peper et al. 1987): *The negative shift of the process output on drug withdrawal signifies the occurrence of antagonistic symptoms with respect to the drug effect and these are consequently in the "direction" of the disorder the drug was intended to counteract* (Kalant et al. 1971). *This implies [...] a worsening of the disorder of the patient after termination of drug treatment.* Apparently, for the body, adaptation to a medicine means a shift in its functioning in the direction of the disease.

decreases during subsequent administrations when the process regulator adapts to the recurrent disturbance. Recall that all parameter settings in the simulations are arbitrary, as are the axes in the figure.

2.4.2 The Open-Loop Gain

The compensatory response only partly compensates the effect of the drug. The extent to which this takes place depends on the capacity of the body to suppress disturbances, which in the model domain is represented by the open-loop gain of the regulation loop. A large open loop gain—as is found in most electronic regulated systems—suppresses disturbances to a large extent. Stability considerations suggest that the open-loop gain in fast biological processes is small (Peper et al. 1987), and the suppression of disturbances only modest. In the example of Fig. 2.10, the open-loop gain is set at 4. This would be a very low figure for a electronic feedback system, but is a common value in physiological regulations.

The open-loop gain in physiological regulations is not fixed but depends on factors such as health, age and fitness (Mitchell et al. 1970; Verveen 1978, 1983; Peper et al. 1987, 1988; Peper 2004a). The open-loop gain determines both the rate of suppression of the drug effect after tolerance has developed and the magnitude of the reactions after withdrawal. This direct link between very different effects forces the organism to make a trade-off between a beneficial and an undesirable effect of the regulation, which may partly explain why the suppression of the drug effect when tolerance has developed tends to be relatively low. Yet another reason why there is a limited suppression of the drug effect in the tolerant situation may be that the organism cannot estimate the exact drug dose at the moment of administration and therefore has to be cautious in opposing the drug effect. If the organism nevertheless overestimates the drug dose, its drug-opposing action may outweigh the drug effect itself, resulting in a paradoxical drug effect: an effect with characteristics opposite to the normal drug effect.

When the time constant of a regulation loop is large, stability becomes less of a factor. In many cases, the open-loop gain of the slow adaptive regulator in physiological processes will therefore be significantly larger than that of the fast adaptive regulator.

2.4.3 Constant Drug Effect

In the simulation of Fig. 2.10, the drug dose has a constant magnitude. In clinical practise, it is not the drug dose but the drug effect that is of primary interest. As the drug effect decreases when tolerance to the drug develops, the dose must be increased to maintain the drug effect at the desired level. In the simulation in Fig. 2.11, the magnitude of the drug dose has been adjusted during the simulation to maintain

Fig. 2.11 The result of a computer simulation showing dose–response relation for constant drug effect. The magnitude of the stimulus has been adjusted during the simulation to maintain a nearly constant effect in the output of the model

Fig. 2.12 Illustration of the consequences of adaptive regulation to a permanent change in level

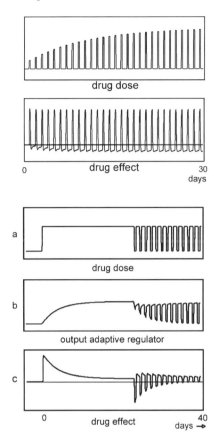

a nearly constant drug effect. After an initial increase, the magnitude of the stimulus settles at a level which yields the desired effect. The relation between the dose and the drug effect in that situation is determined by the open loop gain of the fast adaptive regulator, as explained in Sect. 2.4.2.

2.4.4 Adaptive Regulation

Figure 2.10 demonstrates how the adaptive regulator learns to generate a compensatory response when a drug is administered repeatedly. Figure 2.12 shows its response when a drug is administered permanently. A permanent change in drug level, as shown in the first part of Fig. 2.12a, will result in a permanently changed level of the output of the adaptive regulator (Fig. 2.12b). This level then becomes the new base line for the regulation and is accompanied by a shift in the level of the drug in the bloodstream (Fig. 2.12c). This shift is generally small, as the compensation in slow adaptation is generally large (see Sect. 2.4.2). Interruptions to such a permanent stimulus, shown in the second part of the figure, are now new—negative—stimuli,

Fig. 2.13 A simulation of the effect of a small change in drug dose after tolerance has developed. For a given set of parameters, a 20 percent decrease in dose results in an initial suppression of the drug effect. An increase in dose back to the original value causes an initial large increase in the drug effect

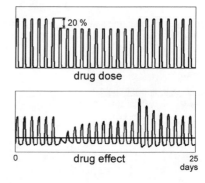

the suppression of which will increase over time similarly to the periodic stimuli shown in Fig. 2.10.

A permanent drug administration is no different from any other permanent change in the environment as was illustrated in Sect. 2.2.2 with an example of the consequences of a permanent change in environmental temperature. Figure 2.12 depicts adaptation to the cold outside when the temperature is substituted for "drug dose". The vertical axes then show an increase in cold and the onset of the signal is the temperature inside. Figure 2.12c then depicts the sensation of cold or warm and Fig. 2.12b the adaptation to the changes in temperature, that is, the compensatory response.

2.4.5 The Effect of Changes in Drug Dose

Because the compensatory response is not based on the actual drug dose but on the dose the subject is accustomed to (see Sect. 2.2.5), the compensatory response will initially not change when the actual dose is changed. The consequence is that a small change in drug dose will have a disproportionately large effect. Figure 2.13 shows a model simulation of the effect of a small change in drug dose after tolerance has developed. For a given set of parameters, a 20 percent decrease in drug dose results in an initial suppression of the drug effect. When the regulation adapts itself to the new situation—it slowly learns to decrease the compensatory response—the magnitude of the drug effect settles at a level reduced proportionally by 20 percent. When the dose is increased to its original magnitude, the drug effect initially increases to approximately twice the normal level.

In Fig. 2.13, with the parameter values selected, a 20% reduction in the dose results in an initial reduction in the drug effect to zero. This implies that at that moment the drug action and the compensatory response are of equal magnitude (S_{digest} and $S_{process}$ in Fig. 2.10). When the dose is reduced by more than 20%, negative reactions occur as the compensatory response then initially exceeds the action of the drug. This is shown in Fig. 2.14, where the dose is reduced to 50%.

Positive reactions to a small increase in drug dose are usually less apparent than negative reactions since the latter may cause a reversal of the symptoms, which is

Fig. 2.14 Effect of reduction in drug dose to 50%

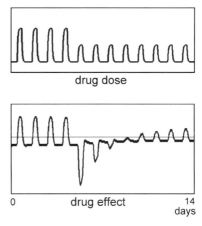

generally unpleasant or undesired, while a positive reaction is of the same nature as the drug effect. The action of many drugs is also subject to an upper limit. Pain medication, for instance, alleviates pain and cannot go beyond no pain. In addition, the effect of a larger dose is often reduced by non-linear transfers in the process. These are not incorporated in the general model presented here.

The large responses to small changes in drug dose are a common feature of the drug effect and are well known in the treatment of addicts. It explains why tapering off the drug dose to prevent negative reactions is such a slow process. A decrease of 10% or less a week is a common value for dependent or addicted subjects as higher values might cause adverse effects.

The disproportionate responses to a change in drug dose in dependence and addiction are not fundamentally different from when only tolerance is present. In dependence, the effect is large because tolerance in dependence is high. When tolerance is lower, as will be the case after a limited number of drug administrations, the effect of a reduction in dose is smaller but the decrease in drug effect may initially still be significantly larger than expected.

2.4.6 The Dose–Response Curve

Existing conceptualisations of the relationship between drug dose and drug effect display fundamental contradictions. It is undisputed that in dependent subjects a reduction in drug dose may generate large reactions. At the same time, the dose–response curve—shown in Fig. 2.15—which postulates that a change in drug dose will produce a proportionate and predictable change in drug effect, is assumed to provide an adequate description of the dose-effect relation. The applicability of the dose–response curve is limited because responses vary widely across subjects (Ramsay and Woods 1997). But it also has other shortcomings.

In standard medical practise, the initial dose of a drug is selected on the basis of the dose–response curve of the drug (curve (a) in Fig. 2.15) and the characteristics

Fig. 2.15 (a) Dose–response curve. (b) Dose–response curve after tolerance has developed

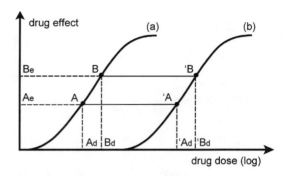

and peculiarities of the patient. In the figure this is assumed to be dose A_d, which has a drug effect A_e. If, after a while, the effect of the dose is not as desired, the dose is adjusted. For instance, if the effect is too small the dose is increased. In curve (a) that would be dose B_d with a drug effect B_e. However, if curve (a) were used to determine the new dose a problem would occur because, during the administration of the drug, tolerance may have developed. The dose–response curve captures an increase in tolerance through a shift to the right to curve (b). A larger dose is required to obtain the same drug effect. In the figure the shift is arbitrarily large, but in reality the shift can also be substantial and dose B_d will be too small to generate the desired effect B_e. If in practise tolerance development can be estimated and the curve is shifted to the right by the measured value, another difficulty arises. Whereas curve (a)—that is, the curve relevant for the first dose—can determine the drug effect values A_e and B_e given the drug dose values A_d and B_d, once tolerance has started to develop, an increase in dose from A_d to $'B_d$ will cause an initial increase in drug effect larger than curve (b) suggests, as was demonstrated in Fig. 2.13. In other words, an increase in the dose of a drug to which tolerance has developed may result in a disproportionately large increase in drug effect. Negative overshoot when the drug dose is decreased will be just as large and both situations may not be without risk to the patient.

The dose–response curve presumes a static relationship between drug dose and drug effect. Yet tolerance development—and thus time—is an important factor in measuring the drug effect. This is demonstrated in the model simulations reported in Fig. 2.16, where the dose and the drug effect are plotted separately against time to illustrate the influence of tolerance development on dose–response curve measurements.

Usually, the dose–response curve is measured by increasing the dose in logarithmic steps. The tolerance which develops during such a measurement distorts the curve. This effect, however, is not very clear in the curve, partly due to the distortion being gradual and partly due to the logarithmic change in dose.[2] When the curve

---

[2]The bend at the bottom of the dose–response curve is largely caused by the logarithmic scale. In a linear process, a linear change in dose will cause a linear change in drug effect, as long as there is no tolerance development (curve (d)). With a linear scale, distortion of the curve due to tolerance development is easily noticed. However, as the dose–response curve is commonly presented using

Fig. 2.16 Simulations with the mathematical model of the relation between dose (**a**) and drug effect, plotted against time to illustrate the influence of tolerance development on the outcome of dose–response curve measurements. The time constant of the tolerance mechanism in the simulations is respectively 7 days (**b**), 30 days (**c**) and 400 days (**d**)

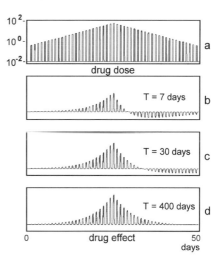

is determined with a decreasing dose, the effect of tolerance development becomes readily apparent. To demonstrate these effects, in Fig. 2.16 the dose is first increased and subsequently decreased (a). In curve (b), which represents the drug effect, a time constant of seven days is chosen for the tolerance process (approximately the time constant used in the simulations shown above and in previous publications on the subject). The effect of the decrease in drug dose is a dramatic shift towards a negative drug effect with symptoms opposite to the normal drug effect. When the time constant is increased to 30 days (c), this effect is still very strong. When the time constant is increased to 400 days (d), the effect has nearly disappeared, leaving a curve where tolerance development does not take place during measurement and the upward- and downward-sloping portions of the curve have a similar shape.

The full implication of the effect of tolerance development in dose–response curve measurements becomes clear during the decrease in drug dose when the decrease in drug action causes the compensatory response to become dominant and the overall drug effect to turn negative. Negative reactions are commonly seen in slow withdrawal when the dose is tapered off too rapidly, a situation comparable to that depicted in the figure. The dose–response curve is naturally measured by increasing the dose, in which case no such reactions are generated. But the distortion of the curve during the increase in dose is significant too, as shown in the figure. In the simulations, doses are administered once a day, over 50 days in total. Simulations with other settings of the model parameters, such as a different maximal dose, fewer stimuli or stimuli with different time intervals give a very similar picture.

The static representation of the relationship between drug dose and drug effect suggested by the dose–response curve cannot be reconciled with the dynamic responses of the organism to changes in drug dose characteristic of the mechanism of

---

a logarithmic dose scale, this has also been adopted here. The saturation in the top of the dose–response curve in Fig. 2.15 is the natural maximal activity of the processes involved. This effect has been left out in the simulation of Fig. 2.16 as it has no relevance to the present subject.

Fig. 2.17 Effect of reduction in drug dose to 10%

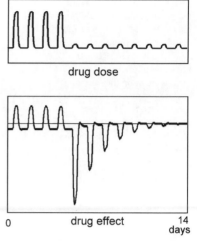

tolerance development. Unless tolerance to a certain drug develops very slowly, tolerance development will distort the curve when the effect of different drug doses is determined in a single subject. Values for the dose–response curve should therefore be determined from the (averaged) responses to single drug administrations measured in different subjects. Even measured in this way, a dose–response curve can only serve one valid purpose: it shows the average relationship between the dose and the *initial* response to a drug.

2.4.7 The Effect of a Further Reduction in the Drug Dose

It was explained above that when the compensatory response exceeds the drug action, negative reactions occur. This was demonstrated in Fig. 2.14 with a reduction in the dose to 50%. When the dose is reduced even more, the net result will be approximately the compensatory response alone, as is shown in Fig. 2.17, where the dose is reduced to 10%. A further reduction in drug dose will give about the same negative effect, as the contribution of any such small dose to the total drug effect becomes negligible.

The negative reactions shown in Fig. 2.17 are not fundamentally different from withdrawal reactions in dependence. In withdrawal, however, reactions occur because environmental cues paired to the drug taking continue to trigger the compensatory mechanism after the drug is withdrawn. When an exogenous substance is taken orally and there are no environmental cues paired to the drug taking, the compensatory mechanism is not triggered when the administration of the drug is stopped and no reactions will occur, as will be discussed in Sect. 2.5.1. When the administration of the drug is continued but the dose is reduced, however, the compensatory mechanism will keep responding at the moments when the drug is administered, as illustrated in Figs. 2.13 and 2.14. When the dose is sharply reduced, yet is still

Fig. 2.18 The drug effect when a small dose is administered at an arbitrary time after the administration of a drug to which tolerance has developed is discontinued

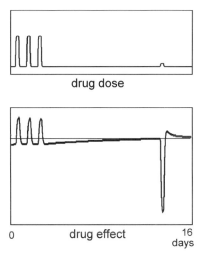

detected by the organism, it is basically not the drug which induces these reactions but the orally acquired information that the drug is present.

Not only oral administrations of small doses can evoke the responses described above, any stimulus able to trigger the compensatory mechanism can cause reactions such as those shown in Fig. 2.17. In other words, the tolerance mechanism will respond, whether it is triggered orally or by environmental cues. But environmental cues are only coupled to drugs which are used regularly whereas a small dose of any drug to which the body has a certain level of tolerance to will trigger a compensatory response. As the oral detection of exogenous substances is a highly sensitive and specialised mechanism, capable of reacting to very small doses, this phenomenon may provide an explanation of such controversial subjects as hormesis and homeopathy.

Hormesis has been defined as a bi-phasic dose–response relationship in which the response at low doses is opposite to the effect at high doses. Examples of opposite effects of drugs (and radiation) at low and high doses can be found abundantly in the literature (Calabrese and Baldwin 2001, 2003; Conolly and Lutz 2004; Ali and Rattan 2006). Hormesis is usually explained by assuming a negative part in the dose–response curve at the low dose end. Homeopathy claims a curative reaction from a small dose of a drug of which high doses cause symptoms similar to those from which the patient is suffering.

In Figs. 2.14 and 2.17, the dose was reduced abruptly. The resulting reactions, however, do not depend on a sudden change in dose but on the difference between the actual dose and the dose to which the organism has developed tolerance. Tolerance to a drug develops slowly and remains present for a long time. Figure 2.18 depicts a model simulation describing what happens when a small dose is administered at an arbitrary time after the administration of a drug to which tolerance exists is discontinued. The figure shows that the small dose evokes a reaction similar to the sudden reduction in dose simulated in Figs. 2.14 and 2.17. The drug dose in the figure of 10% is arbitrary. As the actual dose itself plays only a minor role in the remaining drug effect, any small dose will cause approximately the same reaction as long as the body recognises the drug.

Generally speaking, when there exists tolerance to a substance, the effect of a small dose is limited to triggering the compensatory response, resulting in effects opposite to the normal drug effect. Small doses of a drug apparently separate the compensatory response from the drug effect, which is a peculiar phenomenon. It does not explain the assumed curative effect of small doses in homeopathy. It does show, however, that a small dose of a substance can cause reactions with symptoms opposite to the action of the drug in high doses, a phenomenon that lies at the basis of homeopathy. The small dose mentioned above does not refer to the "infinitesimal dose" or "high potency" homeopathic medicines. On the other hand, the analysis shows that it is not the dose but information about the presence of a substance that triggers the compensatory response.

2.4.8 Sensitisation and Other Paradoxical Effects

Figure 2.13 shows that the fall in drug effect in response to a decrease in dose is followed by a rise in drug effect during subsequent drug administrations. The reduction in drug dose in this figure has been chosen to obtain a large initial reduction in drug effect. However, after tolerance has developed, any reduction in dose will be followed by a rise in drug effect until the organism has readjusted the magnitude of the compensatory response to correspond with the action of the new drug dose. This gradual increase in drug effect may explain cases of sensitisation, a phenomenon whereby the drug effect increases during repeated administrations (Robinson and Berridge 1993; Everitt and Wolf 2002). Figure 2.13 demonstrates the effect of abrupt changes in drug dose. As noted above, tolerance to a drug remains present for a long time. When a drug has not been administered over a certain period but tolerance has remained, or when innate tolerance exists, a dose smaller than the dose to which tolerance exists will result in a similar effect and may also be the origin of other paradoxical drug effects reported in the literature (Heisler and Tecott 2000; Wilens et al. 2003). It should be observed that neither sensitisation nor opposite drug effects necessarily require tolerance to the administered drug as cross tolerance to a related drug may cause similar effects.

Besides the drug dose, the magnitude of the compensatory response also depends on other variables. The capacity of the body to suppress disturbances—the open loop gain of the regulation loop (see Sect. 2.4.2)—is of major importance. The latter parameter is not fixed but depends on the subject's age, state of health and condition. The consequence is that an individual's level of tolerance to a certain drug and the resulting drug effect may appear different in different situations. This may mimic changes in drug dose with all its consequences and may be an additional cause of sensitisation. Rather than a loss of tolerance (Miller 2000) this might then constitute a loss of the organism's ability to express its tolerance.

In addition, the open loop gain may be affected by depressants and stimulants and even by the effect of the drug administration itself. Psychological factors, too, such

Fig. 2.19 Decrease in drug effect after the gain of the regulation loop is increased by 20%

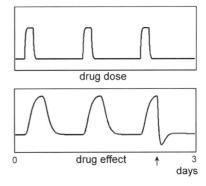

as positive reinforcers may affect the open loop gain, causing changes in the drug effect (Fillmore and Vogel-Sprott 1999; Grattan-Miscio and Vogel-Sprott 2005). Similar to small changes in drug dose, small changes in the open loop gain can have large effects. This is demonstrated in Fig. 2.19, where at the instant indicated by the arrow, the gain of the regulation loop is increased by 20%. There is an instant decrease in the drug effect and even an adverse effect temporarily appears. In the physiological regulation process, the gain is a distributed entity and the speed of change in the drug effect depends on where in the regulation loop a change in gain occurs.

2.5 Practical Significance of the Model

2.5.1 Anticipation and Dependence

When an orally administered drug is taken infrequently, the gustatory detection of the substance will be the main trigger of the compensatory response. When a drug is taken frequently over a longer period, other mechanisms will start to play a role, such as anticipation and the coupling of environmental cues to the taking of the drug. The incorporation of additional information about the drug's presence will change the nature of the mechanism. If a drug is taken infrequently, the effect of not taking the drug will be that the rebound takes its course. When the organism anticipates a drug which, however, is not administered, strong negative reactions can occur.

Figure 2.20 shows a model simulation demonstrating what happens when the administration of a drug is abruptly discontinued after tolerance has developed. When at withdrawal the triggered compensatory action of the adaptive mechanism also ends, the magnitude of the negative shift following withdrawal is comparable to the regular rebound (Fig. 2.20b). Figure 2.20c shows the effect when after withdrawal the adaptive regulator keeps responding, triggered by time factors or environmental cues associated with the administration of the drug. Now, large negative reactions occur at the moment the drug is "expected". In the model, the activation of the compensatory mechanism, independently of the drug's presence, is assumed to

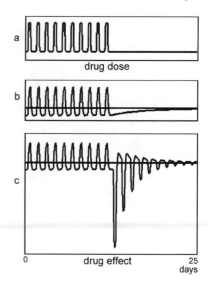

Fig. 2.20 Simulation of the effect of abrupt drug withdrawal in tolerant (**b**) and dependent (**c**) subjects. The drug is administered once a day

be the essential difference between drug tolerance and drug dependence. In reality, this difference is of course much more complex and difficult to define. Even so, in the model domain it provides fundamental insight into the mechanisms playing a role in dependence and addiction. The magnitude of the negative reactions after withdrawal—the magnitude of the compensatory response—is determined by the dose to which the subject is accustomed, the level of tolerance and the capacity of the organism to suppress disturbances to its functioning, that is, the open loop gain in the model.

Compared with the severe reactions in the model to drug withdrawal in a dependent subject, the effect in a tolerant but non-dependent subject is moderate (Fig. 2.20b). Nevertheless, its consequences can be considerable. The negative shift after the termination of drug treatment represents a worsening of the disorder in the patient (see also the note to Fig. 2.10, Sect. 2.4.1). Although this effect will diminish over time as the organism adapts to the new situation, an initial worsening of the symptoms will give the patient a strong incentive to continue drug treatment. In the figure, the reaction declines relatively fast, but the speed of decline is determined for an important part by the slow regulator which can have a long time constant so that the shift may remain for a long time after a drug is withdrawn. Moreover, in the case of a chronic disorder due to a shift in the reference level of a process regulator (Verveen 1978, 1983), it is doubtful whether adaptation to zero drug level will occur at all. A permanent shift in the reference level of a process indicates a certain malfunctioning of the regulation and a negative reaction in the process output to interruption of the stimulus represents a further shift in this reference level (Peper et al. 1987). Consequently, if a chronic disorder is due to a shift in a reference level, the extra shift after a drug treatment has ended might become permanent too and the effect of any drug treatment of limited duration will then be a permanent worsening of the disorder.

Fig. 2.21 Simulation of gradual drug withdrawal

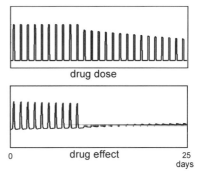

Fig. 2.22 Gradual drug withdrawal, allowing moderate reactions

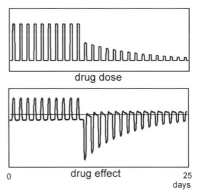

2.5.2 Alternative Protocols for Drug Withdrawal

The large reactions occurring in an addicted subject when a drug is withdrawn, simulated in Fig. 2.20c, are an expression of the high level of tolerance associated with the large dose to which the subject is accustomed. The figure shows that the reactions gradually decrease in time when the body adapts to zero drug level and tolerance to the large dose decreases.

Figure 2.21 shows a simulation of how withdrawal can be achieved in addicted subjects without negative reactions. The dose is initially decreased by 20%, which causes the drug effect to go to zero, as was shown in Fig. 2.13. (The 20% is a result of the parameter values used in the simulation. In practise, this will be different for different drugs and in different subjects.) After this step in drug dose, the dose is gradually tapered off in such a way that the drug effect is kept small. This process is very slow, much slower than when negative reactions are allowed to occur (Fig. 2.20c). The speed of withdrawal can be increased considerably when moderate negative reactions are allowed. This is depicted in Fig. 2.22, where an initial decrease in drug dose of 50% is followed by a fast decrease in the dose of succeeding drug administrations. The reactions in this approach are considerably smaller than with abrupt withdrawal, while the decrease in drug dose is much faster than

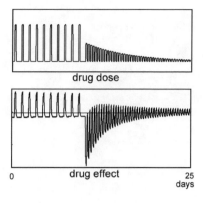

Fig. 2.23 Withdrawal with increased frequency of drug administration

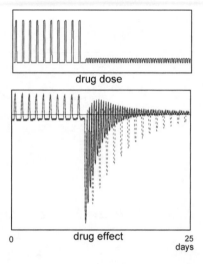

Fig. 2.24 Abrupt drug withdrawal using a small drug dose and an increased frequency of drug administration

is the case in Fig. 2.20. Nonetheless, moderate responses remain for a long time due to what is still a relatively slow decline in tolerance level. As the axes in the figures are arbitrary, the negative reactions in the figures can be interpreted more easily if their magnitude is compared with the positive drug effect in the first part of the figure.

The speed of decline in withdrawal can be increased by administering the drug more frequently. This is demonstrated in Fig. 2.23 where, instead of once a day, the drug is administered three times a day. The negative effect now declines considerably faster than in Fig. 2.22. This method of reducing tolerance can also be used when maximal reactions are allowed in withdrawal. If during drug withdrawal the drug dose is reduced to a low rather than zero value, the reactions become almost as large as in complete withdrawal, depicted in Fig. 2.20. When the small dose is now administered more frequently, the negative effect declines more rapidly. This is demonstrated in Fig. 2.24, where the drug dose is lowered to 10% of the usual dose and the frequency of administration is increased from once a day to three times a day. For comparison, abrupt drug withdrawal—as shown in Fig. 2.20—is represented with a dotted line.

In these simulations of alternative drug withdrawal, the stimulus is obtained by the oral detection of small drug doses. If the drug is not administered orally, this simple means of triggering the compensatory response is not available and other ways have to be investigated to obtain a reliable stimulus. If the drug is administered intravenously, it might be sufficient to inject a diluted sample of the drug itself. If that does not trigger the compensatory response or if the drug is administered in some other way, an unrelated stimulus may be paired with the usual drug administration in the Pavlovian manner, before withdrawal is started. Further research will have to confirm these suggestions and investigate their practical applicability.

Acknowledgements The author would like to thank C.A. Grimbergen, R. Jonges, J. Habraken and I. Jans for their critical support and valuable suggestions.

Appendix

The model is a non-linear, learning feedback system, fully satisfying control theoretical principles. It accepts any form of the stimulus—the drug intake—and describes how the physiological processes involved affect the distribution of the drug through the body and the stability of the regulation loop. The model assumes the development of tolerance to a repeatedly administered drug to be the result of a regulated adaptive process; adaptation to the effect of a drug and adaptation to the interval between drug taking are considered autonomous tolerance processes.

The mathematical model is derived in detail in Peper (2004b). In the present appendix the equations are summarised. A block diagram of the model is shown in Fig. 2.25. For the sake of brevity, the index '(t)' in time signals is omitted.

A.1 The Digestive Tract

The digestive system plays no role in the regulation loop. Drug transport through the digestive tract is modeled as a first order function:

$$S_{\text{digest}} = \int_0^t drug\, dt - \frac{1}{T_{\text{digest}}} \int_0^t S_{\text{digest}}\, dt \tag{1}$$

The input to the block is the drug administration, *drug*. The input signal is integrated to obtain the drug level when it enters the bloodstream, the output of the block S_{digest}. A fraction $1/T_{\text{digest}}$ of the output signal is subtracted from the input to account for the distribution of the drug in the digestive tract. T_{digest} is the time constant of this process.

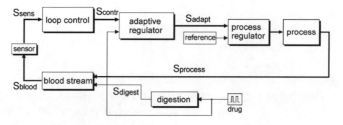

Fig. 2.25 Block diagram of the mathematical model

A.2 The Bloodstream

After digestion, the drug enters the bloodstream where it is dispersed. In the present configuration of the model, the drug and the substance produced by the process are assumed to be identical in composition and consequently add in the bloodstream. The amount of the total substance in the bloodstream will be reduced by the body's metabolism. The processes are modeled by a first order function:

$$S_{blood} = \int_0^t (S_{process} + S_{digest})\, dt - \frac{1}{T_{blood}} \int_0^t S_{blood}\, dt \qquad (2)$$

The input signals—the drug as it moves from the digestive tract into the bloodstream, S_{digest}, and the substance produced by the process, $S_{process}$—are added and integrated, yielding the output of the block, the blood drug level S_{blood}. To account for the body's metabolism, a fraction $1/T_{blood}$ of the output signal is subtracted from the input.

A.3 The Adaptive Regulator

The input signals of the adaptive regulator are the drug administration and the sensor signal, processed by the loop control block. The sensor signal provides the information about the drug effect. The output of the adaptive regulator counteracts the disturbance by lowering the process output during the drug's presence. The adaptive regulator comprises a fast and a slow regulator. The fast regulator consists of the blocks "drug regulator", "interval regulator" and "model estimation". The slow regulator suppresses the slow changes in the input signal, its output being the average of the input signal. As the fast regulator reacts to fast changes only, the output of the slow regulator is subtracted from its input. It is assumed that the body more or less separately develops tolerance to the drug's presence and to the intervals between drug administrations. The fast regulator therefore consists of a regulator which provides the adaptation to the drug's direct effect and a regulator which provides adaptation to the interval between drug taking. The output of the complete adaptive regulator is a combination of signals from its individual components.

The model assumes the body to anticipate the effect of a drug to which it has developed tolerance. This implies that the body has made an estimate of what is

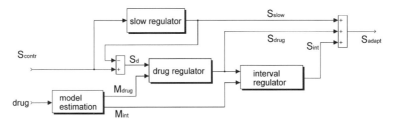

Fig. 2.26 Block diagram of the adaptive regulator

going to happen when the drug is administered: it has a model of it. The organism has also made an estimate of the magnitude of the drug effect at the given state of tolerance development. These two entities are the main factors determining the functioning of the fast regulator: the level of tolerance development and the course of the drug effect.

A.3.1 The Fast Regulator

The fast regulator consists of the blocks "drug regulator", "interval regulator" and "model estimation" (Fig. 2.26). The input signal of the drug regulator S_d is multiplied by M_{drug}, which represents the course of the drug level in the input signal over time. This signal is integrated (1/s) with a time constant T_{drug}, yielding its average. The resulting value is a slowly rising signal, L_{drug}. Multiplying L_{drug} by M_{drug} yields the output signal S_{drug}.

Because of the slow response of the circuit, changes in the input magnitude will be followed only slowly by the output. The speed of change of the output magnitude—representing the slow development of tolerance by the organism—depends on the frequency of occurrence of the drug signal and the amplification of the feedback loop: $1/T_{\text{drug}}$. The relation between the signals is

$$S_{\text{drug}} = M_{\text{drug}} \cdot \frac{1}{T_{\text{drug}}} \int_0^t (S_d - S_{\text{drug}}) \cdot M_{\text{drug}} \, dt \qquad (3)$$

and

$$S_{\text{drug}} = L_{\text{drug}} \cdot M_{\text{drug}} \qquad (4)$$

The input to the interval regulator is obtained when the output signal of the drug regulator—S_{drug}—is subtracted from its top value L_{drug}. The model of the interval is M_{int}.

The relation between the signals in the fast regulator describing the drug's presence is then

$$\begin{aligned} S_{\text{drug}} = & \ M_{\text{drug}} \cdot \frac{1}{T_{\text{drug}}} \int_0^t (S_d - S_{\text{drug}}) \cdot M_{\text{drug}} \, dt \\ & - M_{\text{drug}} \cdot \frac{1}{T_{\text{decline}}} \int_0^t \frac{S_{\text{drug}}}{M_{\text{drug}}} \, dt \end{aligned} \qquad (5)$$

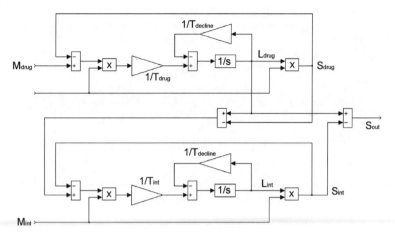

Fig. 2.27 Fast regulator implemented in Simulink

and
$$S_{\text{drug}} = L_{\text{drug}} \cdot M_{\text{drug}} \tag{6}$$

Similarly, the equation describing the interval regulator is

$$S_{\text{int}} = M_{\text{int}} \cdot \frac{1}{T_{\text{int}}} \int_0^t (L_{\text{drug}} - S_{\text{drug}} - S_{\text{int}}) \cdot M_{\text{int}} \, dt$$
$$- M_{\text{int}} \cdot \frac{1}{T_{\text{decline}}} \int_0^t \frac{S_{\text{int}}}{M_{\text{int}}} \, dt \tag{7}$$

and
$$S_{\text{int}} = L_{\text{int}} \cdot M_{\text{int}} \tag{8}$$

The output of the interval regulator is S_{int}. The output signal of the total fast regulator is obtained by subtracting the interval signal from the top level of the drug signal:

$$S_{\text{out}} = L_{\text{drug}} - S_{\text{int}} \tag{9}$$

Figure 2.27 shows the implementation of the fast regulator in the mathematical simulation program Simulink (see Peper 2004b).

A.3.2 Estimation of the Drug Effect in the Adaptive Regulator

As the duration of the drug administration is relatively short in most cases, it may be represented by a short pulse. The model of the course of the drug concentration when it enters the bloodstream—M_{drug}—is then computed by calculating the effect of a pulse with a magnitude of 1 on the digestive tract's transfer function. The input of the interval is acquired when the signal "drug" is subtracted from its top value of 1. Multiplying this signal by the transfer of the digestive tract yields the model of the interval M_{int}:

$$M_{\text{drug}} = \int_0^t drug\, dt - \frac{1}{T_{\text{digest}}} \int_0^t M_{\text{drug}}\, dt \qquad (10)$$

and

$$M_{\text{int}} = \int_0^t (1 - drug)\, dt - \frac{1}{T_{\text{digest}}} \int_0^t M_{\text{int}}\, dt \qquad (11)$$

T_{digest} is the time constant of the digestive system.

A.3.3 The Slow Regulator

The slow regulator models the long term adaptation to the drug effect. In the tolerant state, the slow adaptation causes the magnitude of the negative reaction after the drug effect to depend on the interval between drug administrations: an infrequently taken drug has a small effect during the interval, while a frequently taken drug causes a large rebound. The slow regulator counteracts the disturbance by lowering the level of the process by the average of the drug effect. Its input signal—the sensor signal, processed by the loop control block—provides the information about the drug effect. The average of the input signal is obtained by a low pass filter with a time constant T_{slow}:

$$S_{\text{slow}} = \int_0^t S_{\text{contr}}\, dt - \frac{1}{T_{\text{slow}}} \int_0^t S_{\text{slow}}\, dt \qquad (12)$$

A.4 The Process

The model does not incorporate the characteristics of the process and the process regulator. In a specific model of drug tolerance where the process is included, the effect of the process transfer on loop stability has to be controlled by the loop control block.

A.5 Loop Control

A loop control is an essential element in any regulated system. It incorporates the open loop amplification, which determines the accuracy of the regulation, and it provides the necessary conditions for stable operation of the negative feedback system. For stable operation, the regulation loop has to contain compensation for the effect of superfluous time constants: their effect on the signals in the loop has to be counteracted by circuits with an inverse effect. In the present form of the model, only the effect of the bloodstream on the regulation loop is counteracted as the transfer of the process and its regulator and the transfer function of the sensor are set at unity. The relation between the input and the output of the loop control is

$$S_{\text{sens}} = \int_0^t S_{\text{contr}} \, dt - \frac{1}{T_{\text{blood}}} \int_0^t S_{\text{sens}} \, dt \tag{13}$$

A.6 The Sensor

The sensor transforms the chemical signal S_{blood}—the blood drug level—into the signal S_{sense}. In the present model, this transformation is assumed to be linear and is set at 1. In specific models of physiological processes, this complex mechanism can be described more accurately. Stable operation then requires that the effect of its transfer on loop stability is controlled by the loop control block.

References

Ahmed SH, Kenny PJ, Koob GF, Markou A (2002) Neurobiological evidence for hedonic allostasis associated with escalating cocaine use. Nat Neurosci 5(7):625–626
Ahmed SH, Koob GF (2005) Transition to drug addiction: a negative reinforcement model based on an allostatic decrease in reward function. Psychopharmacology 180(3):473–490
Ali RE, Rattan SIS (2006) Curcumin's biphasic hormetic response on proteasome activity and heat shock protein synthesis in human keratinocytes. Ann NY Acad Sci 1067:394–399
Baker TB, Tiffany ST (1985) Morphine tolerance as habituation. Psychol Rev 92:78–108
Bell D, Griffin AWJ (eds) (1969) Modern control theory and computing. McGraw-Hill, London
Bernard C (1878) Leçons sur les Phénomènes de la Vie Communs aux Animaux et aux Végétaux. Baillière et Fils, Paris
Calabrese EJ, Baldwin LA (2001) The frequency of U-shaped dose–responses in the toxicological literature. Toxicol Sci 62:330–338
Calabrese EJ, Baldwin LA (2003) A general classification of U-shaped dose–response relationships in toxicology and their mechanistic foundations. Hum Exp Toxicol 17:353–364
Cannon WB (1929) Organization for physiological homeostasis. Physiol Rev 9:399–431
Conolly RB, Lutz WK (2004) Nonmonotonic dose–response relationships: mechanistic basis, kinetic modeling, and implications for risk assessment. Toxicol Sci 77:151–157
Deutsch R (1974) Conditioned hypoglycemia: a mechanism for saccharid-induced sensitivity to insulin in the rat. J Comp Physiol Psychol 86:350–358
Dworkin BR (1993) Learning and physiological regulation. University of Chicago Press, Chicago
Everitt BJ, Wolf ME (2002) Psychomotor stimulant addiction: a neural systems perspective. J Neurosci 22:3312–3320
Fillmore MT, Vogel-Sprott M (1999) An alcohol model of impaired inhibitory control and its treatment in humans. Exp Clin Psychopharmacol 7:49–55
Goldstein A, Goldstein DB (1968) Enzyme expansion theory of drug tolerance and physical dependence. In: Wikler A (ed) The addictive states. Research publications association for research nervous and mental disease, vol 46. Williams & Wilkins, Baltimore, p 265
Grattan-Miscio K, Vogel-Sprott M (2005) Effects of alcohol and performance incentives on immediate working memory. Psychopharmacology 81:188–196
Grill HJ, Berridge KC, Ganster DJ (1984) Oral glucose is the prime elicitor of preabsorptive insulin secretion. Am J Physiol 246:R88–R95

Heding LG, Munkgaard Rasmussen S (1975) Human C-peptide in normal and diabetic subjects. Diabetologia 11:201–206

Heisler LK, Tecott LH (2000) A paradoxical locomotor response in serotonin 5-HT2C receptor mutant mice. J Neurosci 20:RC71

Jaffe JH, Sharpless SK (1968) Pharmacological denervation super sensitivity in the central nervous system: a theory of physical dependence. In: Wikler A (ed) The addictive states. Research publications association for research nervous and mental disease, vol 46. Williams & Wilkins, Baltimore, p 226

Kalant H, LeBlanc AE, Gibbins RJ (1971) Tolerance to and dependence on, some non-opiate psychotropic drugs. Pharmacol Rev 23(3):135–191

Kandel ER (1976) Cellular basis of behaviour; an introduction to behavioral neurobiology. Freeman and Comp, San Fransisco

Koob GF, Le Moal M (2001) Drug addiction, dysregulation of reward, and allostasis. Neuropsychopharmacology 24:97–129

Koshland DE (1977) A response regulator model in a simple sensory system. Science 196:1055–1063

Loewy AD, Haxhiu MA (1993) CNS cell groups projecting to pancreatic parasympathetic preganglionic neurons. Brain Res 620:323–330

Martin WR (1968) A homeostatic and redundancy theory of tolerance to and dependence on narcotic analgesics. In: Wikler A (ed) The addictive states. Research publications association for research nervous and mental disease, vol 46. Williams & Wilkins, Baltimore, p 206

Miller CS (2000) Mechanisms of action of addictive stimuli; toxicant-induced loss of tolerance. Addiction 96(1):115–139

Mirel RD, Ginsberg-Fellner F, Horwitz DL, Rayfield EJ (1980) C-peptide reserve in insulin-dependent diabetes: Comparative responses to glucose, glucagon and tabutamide. Diabetologia 19:183–188

Mitchell D, Snellen JW, Atkins AR (1970) Thermoregulation during fever: change of set-point or change of gain. Pflügers Arch 321:393

Peper A, Grimbergen CA, Kraal JW, Engelbart JH (1987) An approach to the modelling of the tolerance mechanism in the drug effect. Part I: The drug effect as a disturbance of regulations. J Theor Biol 127:413–426

Peper A, Grimbergen CA, Kraal JW, Engelbart JH (1988) An approach to the modelling of the tolerance mechanism in the drug effect. Part II: On the implications of compensatory regulations. J Theor Biol 132:29–41

Peper A, Grimbergen CA et al (1998) A mathematical model of the hsp70 regulation in the cell. Int J Hypertherm 14(1):97–124

Peper A, Grimbergen CA (1999) Preliminary results of simulations with an improved mathematical model of drug tolerance. J Theor Biol 199:119–123

Peper A (2004a) A theory of drug tolerance and dependence I: A conceptual analysis. J Theor Biol 229:477–490

Peper A (2004b) A theory of drug tolerance and dependence II: The mathematical model. J Theor Biol 229:491–500

Peper A (2009a) Aspects of the relationship between drug dose and drug effect. Dose Response 7(2):172–192

Peper A (2009b) Intermittent adaptation. A theory of drug tolerance, dependence and addiction. Pharmacopsychiatry 42(Suppl 1):S129–S143

Poulos CX, Cappell H (1991) Homeostatic theory of drug tolerance: a general model of physiological adaptation. Psychol Rev 98:390–408

Ramsay DS, Woods SC (1997) Biological consequences of drug administration: implications for acute and chronic tolerance. Psychol Rev 104:170–193

Rescorla RA, Wagner AR (1972) A theory of Pavlovian conditioning: variations in the effectiveness of reinforcement and non-reinforcement. In: Black AH, Prokasy WF (eds) Classical conditioning II: Current research and theory. Appleton-Century-Crofts, New York, pp 64–99

Robinson TE, Berridge KC (1993) The neural basis of drug craving: an incentive-sensitisation theory of addiction. Brains Res Rev 18:247–291

Schulkin J (2003) Rethinking homeostasis: allostatic regulation in physiology and pathophysiology. MIT Press, Cambridge
Siegel S (1975) Evidence from rats that morphine tolerance is a learned response. J Comp Physiol Psychol 89:498–506
Siegel S (1996) Learning and homeostasis. Integr Physiol Behav Sci 31(2):189
Siegel S (1999) Drug anticipation and drug addiction. The 1998 H. David Archibald lecture. Addiction 94(8):1113–1124
Siegel S, Allan LG (1998) Learning and homeostasis: drug addiction and the McCollough effect. Psychol Bull 124(2):230–239
Siegel S, Hinson RE, Krank MD, McCully J (1982) Heroin "overdose" death: contribution of drug-associated environmental cues. Science 216:436–437
Snyder SH (1977) Opiate receptors and internal opiates. Sci Am 236:44–56
Solomon RL, Corbit JD (1973) An opponent-process theory of motivation II: Cigarette addiction. J Abnorm Psychol 81:158–171
Solomon RL, Corbit JD (1974) An opponent-process theory of motivation. I: Temporal dynamics of affect. Psychol Rev 81:119–145
Solomon RL (1977) An opponent-process theory of acquired motivation: the affective dynamics of addiction. In: Maser JD, Seligman MEP (eds) Psychopathology: experimental models. Freeman, San Francisco, pp 66–103
Solomon RL (1980) The opponent-process theory of acquired motivation: the costs of pleasure and the benefits of pain. Am Psychol 35:691–712
Steffens AB (1976) Influence of the oral cavity on insulin release in the rat. Am J Physiol 230:1411–1415
Sterling P, Eyer J (1988) Allostasis: a new paradigm to explain arousal pathology. In: Fisher S, Reason J (eds) Handbook of life stress, cognition and health. Wiley, New York, pp 629–649
Sterling P (2004) Principles of allostasis: optimal design, predictive regulation, pathophysiology and rational therapeutics. In: Schulkin J (ed) Allostasis homeostasis and the costs of adaptation. Cambridge University Press, Cambridge
Thorpe WH (1956) Learning and instinct in animals. Methuen and Co, London
Tiffany ST, Baker TB (1981) Morphine tolerance in rats: Congruence with a Pavlovian paradigm. J Comp Physiol Psychol 95:747–762
Tiffany ST, Maude-Griffin PM (1988) Tolerance to morphine in the rat: associative and nonassociative effects. Behav Neurosci 102(4):534–543
Toates FM (1979) Homeostasis and drinking. Behav Brain Sci 2:95–136
Verveen AA (1978) Silent endocrine tumors. A steady-state analysis of the effects of changes in cell number for biological feedback systems. Biol Cybern 31:49
Verveen AA (1983) Theory of diseases of steady-state proportional control systems. Biol Cybern 47:25
von Bertalanffi L (1949) Zu einer allgemeinen Systemlehre. Biol Gen 195:114–129
von Bertalanffi L (1950) An outline of general systems theory. Br J Philos Sci 1:139–164
Wagner AR (1978) Expectancies and the primary of STM. In: Hulse S, Fowler H, Honig WK (eds) Cognitive processes in animal behaviour. Erlbaum, Hillsdale, pp 177–209
Wagner AR (1981) SOP: a model of automatic memory processing in animal behaviour. In: Spear NE, Miller RR (eds) Habituation: perspectives from child development, animal behaviour and neurophysiology. Erlbaum, Hillsdale
Wiegant FAC, Spieker N, van Wijk R (1998) Stressor-specific enhancement of hsp induction by low doses of stressors in conditions of self- and cross-sensitisation. Toxicology 127(1–3):107–119
Wiener N (1948) Cybernetics: or control and communication in the animal and the machine. Wiley, New York
Wilens TE, Biederman J, Kwon A, Chase R et al (2003) A systematic chart review of the nature of psychiatric adverse events in children and adolescents treated with selective serotonin reuptake inhibitors. J Child Adolesc Psychopharmacol 13(2):143–152

Chapter 3
Control Theory and Addictive Behavior

David B. Newlin, Phillip A. Regalia, Thomas I. Seidman, and Georgiy Bobashev

Abstract Control theory provides a powerful conceptual framework and mathematical armamentarium for modeling addictive behavior. It is particularly appropriate for repetitive, rhythmic behavior that occurs over time, such as drug use. We reframe seven selected theories of addictive behavior in control theoretic terms (heroin addiction model, opponent process theory, respondent conditioning, evolutionary theory, instrumental conditioning, incentive sensitization, and autoshaping) and provide examples of quantitative simulations for two of these models (opponent process theory and instrumental conditioning). This paper discusses theories of addiction to lay the foundation for control theoretic analyses of drug addiction phenomena, but does not review the empirical evidence for or against any particular model. These seven addiction models are then discussed in relation to the addictive phenomena for which they attempt to account and specific aspects of their feedback systems.

3.1 Control Theory and Addiction

Addictive behavior occurs rhythmically over time in what is often a roughly cyclic process (Bobashev et al. 2007). This form of repetitive behavior is frequently the signature of a controlled, regulated system with delayed feedback (Ahmed et al. 2007). For example, a thermostatically controlled environment with a heating system produces such a time series in its output—in this case, rhythmic oscillations in room temperature. There is an inherent delay between thermostatic actuation and the delivery of heat to the room. We explore the application of control theory (see Carver and Scheier 1990; Johnson et al. 2006; Warren 2006) to addictive behavior on an exclusively theoretical level. In this discussion, explicit control theoretic models of seven psychobiological theories of addictive behavior are developed. These include a heroin addiction model, opponent process theory, respondent conditioning, evolutionary psychological theory, instrumental conditioning, incentive-sensitization theory, and autoshaping. These theories were chosen to represent a broad range of models of addiction, from traditional learning theory to evolutionary theory, but they are far from exhaustive.

Control theory has been considered under a variety of labels, including (but not limited to) set point theory, feedback models, cybernetics, regulator theory, and systems theory. We use the term "control theory," noting the vast mathematical and engineering literature in which these ideas have been developed more fully.

We argue that formal control theory increases understanding of how these theories model the disparate behavioral, neurochemical and pharmacodynamic aspects of drug use.[1]

Control theory is particularly valuable because it facilitates comparison of the structural similarities and differences among competing theories. We provide control theoretic diagrams of each model; control theorists can easily convert these diagrammatic representations into mathematical formulations. We chose this intermediate step (diagrams) for all but two of these models because they are both conceptually intuitive and lead rather directly to control theoretic implementations. Simulations of the outputs of two of these controlled systems, opponent process theory (or respondent drug conditioning, which is structurally similar) and incentive-sensitization theory, are illustrated with tracings that represent different values for the parameters of the models.

The significance of control theory in this discussion is as a "way of thinking" about addiction issues and a collection of examples exhibiting diverse behavior. In fact, most control theory is purposeful in the sense that people specifically design systems using control theoretic principles to solve concrete problems. On the other hand, we can consider these mechanisms (and the underlying physiology) as evolving over time with the usual teleology defined by natural selection processes. More generally, one has dynamical system models (which also serve as a "way of thinking" and collection of examples). It might only be the evolutionary (selection) of ideas which would emphasize optimization and stabilization.

---

[1] We do not model pharmacokinetic functions in this discussion, although we recognize that physiologically based pharmacokinetic modeling (e.g., Umulis et al. 2005) is needed to complete the modeling process. For simplicity, drug metabolism and excretion is modeled here simply as "time elapsed since drug administration."

3.1.1 Scope

This discussion is an initial application of control theory to addictive behavior in ways that promote its understanding. It is a selective review of *theories* of addiction rather than the empirical results that may be used to verify or reject these theories. At the present time, we are modeling theories rather than empirical data. The adequacy of these models in terms of their congruence with experimental results is far beyond the scope of this discussion, and is beyond that of any single paper. For example, Petraitis et al. (1995) reviewed theories of adolescent substance abuse in terms of their congruence with each other and their theoretical differences, but did not review empirical evidence related to these theories. This discussion lays the foundation for future research in which control theoretic implementations may be key to modeling empirical data and, in turn, the degree to which these data conform to theoretical prediction.

Our decision to avoid discussion of the vast empirical literatures relevant to these theories should *not* be taken as promotion or acceptance of these models. Far from it. As we shall see, many models are mutually exclusive, particularly as they collide with each other in terms of empirical results of real-world experimentation. Our discussion may guide control theoretic research on these theories that is data-driven, and may highlight ways in which models are similar or different, but does not attempt to evaluate these theories in terms of their congruence with existing empirical data.

In our discussion, we emphasize that drug self-administration is neither necessary nor sufficient for drug addiction, depending of course on one's definition of "addiction" in human and nonhuman animals. Therefore, a model of drug self-administration, though highly relevant to addictive behavior, is not by itself a theory of addiction. For example, Deroche-Gamonet et al. (2004) tackled this issue in studies of rats that were trained to self-administer cocaine. They noted that the majority of human drug users and drug self-administering rodents are *not* addicted to the drug. Deroche-Gamonet et al. (2004) adopted human-like criteria for addiction in their study of rats: (1) difficulty stopping or limiting drug self-administration, (2) extreme motivation to use the drug, and (3) continued use despite harmful consequences. Note that the classic criteria of drug tolerance and withdrawal effects upon cessation were not included in this definition of addiction, and that it would be possible to have an addicted rat that had prolonged passive administration of the drug rather than self-administering it—that is, drug self-administration may be neither necessary nor sufficient for addiction.

We also seek to present these theories in a manner that is helpful for comparing and contrasting their formal characteristics. Some problems with individual theories become evident when they are formalized in terms of their structural components. Control theory may also lead ultimately to prediction of phenomena that are difficult to envision or model without explicit, formal control theory modeling.

Therefore, the purposes and goals of this discussion are to (1) present capsule descriptions of various homeostatic and nonhomeostatic models of addictive behavior, (2) reframe them in control theoretic terms, (3) implement two of the models so that simulations can be constructed, and (4) compare and contrast these theories in relation to control theoretic analyses.

3.2 Key Concepts

3.2.1 Feedback

Feedback refers to modifying the driving forces of a dynamical system based on measured responses, in order to manipulate the dynamic behavior of the system. Familiar examples are found in economics (e.g., adjusting the prime rate according to economic indicators), navigation (firing rocket thrusters to correct for trajectory variations of a spacecraft), temperature control (turning a heater on when the temperature drops below a critical value, and turning if off when the desired temperature is reached) and even gaming (adjusting one's wager based on perceptions of another player's hand). In these examples, "feedback" represents a corrective action that is a function of the measured or observed behavior of the system. A "dynamic" system is simply one that changes over time, such as drug self-administration, and a "system" is a set of interacting variables that, in the case of feedback systems, are interconnected and guided by feedback from the system's measured output. If feedback functions to regulate the system, as in the examples above, then control theory is a useful conceptual and mathematical approach for describing, modeling, and controlling that dynamic system.

Feedback describes the situation when the signal from a past event influences the same event in the present or future. When this past event is part of an interlocking network of causes and effects that forms a system, then the system represent a feedback loop. In mathematical terms, with feedback the input signals or parameters of the system are changing (adapting) in relation to the observed or perceived response.

The following is a simple mathematical example. A more detailed mathematical treatment is in the Simulations section.

Let response x at time $t + \Delta t$ be positively and linearly dependent on some input $y(t)$ (i.e., the larger the $y(t)$ the larger becomes the value of x after an interval Δt).

$$x(t + \Delta t) = ay(t) + b, \quad \text{where } a \text{ and } b \text{ are positive parameters.} \quad (3.1)$$

Linear feedback will occur if $y(t)$ is itself linearly dependent on the output $x(t)$ (e.g., $y(t)$ can have a negative relationship with $x(t)$):

$$y(t) = -cx(t) + d, \quad \text{where } c \text{ and } d \text{ are positive parameters.} \quad (3.2)$$

Evaluating Eq. (3.1) in relation to (3.2) yields an iterative equation for $x(t + \Delta t)$, that is,

$$x(t + \Delta t) = -acx(t) + ad + b. \quad (3.3)$$

This equation leads to a steady state solution when $x(t + \Delta t) = x(t) = (ad + b)/(1 + ac)$. This solution is stable when the product ac is positive.

Feedback is ubiquitous in nature and in human behavior. It is generally of two types: enhancing the behavior of the system (positive feedback) and attenuating it (negative feedback). Positive feedback tends to support the beginning of a process or to prevent the process from becoming extinct. For example, the use of a food treat in animal training is a typical example of positive feedback. As the animal exhibits more of the targeted behavior, the more treats it receives (i.e., instrumental conditioning). A problem with positive feedback is that if this process were not countered by resource limitation or negative feedback, the process would build exponentially into the future. For example, if there were no negative feedback (e.g., satiety) built into the organismic system, then humans with unlimited access to food would eat continuously until they died. In other words, the system is unstable and explosive.

Negative feedback in physiological processes gives rise to a variety of stable dynamical processes. In a simple case, a combination of positive or negative feedback can lead to a steady state. For example, if the goal of reaching a destination in the shortest time possible leads us to drive as fast as possible, then safety and speed limit considerations (negative feedback) settle the driver to a stable compromise. If the responses of the positive and negative feedback are slow or delayed, the process could exhibit periodic or semiperiodic behavior (i.e., biological cycles). An example of these cycles is feeding in humans (e.g., breakfast, lunch, and dinner) that are semi-periodic responses to positive (food reward) and negative (satiety) feedback, among other limiting and entraining factors. In nature, the dynamic behavior of physiological systems with feedback have tended to evolve such that they are relatively stable and lead either to homeostasis or semi-periodicity, often with slow secular trends such as maturation or—in our case—the development of addiction.

3.2.2 Feedback Systems Gone Awry

The guiding principle behind this work is the idea that normally adaptive feedback systems in the brain become ensnared by highly potent drugs that were not encountered in our deep ancestral past, or at least not with the potency and abundance they now have (e.g., Nesse and Berridge 1997; Redish et al. 2008). This view is made explicit in the evolutionary psychology model of addiction (see SPFit theory, below), but applies as well to all seven of the theories presented herein. By "normally adaptive" we mean these feedback systems are fundamental to mammalian functioning, such as those that support respondent or instrumental conditioning. They are adaptive in the sense that they promote survival and reproduction more often than not. This view speaks to the relevance and importance of control theory modeling for behavioral science in general. The ways in which these feedback systems go awry with drugs of abuse may be very informative as to their operating characteristics. Moreover, we hope that better understanding the fundamental organization and modes of operation of these feedback systems in addiction will be useful ultimately in preventing and treating drug abuse and dependence.

Fig. 3.1 Quasi-linear operating range of some common functions

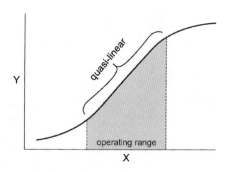

3.2.3 Linear and Non-linear Dynamical Systems

A linear dynamical model (Chen 1999; Kailath 1980) is one in which the relations between input and output are linear (i.e., change in the output is uniformly proportional to change in the input). Uniformity means that the proportionality coefficient is the same for the entire range of inputs and outputs. Graphically, this relationship is expressed as a straight line. However, many natural responses show a more complex relationship between input and output. For example, the pharmacodynamic dose-response relationship for many drugs is disproportionately small at very low doses, and either reaches saturation or there is some catastrophic change in state (e.g., toxicity or mortality) at very high doses. This common sigmoidal dose-response function often affords a quasi-linear range through the moderate dosages in which the output is approximately proportional to the input. These relationships between input X (e.g., dose) and output Y (e.g., pharmacodynamic response) are illustrated in Fig. 3.1. There are an infinite number of nonlinear response shapes. Some could be linearized (i.e., approximated with a straight line), others described explicitly by a smooth curve, others as a collection of linear relationships (piece-wise linear), but some can include disconnected "jumps" and "brakes." The exact formalism for dealing with these functional forms is beyond the scope of the paper. We emphasize that while most processes are nonlinear by nature, the analysis of simpler, linearization models can be very useful to capture the essence of the system's regulatory mechanisms.

This discussion generally concerns single variables in the control theoretic models. In the long term, vectors of variables rather than single variables will be the focus of inquiry as the ability to measure from multiple sensors becomes more commonplace, and the theoretical models become multicomponent, just as in other control theoretic applications (such as electronics or physical systems). This generalization will not be difficult from a control theoretic perspective since the procedures for modeling multicomponent systems have been developed fully.

3.2.4 Feedback Delay

Most of the addiction models we present have variable delays in their feedback loops. These delays can be due to a host of factors specific to the drug, the person, constraints of the environment, or the interactions among these. One result of these variable delays is that there can be positive and negative swings in the response of the system that the individual may experience as euphoria, craving, acute withdrawal, etc., depending on the model. This is true even with homeostatic models that "seek" a neutral, quiescent state. A second result, related to the first, is that these systems may exhibit semi-periodic oscillatory behavior, particularly if the rate of drug self-administration is relatively constant or if the feedback delay is similar each time that the drug is taken. We noted this system behavior in the example of the temperature-controlled room. This rhythmic, oscillatory behavior would appear to be characteristic of addiction as well.

3.2.5 Control Theoretic Diagrams

One may be tempted to view a formal control theory diagram as a sequence of operations (i.e., from "*a*" to "*b*" to "*c*" to "*d*" as a function of time). Although this sequential functioning is possible, in most controlled systems to understand and analyze the system one must view many operations as occurring simultaneously in time and fully interdependently. In other words, it functions as a *system* rather than simply the amalgamation of effects of "*a*" on "*b*" and "*b*" on "*c*", etc. This is a major shift in thinking away from reductionist science in which one deconstructs systems to study isolated variables, such as the effect of "*a*" on "*b*" while controlling "*c*."

3.3 A Brief Introduction to Control Theory

We are concerned here with a branch of control theory called "cybernetics" or "regulator theory"—typically stabilization in the neighborhood of a set point (desired state). This has two related aspects: design and analysis. The mathematical theory, going back to Airy (1840) and Maxwell (1868), has shown what phenomena can occur in various dynamic contexts and can therefore improve design characteristics. Our present use of regulator theory is intended to be intuitive, in the sense of reverse engineering (see below) mechanisms producing phenomena similar to those observed in addiction. This modeling can be relevant ultimately to causal understanding and possible intervention in drug addiction. In evolutionary terms, it is meaningless to speak of "optimal control," which is often the goal of control theoretic analyses, but we occasionally refer to a set point as "optimal" by analogy with control design methodology.

Wiener (1948) introduced the term "cybernetics" based on the metaphor of steering a ship: the pilot (steersman; from the classical Greek word, *kybernetes*) observes the current state of the ship's motion and, on noticing any deviation from the desired course, adjusts the steering oar (control mechanism) to make the ship move back to course. This sequence:

$$\text{state} \rightarrow \text{observation} \rightarrow \text{change of control} \rightarrow \text{correction of state}$$

represents a "feedback loop." Since it is necessary for stabilization that the resultant state correction be opposite directionally to the initiating deviation, this is an example of "negative feedback." The action of a competent pilot is then a paradigm of successful homeostatic feedback, smoothly maintaining a stable course despite environmental perturbations. Even in this simple setting we recognize the oversimplification in omitting consideration of such factors as rower fatigue, wind and currents, and obstacles to be avoided, which would intervene in a more complete treatment. This emphasizes, to paraphrase Hamming (1962), that "The purpose of modeling is insight, not formulas," or, as Albert Einstein is thought to have said, "Our theories should be as simple as possible, but no simpler."

Another familiar example is riding a bicycle. For a competent bicyclist, maintaining balance is unconscious; one reflexively shifts one's weight and steering proportionally to compensate for almost imperceptible changes in the bicycle's balance so the underlying feedback process is effectively invisible to the observer. This process is, however, much more noticeable for the beginner, who must provide the feedback deliberately; rather than inducing a rapid smooth response, a shift in the bicycle's balance has time to grow before reaching a perceptual threshold and only after this delay producing the response. There are several important inferences that can be drawn from this example:

- The simplest design for feedback mechanisms is to use control changes (approximately) proportional to deviation in the current observation, but directed oppositely in effect. Violent responses to perceived deviations (i.e., "overcorrection") can lead to extreme oscillatory behavior.
- Even when a feedback control mechanism operates appropriately, delay in its implementation can cause oscillations. This may be actually an objective, such as in a drug abuser who desires "overshoot" in their homeostatic system so that they experience drug-induced effects (e.g., "euphoria") before compensatory (negative feedback) processes attenuate or negate the drug effect.
- Alternatively, oscillations might signal overcompensation, amplifying the deviation each time in the opposite direction (unless constrained by training wheels in the bicycling example), so these corrections effectively become a kind of positive feedback, leading to catastrophic failure.
- The most important inference, perhaps, is that a control mechanism is often observed most clearly when it is not functioning properly, such as in drug addiction.

This bicycling model is illustrated in Fig. 3.2. The goal of the bicyclist is to follow a certain path, and deviations from that goal lead to corrective negative feedback. These deviations from the desired direction are sensed by a sensori-motor executive within the rider that is informed by an internal model of bicycle controls and control actions, an internal model developed as the bicyclist first learned to ride.

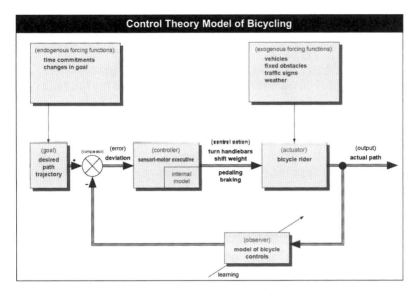

Fig. 3.2 Control theoretic example of bicycling as a negative feedback system

In this graphic (Fig. 3.2), the negative feedback arm is controlled by an "observer" component that has an internal model of bicycling. The "observer" is a control theoretic construct invoked when one component has information about the inner workings of other components in the control system. In this sense, the observer is an intelligent component that can correct error based on observation of other structural components. This does not imply that the intelligent observer consciously controls the bicycle. In fact, the bicycle rider may fall if he or she attempts to consciously control the system through verbal thought and planned adjustments to remain vertical, just as inexperienced riders often fall before their internal model of bicycling has developed fully.

Once learned, the experienced bicyclist is unlikely to forget how to ride, and will relearn efficient bicycling very quickly even after years of disuse. This is analogous to the addicted individual who has been abstinent for many years, but who returns to an addicted state very rapidly after reinitiating drug use.

Another familiar example of control feedback is the use of a thermostat. The temperature measured at a single point in space or time determines whether the furnace is turned on or off. If the furnace is on, then the resultant heat reduces error and the thermostat then shuts off the furnace. In this case, the change in control action is binary (on or off) and does not depend on the amount of deviation from the threshold setting; therefore, this is not a linear dynamical system because the feedback is not proportional to the input. Several additional inferences can be drawn:

- While many feedback mechanisms have control changes (approximately) proportional to the current deviation in the observation, there are other possibilities (e.g., hybrid control [involving thresholds, Goebel et al. 2009], nonlinear dependence, or delayed action).

- The observation used in the feedback loop is not necessarily the state of interest. In this case, the temperature at a single point in space is used as surrogate for the entire distribution of temperatures in the room or building.

In this example of temperature regulation, we do not need to "reverse engineer" (see below) the system because we already know precisely how a thermostat works. The negative feedback may be (strikingly) compared to a similar room in which the thermostat was miswired such that it actuated the furnace when the temperature reached *above* (rather than below) the set point. This would be a positive feedback system in which feedback increased error rather than decreased it. The result would be runaway heating, and a very hot room! We raise this issue because some of the addiction models that are discussed (below) employ positive feedback, with or without parallel negative feedback to limit the inherently runaway escalation of positive feedback systems.

To pursue this example further, suppose that a small fire were introduced into a temperature-controlled room in which the heating-cooling systems were working properly. One first might predict that the additional heat would lead to an increase in the ambient temperature of the room, and indeed, this would be the case if the heating/cooling system were turned off. However, because temperature is the controlled variable in this example, we would expect little or no change in ambient temperature when the additional heating source is introduced, particularly if the thermostatic control is regulated tightly. Note too that there would likely be changes in other variables, such as the humidity of the air or the particulate levels in the room, because these variables are not part of the control system.

A final example is the delivery of anesthesia during surgery. The goal of anesthesia is to reduce pain and motor movement, but the model of state evolution (determining the anesthesia level in the body) must include such considerations as drug metabolism, localization, and heart and lung function. However, the sensor data used by the anesthetist to control the depth of anesthesia—defining the negative feedback loop—are typically just blood pressure and oxygenation. These variables serve as surrogates for the actual internal values of a full state description.

The dynamical systems (and consequent control theoretic models) for biological settings are typically complicated, with many interacting variables. We will use the simple single-variable examples to develop intuitive understanding of these systems, but recognize that the actual complexity of these systems requires deeper mathematical tools of multivariate analysis to allow full treatment of these problems.[2]

---

[2] From Wiener's (1948) introduction to *Cybernetics*: "The mathematician need not have the skill to conduct a physiological experiment, but he must have the skill to understand one, to criticize one, and to suggest one. The physiologist need not be able to prove a certain mathematical theorem, but he must be able to grasp its physiological significance and to tell the mathematician for what he should look."

3.3.1 Applying Control Theory

Homeostatic theories are modeled well in control theoretic terms. Control theory describes the behavior of processes in interconnected systems that exert mutual influences over each other, especially closed-loop feedback systems. Mathematical laws may be invoked to deduce equilibrium or set points, if any exist. The set point of a system, typically its input or "goal-state," is a level about which the output tends to vary or oscillate. The system can be described mathematically in terms of its external influences and the internal descriptions of how individual components in the system behave and interact.

In this case, much is known about the specific influences of individual components (such as the effects of drug-related stimuli) on other components in the system (such as craving or drug self-administration). These relations are studied typically as isolated influences (i.e., the effect of one variable on another). At the same time, very little is known about the mathematical properties of these models as closed-loop feedback systems in freely behaving animals or people.

3.3.2 Modern History

Modern mathematical descriptions of feedback systems are usually traced to Nyquist (1932), Lurie and Enright (2000), whose feedback stability criterion was critical to the design of vacuum tube amplifiers in the early days of telephony. The space age ushered in the need for more advanced internal models of feedback systems and stability criteria, with major progress in this direction usually credited to Kalman (1963) who favored an internal or "state-space" viewpoint. This refined approach culminated in the mature field of optimal filtering in which feedback laws are designed to minimize estimation errors between an observed phenomenon and its predicted behavior based on a self-adjusting mathematical model (Anderson and Moore 1979). Subsequent trends incorporate the effects of uncertainty in modeling and measurements (Doyle et al. 1989; Francis 1988), and the influence of nonlinear phenomena (Isidori 1989).

Control theory carries with it a powerful armamentarium of conceptual, analytical, mathematical, and simulation tools for studying regulated systems. This is because it has been fully developed over many decades, in part due to its broad applicability to a wide range of engineering and other problems. There is considerable potential for developing control theoretic analyses of addictive behavior and its neurobiology in future research (e.g., DePetrillo et al. 1999; Ehlers 1995; Ehlers et al. 1998; Hufford et al. 2003). Modern control theory emphasizes settings with multicomponent sensors and responses and even distributed parameter models; we simplify here by avoiding such complications for this stage of the development of control theoretic modeling in addictions research.

Control systems generally exhibit a wide variety of solutions that can be classified in several types. Usually the solutions are designed by the control theorist to be bounded (i.e., occurring within certain predefined limits). Unbounded solutions could lead to system destruction (e.g., a thermostat with only positive feedback would not result in infinitely high room temperature but rather would self-destruct). Common bounded solutions are steady states, periodic behavior, and chaos regimes, each of which could be perturbed by stochastic noise. Although stochastic and chaotic solutions often look similarly "random" there is a major distinction between them. A chaotic solution, because of the underlying determinism, allows short term predictions but does not usually permit long term predictions because small initial perturbations can lead to large divergences in the future. Nonlinearity is necessary but not sufficient for chaotic systems. Conversely, stochastic systems explicitly assume the presence of noise and thus precise short term predictions are not possible, although long term average trends can often be predictable. Stochasticity can be present in any system, linear or nonlinear. Noise simply represents the impact of numerous factors for which the system does not account explicitly.

This discussion focuses on more conventional control systems in which the input and output variables and feedback are continuous and there are no thresholds. Hybrid control systems (see Goebel et al. 2009) represent a second class of control theoretic models in which there are both continuous and binary (nonlinear) variables. Hybrid systems generally have a threshold for control activation (such as a minimum sensed temperature in a thermostat) and the control action is binary (either "on" or "off," like a furnace in a heating system). None of the addiction models discussed here was conceived and constructed in a manner akin to hybrid control systems,[3] although future modeling may lead to hybrid model applications.

3.3.3 Reverse Engineering

In control theoretic terms, inferring models of addiction from measured physiology and behavior represents a classic problem in reverse engineering (Csete and Doyle 2002; Eisenberg 2007). We seek to proceed from the observable outputs of a controlled system to the psychobiological processes that govern them. In a sense, this is the work of much of integrative neuroscience (Aarmodt 2006). It may be contrasted with the typical situation in engineering in which one designs and builds a mechanical or electronic device with known governing principles in order to achieve specific outputs. In the case of drug addiction, we have a complex biological machine that is malfunctioning (at least in relation to societal values), and we attempt to determine what characteristics of the controlling system might account for this problematic behavior. Control theory provides an extensive body of knowledge about regulated systems that may prove useful in describing the systems involved. If the control

---

[3] If self-administration were at a constant dose of the drug, then one might construe several addiction models herein as hybrid control models because the actual self-administration appeared binary (either take the drug or not). However, using a longer time frame, such as weeks or months, the *rate* and *dose* can be viewed as continuous measures, making hybrid modeling unnecessary.

system were well understood, it would likely have important implications for preventing and treating drug addiction.

In this discussion, we focus on the "goal state" of the system as the critical input, although we note other inputs from exogenous (e.g., drug-associated stimuli) and endogenous (e.g., time since last use of the drug) influences. The term "goal state" is used here to refer to a characteristic of the system, determined in many cases by currently unknown or controversial neurobiological mechanisms, rather than as a design characteristic. We suggest that evolutionary selective forces have shaped these neurophysiological systems in a manner that has led to systems that appear to "seek" certain states, such as a neutral affective state (e.g., Solomon 1980). This clearly does not imply active design as in other applications in which set points and other characteristics are deliberately engineered.

The "state" of the system may be defined as the set of descriptive characteristics at a certain time point that allows one to predict the future behavior of the system (aside from varying external influences). The system output in these examples is the measured response.

3.4 Characteristics of Control Theoretic Systems

Control theory concerns the relationships among the inputs and responses in a dynamical system. The term "control" is key and implies four components: (1) a desired path or goal, (2) observed system output, (3) defined error which characterizes the difference between the desired path and the system output, and (4) feedback that changes the system input in order to correct error. Feedback modeling distinguishes control theory from other disciplines that study dynamical systems. Control theory solves "direct" or "inverse" problems. In the direct problem, the intent is to predict the response to the input signal given a particular system. In solving an inverse problem, control theory aims to define or construct a system that would behave as prescribed or observed. In the current discussion, we seek to lay the foundation for using control theoretic models to reverse engineer addiction phenomena.

Control theory could be applied to "systems theory" with which it shares many components: specialization (dividing the system into components), grouping (providing the optimal level of detail for the purpose), synthesis (identifying interactions between the components), and aggregate behavior (which might evolve in time) of multi-element systems, such as a flock of geese flying in "V" formation. Control theory originated in mechanistic representations of natural processes and in engineering, and thus offers a well developed set of mathematical and conceptual tools that can be applied to the analysis of biological systems. For some applications, like physiological modeling, the use of control systems is inherently applicable because many processes in nature have evolved in a manner that maintains stability and sustainability. Other applications, such as drug addiction, involve a wide range of psychobiological processes spanning physiology, neurobiology, social interactions, and interactions with the environment. This breadth of processes makes the rigorous application of control theory tools more challenging. In this discussion, we address this challenge by applying control theory to demonstrate that, although there are different and competing theories of addiction, there is a possible unifying approach to describe them with the same classes of feedback models.

The simplest scientific models have the form of cause and effect, but the world is often more complicated than this and more sophisticated models are required. These models take the form of associations of variables without a clear cut distinction between "independent" and "dependent" variables. We noted above that much of psychopharmacology and neurobiology involves studying the effect of one variable on another. In testing hypotheses researchers usually seek to determine the causal effect of the independent on the dependent variable. However, when feedback loops control the process, causal relationships are in both directions and thus it is more accurate to describe this as an *association* of the independent and dependent variables. An association between variables X and Y could be the result of a variety of different combinations of causal relationships of X on Y and Y on X. Control theory methods describe these relationships in a unified way that considers both direct and indirect causal and dynamic pathways.

Very misleading interpretations can occur if the researcher is dealing with a controlled system. A probe that has dramatic effects *in vitro* can have no apparent effect *in vivo*. In the latter case, there may be intact feedback loops that drastically limit the effects of the probe. The same probe may lead to opposite effects *in vivo* if it boosts the negative feedback arm of the controlled system. Or, as in the temperature-controlled room example above, a probe may alter only variables other than the target variable because the latter is under regulatory control. Enhanced understanding of misleading results such as these is one of the values of integrative neuroscience.

3.5 Homeostatic Theories of Addiction

Many theories of drug addiction are explicitly (Koob and Le Moal 1997; Poulos and Cappell 1991) or implicitly homeostatic. Homeostasis (Cannon 1929) is a characteristic of biological organisms in which feedback systems maintain a relatively constant internal milieu despite external and internal perturbations that can be intense. This means they assume a process by which the organism (whether human or nonhuman animal) "seeks" a "normal" homeostatic balance. Deviations away from this quiescent state are sensed as requiring corrective action, including taking a drug of abuse. Such feedback is relatively ubiquitous in biological organisms. Nonhomeostatic systems do not have this self-righting characteristic, although they may have feedback systems in place.

3.5.1 Heroin Addiction Model

The classic homeostatic model of heroin addiction was developed by Himmelsbach (1943). Among individuals who were addicted to heroin, he found that ceasing to

Fig. 3.3 Himmelsbach's (1943) negative feedback model of heroin addiction. Note the integration of highly disparate elements in one system, a hallmark of control theory implementations

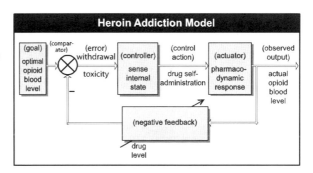

use the drug led to a well-defined heroin withdrawal syndrome consisting of dysphoria, nausea, tearing, yawning, diarrhea, sweating, etc. Implicit in his model was that addicts required heroin to feel "normal" or in homeostatic balance. Individuals who were addicted to heroin would therefore take the drug to stave off withdrawal sickness or to alleviate it once the syndrome had developed: if heroin were not taken, the system would deviate strongly from its neutral state. This deviation (or "error" signal) would be manifest as the heroin withdrawal syndrome. We add that deviation or error in the opposite direction would be system "overshoot" or euphoria in this case. This subjectively rewarding state could result from more heroin than is needed to achieve a neutral or homeostatic state.

Himmelsbach's (1943) classic model of heroin addiction is illustrated in a conventional control theory diagram in Fig. 3.3. The power of control theory is that it integrates a number of highly disparate elements in one functioning system. In this case, the constituents are neurochemical (the input to the system, or goal-state, and the output or actual state of the system), behavioral (the control action), and pharmacodynamic (the response). The diagram models how these elements operate together as a controlled system, in this case instigating either heroin use or the opioid withdrawal syndrome. In this simple negative feedback system, heroin self-administration counteracts drug craving (conceptualized here as an error signal in the system) and also prevents or ameliorates the withdrawal syndrome. Note that the negative feedback is related to the level of the drug in the system, which is itself a function of the dose that is self-administered and the duration of time after use (i.e., metabolism and excretion). If the behavior (heroin self-administration) were not executed (i.e., if it became an "open-loop" system, then craving for heroin and the opioid withdrawal syndrome would ensue).

Several features of the heroin addiction model schematized in Fig. 3.3 are specific to control theory. First, the comparator in this case is a summing or integrating function that compares the set point to the output (actual level) of the system. It calculates error (deviation of the output from the input) that can be positive (withdrawal) or negative (toxicity), depending on whether the actual opioid blood level is too low or too high, respectively, relative to the optimal level for the addicted individual. The feedback from the pharmacodynamic response to the heroin is negative because it *reduces* positive error (i.e., it counteracts the heroin withdrawal syndrome) by moving the system in the direction of higher opioid blood levels. Note that this simple negative feedback system is analogous to the example above of a temperature controlled room with a thermostatically controlled heating system.

3.5.2 Control Theoretic Considerations

The heroin addiction model is a fairly simple negative feedback model, with a set point of the optimum blood heroin level for an individual addict. It is remarkably stable, predicting a relatively constant rate of heroin self-administration behavior (assuming equivalent dosings). Interruption of that behavior, or the administration of an opioid antagonist, drives the system into the classic opioid withdrawal syndrome. This is a model of addiction based on the key concept of the withdrawal syndrome; it is not merely a model of drug self-administration.

This simple—but powerful—homeostatic model of opioid addiction (Himmelsbach 1943) in Fig. 3.3 was prominent for many decades. In fact, it is still the dominant model (e.g., Baker et al. 2004) for most drugs in addition to opioids, and among many researchers who are not familiar with recent theoretical developments in the addictions field. It also instituted homeostatic approaches to addiction that persist today.

3.5.3 Opponent Process Theory

Opponent process theory (Solomon 1980; Solomon and Corbit 1973) has had a profound influence on the study of motivation, including that of substance use disorders. The theory posits that intense affective stimulation elicits a primary process, termed the "a" process, that may be affectively positive or negative, but not neutral. The "a" process is relatively constant in both its short- and long-term response. Its duration is normally assumed to be limited by the length of time that the eliciting stimulus is present. Therefore, it is often modeled as a quasi-square wave, and that wave is the same each time it is elicited. It is also proposed that the "a" process automatically (nonassociatively) engenders a "b" process that is subjectively and physiologically opposite in direction to the "a" process. The "b" process is sluggish. It is recruited slowly (delayed in time), terminates slowly, and extends in time beyond the end of the "a" process. Moreover, the "b" process is strengthened each time it is elicited by the "a" process so that it grows in magnitude and duration each time it is manifest. The "b" process also moves forward in time each time that it is elicited by the "a" process, and can come to partially precede the "a" process.

Importantly, since the "a" and "b" processes are opposite in direction, the "b" process partially or fully cancels the "a" process. This cancellation reflects a primary assumption of opponent process theory (and other theories, below) that different responses combine additively. Because the "b" process is slow, it leads to a subjective and physiological response immediately after the "a" process ends that is opposite

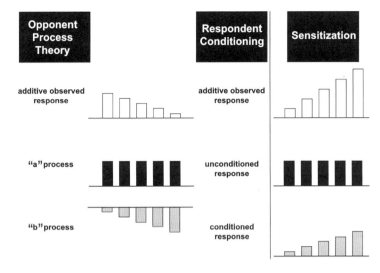

Fig. 3.4 Opponent process theory. This figure schematizes the assumptions of opponent process theory and respondent conditioning: (1) a growing counter-response with each elicitation of the drug effect, and (2) additivity. In the *left panel*, the X axis has separate elicitations of the affective stimulus (in this case, a drug effect) over time, and the Y axis shows the magnitude of the responses. In the *right panel* are the same assumptions, although the secondary response is in the *same* direction as the primary drug response. Opponent process theory and the respondent conditioning model (*left panel*) account well for drug tolerance, but cannot account easily for drug sensitization unless one makes the secondary assumption (*right panel*) that the "b" process or conditioned response can be in the same direction as the drug, resulting in sensitization rather than tolerance

in direction to the initial response as the b process wanes. This produces a biphasic drug response in which the early response to the drug is qualitatively different from the late response to the drug—that is, the early and late components of the acute drug response are opposite in direction.

The result of these temporal dynamics is a homeostatic model in which intense affective responses ("a" processes) are diminished following repeated elicitations. The organism "seeks" and generally achieves a neutral affective state that is less perturbed by strong stimuli or intense emotions and returns relatively quickly to a quiescent state. These relations are illustrated in Fig. 3.4 in which repeated elicitations of the "a" process (or unconditioned response in respondent conditioning, see below) are countered by an opposite-direction "b" process that grows with each elicitation. The combined effect is a diminishing response. Note that if we simply assume that there are certain conditions in which the "b" process is in the same direction as the "a" process, then the result is sensitization of the combined response.

The application of opponent process theory to drug addiction is clear, and was envisioned from inception of the theory (Solomon and Corbit 1973). The "a" process is the primary response to the drug of abuse and the "b" process partially cancels the "a" process. This produces drug tolerance as the drug response diminishes with repeated use at the same dose. In addition, gradual recruitment and strengthening of

Fig. 3.5 Control theory implementation. The control theory implementation of opponent process theory has both negative (the "a" process) and positive (the "b" process) feedback arms. Forcing functions, whether external to the human or nonhuman animal (exogenous) or internal states or processes (endogenous) impinge on the system and affect the goal-state or drive control actions

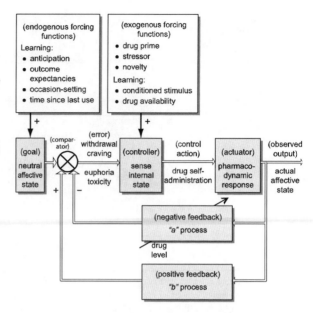

the "b" process leads to the drug withdrawal syndrome, which is generally opposite in direction to the initial response to the drug (Siegel 1983), as the "b" process is expressed in the absence of the "a" process.

At its core, opponent process theory is nonassociative. The elicitation and recruitment of the "b" process occurs reflexively without any learning having taken place. Having said that, Solomon (1980) also allowed for respondent (Pavlovian) conditioning of both the "a" and "b" processes, particularly the "b" process. So, for example, an environmental cue that has reliably predicted the drug in the past may elicit a withdrawal-like "b" process (conditioned withdrawal). This additional complexity makes the theory more powerful because it can "explain" many psychological phenomena. However this is at the expense of parsimony and leads to greater difficulty in designing specific and unique empirical tests of the theory.

Opponent process theory may be diagrammed in control theory terms as in Fig. 3.5. Note there is both negative feedback (as in the Heroin Addiction Model, Fig. 3.3) and positive feedback in the form of a drug-opposite "b" process. The latter is positive feedback because it increases error, potentially subserving craving and drug withdrawal. The output of the system is determined by the relative magnitude or "gain" of the negative and positive feedback loops (assuming they combine additively).

We introduce components ("forcing functions" in control theory terms) that are external to the model, but affect the "goal-state" of the system and the actions of the "controller." This has many effects, including allowing influences of the state of the organism (such as anticipation of the drug) and those of triggering stimuli (such as drug-related cues or stressful, noxious stimuli). It also provides mechanisms for the influence of learning, such as conditioning of the "b" process. It is important to note that all of these forcing functions are positive in the sense that they drive the

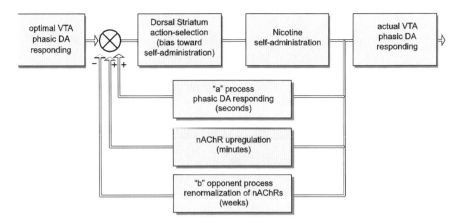

Fig. 3.6 Opponent process model of nicotine addition (Gutkin et al. 2006). Schematic of Gutkin et al.'s (2006) opponent process model of nicotine. Note that the "a" process is in a positive feedback loop, while the "b" process represents negative feedback. Ventral tegmental (VTA) phasic dopamine signaling drives the system. Unlike Solomon's (1980) original opponent process theory (Fig. 3.5), this relatively complex system is very specific in its hypothesized neurobiological mechanisms

system toward greater drug self-administration. None of the forcing functions introduce limits to the system. The only limiting factors are the negative feedback loop from the "a" process, and the implication that "toxicity" from drug levels that are too high may limit self-administration. We model mathematically a version of the system schematized in Fig. 3.5 in the Simulations section. Opponent process theory is clearly a model of addiction, based largely on classic indices of the addictive process such as tolerance, withdrawal and craving, etc.

3.5.3.1 Gutkin et al. (2006)'s Opponent Process Model of Nicotine Addiction

Gutkin et al. (2006) presented a neurocomputational model of nicotine dependence with opponent process aspects. Their model is schematized in Fig. 3.6. This relatively complex system is driven in turn by nicotine, nicotinic acetylcholinergic receptors (nAChRs), particularly the beta-2 subunit, and dopamine; it is modulated by gabaergic and glutamatergic influences. The model has two compartments, an action-selection compartment that is roughly equated with dorsal striatal function, and a ventral striatal dopamine compartment that drives the system. This model is of particular interest because the feedback loops, whether positive or negative, have different time scales—from seconds to minutes to weeks. The primary positive feedback (the "a" process) represents phasic (seconds) dopamine signaling, which is closely related to the input and output of the system. This signaling is driven (positive feedback) over seconds by nAChRs activation, followed by upregulation over minutes. Finally, an opponent "b" process consisting of renormalization over weeks of the nAChRs represents a negative feedback loop that stabilizes the system.

The actuator of this system is an action-selection compartment that biases action toward or away from self-administration of nicotine depending on weighting functions. In turn, the system is driven by dopamine signaling that functions to influence the weighting of action-selection. This signaling also provides positive feedback, whose action outcome evaluation (similar to "reward") supports acquisition, maintenance, and reacquisition of nicotine self-administration.

The detailed specification of this neurocomputational model in terms of neurobiological mechanisms, differential equations, and experimental tests makes it unusual and of particular interest. Although it was not presented originally in control theoretic terms, it is easily specified in relation to control theory.

3.5.4 Respondent Conditioning

In a separate development during the same period of time as Solomon's (Solomon and Corbit 1973) theorizing, Siegel (1975) provided empirical evidence for a respondent (Pavlovian) conditioning model of morphine tolerance. His theory was explicitly and entirely associative. Wikler (1973) had theorized earlier concerning drug conditioning (as had Pavlov 1927), but Siegel's empirical work was more compelling, at least in demonstrating environmental specificity of tolerance to morphine.

Siegel (1975) proposed that with repeated elicitations of the drug effect (the unconditioned response[4] to the drug) in the presence of drug-predictive environmental stimuli (conditioned stimuli), respondent learning occurs such that a compensatory conditioned response develops that is opposite in direction to the unconditioned response. Since the conditioned and unconditioned responses are assumed to combine additively, there is partial cancellation, producing morphine tolerance. For example, morphine causes analgesia to painful stimuli, so this model predicts an hyperalgesic (increased pain sensitivity) compensatory conditioned response when elicited by drug-predictive stimuli in the absence of morphine. Siegel's (1975) respondent conditioning model is illustrated in control theory terms in Fig. 3.7. Note the strong structural similarity to Solomon's (1980) opponent process theory, even though Siegel's model is associative and Solomon's is nonassociative. In fact, the two control theory diagrams (Figs. 3.5 and 3.7) are virtually identical.

---

[4]In its original formulation (Siegel 1975), Siegel's associative model of morphine tolerance assumed that the unconditioned response was the drug effect, itself. However, this was revised (Eikelboom and Stewart 1982) to acknowledge that the drug effect (analgesia) is actually the unconditioned stimulus, and the adaptive, counter-directional response to the drug effect is the unconditioned response. In this sense, the conditioned and unconditioned responses are actually in the same direction, in this case hyperalgesia. For the purposes of this discussion, we will revert to Siegel's terminology, keeping in mind that these psychobiological systems are themselves controlled systems for which the issue of whether the response is part of an afferent or efferent response can be critical (Eikelboom and Stewart 1982).

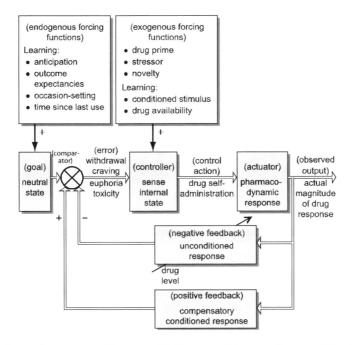

Fig. 3.7 Respondent drug conditioning model. The control theoretic diagram of Siegel's (1975) respondent (Pavlovian) model of drug conditioning, is structurally similar to that of opponent process theory (Fig. 3.5). However, it is an associative model

This structural equivalency of two theories that were developed independently requires some comment. First, control theoretic modeling has the virtue of making this similarity more explicit. These structural similarities may not seem as surprising if one views the function of the theorist to reverse-engineer addictive processes in which many of the same phenomena are apparent (i.e., tolerance, withdrawal syndrome, a shift from euphoria to discomfort, etc.). However, it raises the questions of whether control theoretic modeling has missed critical differences (such as associativity) between models (and phenomena) or whether these mathematical models are unique to a given theoretical model? This is confused by Solomon's (1980) assertion that the "a" and "b" processes are conditionable, in which case the theories are themselves indistinguishable other than the fact that opponent process theory is couched in terms of hedonics while the respondent conditioning model concerns primarily response magnitude. We note below that there are an infinite number of linear mathematical models that can account for a given set of inputs and outputs. Therefore, these issues are unlikely to be resolved because of ambiguities in the theoretical models, but it is clear that control theoretic modeling focuses and clarifies the issues, particularly when implementing the models in simulations.

3.5.5 Control Theoretic Considerations

Opponent process theory and respondent conditioning models do not afford simple set points, as was found with Himmelsbach's (1943) heroin addiction model. Instead, there is periodic or semi-periodic behavior of the system, with relatively stable equilibrium functions that vary consistently around the neighborhood of (but not on) average values. This average value could be taken as a kind of set point in the sense that the system "seeks" this point, but varies around it rather than settling on it. There is a degree of stability in the oscillating functions. These considerations apply equally to both opponent process theory and respondent conditioning models since they are structurally equivalent (Figs. 3.5 and 3.7).

3.5.6 Problems with Homeostatic Models

In terms of reverse-engineering the addictive process, drug use disorders are characterized by strong acquired motivation. After all, these drugs are *self*-administered in an apparently compulsive, driven manner. We need control theoretic models that are strongly self-sustaining and that capture the "drive" in drug addiction.

In our discussion of opponent process theory and the respondent model of drug conditioning, one might view the human or nonhuman animal as a passive recipient of powerful affective stimuli and Pavlovian learning processes (so-called "stamping-in"). These opponent or compensatory processes might seem to be fully automatic with the organism passively "seeking" a neutral state. In contrast, several theorists have recast drug conditioning findings in terms of a more active organism. In this case, the nonhuman animal or human has critical behavioral demands, senses perturbations that may be biologically relevant, and responds actively to counteract drug effects that contribute to these demands or obstruct those behavioral agenda. After all, the dog in Pavlov's (1927) classic experiments was physically restrained so that its skeletal behavior could not be fully observed. At the same time that the dog salivated to the bell signaling imminent meat, a host of other physiological and behavioral phenomena were occurring concurrently that were not measured or that were prevented. In other words, a casual observer of Pavlov's experiments might see an active dog that was highly motivated to cope with restraint, and to anticipate the meal that was coming!

In terms of considering the organism as more than a passive recipient of "stamping-in" processes, Poulos and Cappell (1991) emphasized in their general homeostatic theory of physiological adaptation (including drug tolerance) that the disturbance in homeostasis caused by a drug (or other stimulus) must be *sensed* (but not necessarily consciously) to be effective. Moreover, this biological perturbation is not exclusively an internal regulatory function, but interacts with the organism's behavior and environment. For example, they argue that "an adaptation to a disruption in hunger (anorexia) will not occur in the absence of a behavioral interaction with food." Poulos and Cappell's (1991) elaboration of homeostatic theory places the organism squarely within the behavioral demands of its environmental milieu.

3.5.7 Cognitive Theories of Addiction

Purely cognitive theories of addiction (e.g., Cox and Klinger 1988; Tiffany 1990), though certainly important, are not discussed further because they are not easily rendered in control theoretic terms. It may be without coincidence that this class of theories lends itself poorly to characterizing motivation and drive, which have been more typically the province of models of physiological homeostasis (i.e., "drive-reduction") and behavioral learning theories.

3.5.8 Internal Model

The cleft between informational and motivational theories of addiction is not complete. Feedback models can, and sometimes do, incorporate cognitive components. Peper's conceptual (2004a) and mathematical (2004b) account of drug tolerance is somewhat different from those discussed above. He assumed that the human or nonhuman animal has an internal model of the expected action of the drug, and homeostatic adaptive responses counter it in order to maintain optimum functioning. This internal representation may be present even on first administration in a drug-naïve organism because drugs of abuse mimic endogenous biochemicals that are "known" to the brain's regulatory systems. However, the dose of the drug is not known to the subject, and many self-administrations may be needed to calibrate this internal model so that it can effectively cope with the drug response, resulting in profound tolerance. Moreover, abrupt cessation of drug intake will result in the withdrawal syndrome in a classic rebound response.

Peper's (2004a) theory took an explicitly informational approach to the regulation of drug responses. His internal representation of the drug effect echoed Baker and Tiffany's (1985) cognitive-habituation model of tolerance, although Peper's model was homeostatic while Baker and Tiffany's was not. More importantly, Peper (2004a) emphasized that the "goal" of the organism is to maintain optimal levels of functioning despite drug disturbances, implying a more active animal or human than in other homeostatic models.

Another problem with homeostatic models, besides their inherently passive nature, is they beg the question of what is the neurobiological mechanism controlling the set point or goal-state of the system? This is a critical element in a homeostatic system, like the thermostat in a building with an automated heating/cooling system. At the present time, this biological controlling mechanism is usually inferred rather than measured directly. The strength of this inference depends on indirect evidence that deviations away from the hypothetical set point are sensed and acted upon by neurobiological processes that can be measured, at least to some extent. Koob and Le Moal (2006), Di Chiara and Imperato (1988), Robinson and Berridge (1993, 2003), and Gutkin et al. (2006) have been the most specific regarding the neurobiology of these regulatory mechanisms.

3.5.9 Direction of the Conditioned Response

Theories of addiction have made much of the issue of whether the conditioned response to drug cues is in the same direction as the unconditioned drug effect or opposite in direction. Opponent process theory (Solomon 1980) and Siegel's (1975) respondent conditioning model of drug tolerance assume counter-directional cued responses, while Stewart et al.'s (1984) model and Robinson and Berridge's (1993) incentive-sensitization model propose that the conditioned response is drug isodirectional. As we shall see, the direction of the conditioned response is critical to the behavior of control theoretic models. Moreover, this directionality, often expressed in terms of whether the conditioned response is in a negative or positive feedback loop, helps determine how the model deals with issues of motivation and drive.

3.6 Non-homeostatic Models of Addictive Behavior

3.6.1 Instrumental Conditioning

One of the great ironies of addiction is that individuals demonstrate relentless use of a drug they often report gives them little or no pleasure (Robinson and Berridge 1993). This presents a real problem for instrumental conditioning theories of addiction (Pickens and Thompson 1968), which are in some ways the dominant models of addictive behavior in current thinking.

McSweeney et al. (2005) proposed a disarmingly simple—yet powerful—behavioral theory of chronic drug use that is based on instrumental conditioning (Skinner 1938). It is a modern update of the original instrumental models of drug use that simply observed that drugs of abuse serve as primary reinforcers (a radical notion at the time). That is, they increase the likelihood of immediately preceding behavior, specifically the instrumental response that instigated delivery of the drug. McSweeney et al. (2005) argued that the instrumental response is strongly modulated, initially by sensitization, then by habituation to the reinforcer. Drug sensitization, a well-documented phenomenon, increases the rate and strength of instrumental responding, while habituation, an even more thoroughly studied process, decreases it.

McSweeney et al. (2005) use the terms sensitization and habituation somewhat differently than other models discussed here (except Robinson and Berridge 1993). They use them to refer to the increased valuation or devaluation, respectively, of the drug reinforcer (i.e., its reward value), and therefore, the rate of self-administration; other theorists have used the terms sensitization and tolerance to refer to the magnitude of the drug response.

This is an entirely behavioral theory. It does not draw in the least on the massive literature concerning the neurobiology of drug use and addiction. For example, McSweeney et al. (2005) noted the model cannot account for the drug withdrawal syndrome. Biological processes external to the model are needed to model this and

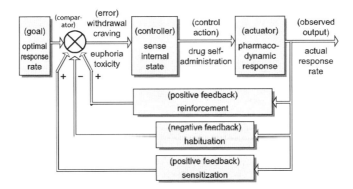

Fig. 3.8 Instrumental drug conditioning model. McSweeney et al.'s (2005) instrumental (Skinnerian) conditioning model. The control theoretic implementation has both negative (habituation) and positive (sensitization) feedback arms, although it is a theoretically different model from those in Fig. 3.5 (opponent process theory) and Fig. 3.7 (respondent conditioning). In instrumental conditioning, the reinforcer (the drug effect) is contingent on the response (drug self-administration). Note the reversal of direction of error relative to Figs. 3.3, 3.5, and 3.7

other phenomena that many other theories are intended to address. Technically, it is not a theory of drug addiction. Constructs such as incentive motivation and craving are also external or irrelevant to the model.

McSweeney et al.'s (2005) theory is schematized in Fig. 3.8. Note first that, like instrumental conditioning, itself, this is a positive feedback model. Drug reinforcement simply increases the likelihood of the instrumental response, such as lever-pressing or pecking a key (Skinner 1938). Positive feedback systems such as this are explosive because there is no inherent negative feedback to limit growth of the behavior over time; as such, they are not realizable in nature, at least not without some other process to limit the system's behavior. Initial sensitization increases this growth curve as it enhances the reward value of the drug reinforcer. Responding is ultimately limited by habituation processes that devalue the reinforcer and limit its efficacy. This model predicts a bitonic curve: sensitization and increased rates of responding followed by habituation and decreased rates. McSweeney et al. (2005) did not explain why drug sensitization generally precedes habituation other than the empirical observation that this sequence is dominant in the literature on nondrug stimuli. In terms of biphasic responses, they emphasized that low drug doses often engender sensitized rates of responding while high doses promote habituation.

3.6.2 Control Theoretic Considerations

McSweeney et al.'s (2005) model of instrumental drug conditioning has both positive and negative feedback. It does have set points or "set responses," but they vary over time with habituation and sensitization processes. The presence of positive feedback injects "drive" into this system that many purely negative feedback systems do not exhibit, while the negative feedback affords stability to the system.

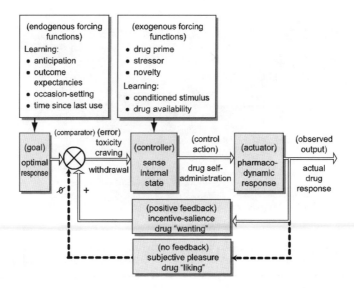

Fig. 3.9 Incentive-sensitization theory. Incentive-sensitization theory is modeled as a positive feedback system (i.e., no negative feedback), which can be problematic because it is inherently explosive (see text). Note that the direction of error has been reversed from that of previous models (Figs. 3.9 and 3.13)

3.6.3 Incentive-Sensitization Model

Homeostatic models of addiction do not account easily for drug sensitization. This phenomenon, when the response to the drug given repeatedly on different occasions at the same dose increases rather than decreases, is well established. In fact, some authors (e.g., Peper 2004a, 2004b; Poulos and Cappell 1991) simply stated their model does not apply to sensitization. However, sensitization has played a central role in some theories of addiction.

Robinson and Berridge's (1993, 2003) incentive-sensitization theory of addiction proposed that the incentive-salience of drugs and drug stimuli becomes sensitized such that the organism comes to "want" or crave the drug more than they "like" it. It is modeled graphically in Fig. 3.9. Note that drug wanting increases due to positive feedback at the same time that drug liking is relatively constant because it is not part of a feedback loop. This is not a homeostatic theory of addiction because it does not "seek" a neutral state, but it is instead a learning theory (Bolles 1972) based on reward and plasticity in the neural systems that control drug-seeking behavior. It is analogous in some ways to Stewart et al.'s (1984) argument that addictive behavior is driven by conditioning of a drug-like (rather than drug-opposite) appetitive response to drug cues. These theories of drug-seeking behavior stand in direct contrast to most homeostatic models, particularly those that emphasize negative reinforcement (prevention or reversal of aversive affective states such as the drug withdrawal syndrome or stress).

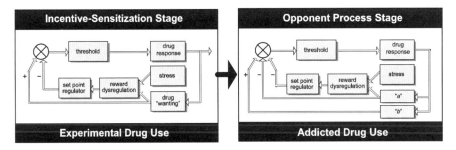

Fig. 3.10 Koob and Le Moal's model. Koob and Le Moal's (1997) model involves two stages, with incentive-sensitization processes early in addiction, and opponent processes (negative feedback) in the later stages of addiction. It is also a stress allostasis model (see text)

Robinson and Berridge's (1993, 2003) model is schematized in Fig. 3.9 as an exclusively positive feedback system. This makes it very different from the theories discussed above (Figs. 3.3, 3.5, 3.6, and 3.7). It is essential to observe in Fig. 3.9 and in other positive feedback models (Figs. 3.9 and 3.13) that the direction of error has been reversed from that of the previous models; both craving and toxicity are schematized as positive error. Incentive-salience increases (sensitizes) craving and drug "wanting" in a positive feedback loop. Because positive feedback systems are rarely manifest in nature (since they are inherently explosive), it raises the question of what (negative feedback) processes may ultimately limit the growth of drug self-administration. The authors (Robinson and Berridge 1993) are silent on this issue.

Having written that, it is not difficult to integrate drug sensitization phenomena into control theory models. We simply assume that the conditioned response is in the same direction as the unconditioned response and they combine in a constructive (positive) way, if not necessarily additively. This is illustrated in Fig. 3.4, right panel. Note that this produces a positive feedback system rather than negative feedback, much like the models of Stewart et al. (1984) and Robinson and Berridge (1993; see Fig. 3.10).

Most theories of addiction implicitly assume that the same biological or learning processes that control initial, experimental drug use also determine addictive, dependent use. Although several theories emphasize change over time with repeated drug use on many occasions, the structural models that we have diagrammed in formal control theory terms remain the same in different phases of drug use and abuse. There may be growth of a process, such as the "b" process, compensatory conditioned response, drug "wanting," and either respondent or instrumental learning functions, but the feedback loops and other structural characteristics do not change. This speaks to the ambitiousness and parsimoniousness of these theories.

Koob and Le Moal (1997, 2006) proposed an ambitious model that attempts to integrate many of the structural components in the theoretical models above. Their theory involves primarily negative feedback that subserves negative reinforcement (prevention or alleviation of aversive states such as the drug withdrawal system or stress). However, it incorporates "allostasis" (Goldstein and McEwen 2002) in which acute and chronic stress to the organism changes the set point or goal-state of the system over time. This structural change is important because it suggests mechanisms (i.e., allostatic load) by which the set point of the system may change from "neutral" or a zero blood drug level in the drug-naïve person or nonhuman animal to a positive drug level. Note that in Koob and Le Moal's (1997, 2006) model, allostasis affects the set point of the system rather than the unconditioned response to the drug, or even the conditioned response to drug cues. This model is illustrated in Fig. 3.10.

Although the theory of stress allostasis has been criticized (Day 2005) for being unnecessary and a misbegotten take on homeostatic theory, the concept of allostasis is in current usage. Koob and Le Moal (1997, 2006) argue that the role of stress in the development and maintenance of addictive behavior is critical, but this is either not incorporated into most homeostatic or other theories of addiction, or is peripheral to these models. Their inclusion of stress into the basic structure of a homeostatic model appears to strengthen the generalizability of the approach.

3.6.4 Control Theoretic Considerations

The early stages of drug use in Koob and Le Moal's (1997) model are viewed as a positive feedback system with no set point or stability. However, asymptotic drug use can yield a set point if the composite influence of both positive and negative feedback is a net negative. In their later writings (Koob and Le Moal 2008a, 2008b), they have emphasized negative feedback in the sense of avoiding the drug withdrawal syndrome. This would tend to stabilize the system, and would admit a set point.

3.6.5 Autoshaping

Autoshaping is simply respondent conditioning in which the conditioned response is skeletal approach toward the conditioned stimulus rather than an autonomic or visceral response (Hearst and Jenkins 1974). Therefore, it has elements of both instrumental (skeletal responding) and respondent conditioning (the stimulus–stimulus contingency). In the classic autoshaping procedure, the conditioned stimulus is presented immediately prior to the unconditioned stimulus (in this case, some reward such as an injection of a drug of abuse). This stimulus-stimulus pairing does not depend in any way on the animal's response. Despite this lack of stimulus-response contingency, the animal approaches and contacts the conditioned stimulus (e.g., Brown and Jenkins 1968). For example, if the conditioned stimulus is the introduction of a lever into the cage and the unconditioned stimuli is an injection of cocaine, the animal will approach the lever each time and eventually press it. This has been used in some cases as a standardized, "hands off" procedure to induce drug self-administration (Carroll and Lac 1993) because the approach behavior elicited

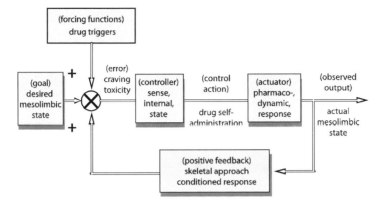

Fig. 3.11 Autoshaping. The autoshaping model of addiction proposes a skeletal approach conditioned response toward drug cues and drugs. This is a positive feedback model

in this way is remarkably durable and resistant to extinction. In fact, the animal will approach the conditioned stimulus even when it is prevented from receiving the unconditioned reward. This has been taken as evidence that the learning is Pavlovian rather than instrumental, and at least in this case, the respondent conditioning is more potent than instrumental conditioning (!) (Hearst and Jenkins 1974).

Newlin (1992) proposed an autoshaping model of human drug craving. Stimuli that have been paired repeatedly with drug effects, such as procurement procedures, commercials for the drug (e.g., tobacco advertisements), alcohol drinking buddies, or injection paraphernalia, elicit through autoshaping procedures a psychological orientation toward drug triggers and approach toward the drug and stimuli associated with it that we refer to as "craving." This autoshaping model is illustrated in Fig. 3.11. This is a positive feedback model because the approach behavior increases both craving and drug use. Autoshaping was proposed (Newlin 1992) as a model of craving and drug addiction because it might account for the remarkable resistance to extinction of drug self-administration behavior and craving for drugs.

Although the autoshaping model (Fig. 3.11) employs a form of respondent conditioning, it differs in important ways from both opponent process theory (Fig. 3.5) and Siegel's (1975) respondent conditioning model of tolerance (Fig. 3.7). First, the autoshaping model employs only positive feedback; limiting factors, such as satiety or toxicity of the drug are external to the model. The drug use and craving will increase progressively in this model without end (i.e., it is explosive, or a "vicious cycle"). Therefore, it accounts well for acquisition of drug use, intense bouts of drug use (bingeing), relapse, sensitization and craving, but does not reach asymptote.

3.6.6 Control Theoretic Considerations

This positive feedback model does not admit a set point. The instability of this system raises questions about whether additional control components (such as negative

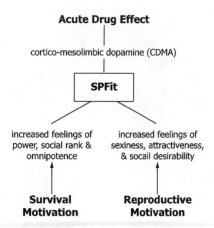

Fig. 3.12 Foundations of SPFit. Foundations of SPFit (self-perceived fitness) theory in survival and reproductive motivation

feedback) are needed for it to be realizable. This problem is inherent to some extent in nonhomeostatic models of addiction. Control theoretic analyses highlight these "design" problems. Although the autoshaping model of addiction (Newlin 1992) does not account well for the classic criteria of tolerance or withdrawal, it does cover many other human-like criteria of dependence (such as craving, narrowing of behavioral repertoire on drug use, drug self-administration, etc.).

3.6.7 Evolutionary Psychological Theory

Negative feedback models may have difficulties with motivation and "drive." A final model of addiction is derived from evolutionary theory and evolutionary psychology. In 1997, Nesse and Berridge suggested that "drugs of abuse create a signal in the brain that indicates, falsely, the arrival of a huge fitness benefit." Newlin (2002, 2007) developed an evolutionary model of addiction, Self-Perceived Fitness (SPFit) theory, that expands upon this premise. The essence of SPFit theory is that humans (and perhaps some other mammals) seek to increase or protect their self-perceived sense of being empowered and sexually/personally attractive. Power enhancement derives from survival motivation because a more powerful individual is viewed as more likely to survive. The desire to be sexually and personally attractive is related to reproductive motivation because the attractive person may be better able to attract and keep a mate for reproduction. In this sense, SPFit theory proposes that people strive to increase their Darwinian fitness (Darwin 1871), and *self-perception of this fitness* is vitally important to their behavior, cognitions, and emotional functioning. These theoretical views are illustrated in Fig. 3.12. Survival and reproductive motivation drive the acute response to abused drugs in which SPFit is artificially, and in many cases, dramatically elevated.

Importantly, SPFit theory proposes that drugs of abuse artificially inflate the person's feelings of powerfulness and being attractive because they directly or indirectly affect functioning of the cortico-mesolimbic dopamine (CMDA) system that has been studied so intensively in neurobiology and neuropharmacology. SPFit theory views the CMDA as the primary substrate for survival and reproductive motivation. The CMDA is viewed in most theories as a "reward center," "reward pathway," or "reward circuitry" (the latter terms are retrenchments of sorts, given more recent evidence concerning the complexity of this system). SPFit theory proposes instead that the CMDA is a survival and reproductive fitness motivation system rather than a reward system in the brain. Newlin (2002, 2007) noted that aversive or noxious stimuli, novelty, and other non-reward stimuli also increase activity in the CMDA, which argues against the common view that this is a reward system. This evidence is consistent, though, with the view that it is a survival motivation system because aversive or novel stimuli would be expected to activate a survival motivation system along with "reward" stimuli. While it may be argued that the entire body and brain are honed sharply by evolutionary forces to promote survival and reproductive functions, most bodily organs and brain systems are not involved directly in *motivation*.

An important way in which SPFit differs from models of addiction discussed above is that it does not make use of constructs such as "reward," "reinforcement," or "euphoria," although it does include drug craving in its model. Evolutionary psychology derives from a very different intellectual tradition than that of behavioral learning theory.

The control theoretic illustration of SPFit theory appears in Fig. 3.13. The control theory models in Fig. 3.2 (bicycle riding) and Fig. 3.13 (SPFit) can be highly adaptive and at least semi-automatic. Bicycle riding is a skill and a habit that is highly resistant to unlearning. In a similar manner, promotion of SPFit and protecting it against environmental threats represents semi-automatic or automatic functioning that is usually highly adaptive, except in the case of drug abuse and addiction. In Fig. 3.13, the goal that the system seeks is a desired mesolimbic state. The actual mesolimbic state is affected rather directly by drugs of abuse. A prefrontal executive, with an internal model of the drug's anticipated and actual effects, senses deviations (error) in the mesolimbic state.

There is a fundamental distinction between the prefrontal executive and the observer (SPFit). The executive senses error and executes motor, affective, and cognitive operations that support SPFit, and in this case, promote drug use. In contrast, the observer is privy to information about the state of other components in the model (through imperfect self-observation) and *provides corrective feedback* to the prefrontal executive. The drug effect inflates SPFit in a way that would appear to promote Darwinian fitness, but in the case of drug use, it actually impairs fitness. In the bicycling example, the prefrontal executive executes motor sequences for riding the bicycle, and the observer provides necessary, and generally accurate (in the experienced rider), feedback to guide the executive's actions.

3.6.8 Control Theoretic Considerations

SPFit theory is a positive feedback model, although there are social constraints that limit SPFit in a negative feedback loop. The operating characteristics of this model

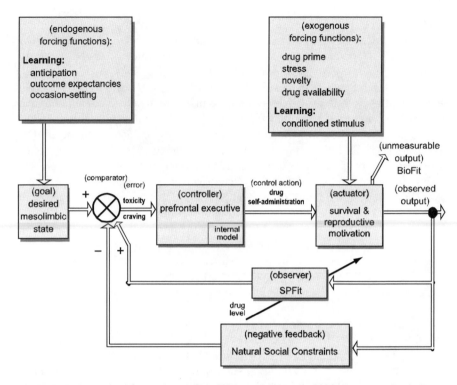

Fig. 3.13 Self-perceived fitness theory. This SPFit model (Newlin 2002) is structurally similar to the bicycling example (Fig. 3.2) of a control model with an "observer" (see text), although it is a positive feedback model. It "seeks" higher levels of SPFit rather than a fixed level (i.e., there is no set point). There are multiple social environmental constraints that "weigh down" SPFit in a negative feedback loop, including the short and long term deleterious effects of the drug that is used. "BioFit" is an unmeasurable output of the system that corresponds roughly to classic Darwinian fitness, while SPFit is the self-perception of that fitness

may be compared to those of the heroin addiction model. The heroin addict titrates the amount of heroin taken to "hover around" or to "seek" an optimum blood level of the drug. In contrast, individuals generally strive to increase (or to protect) their SPFit within social environmental constraints and available resources. Therefore, SPFit theory does not admit a set point. In this way, SPFit theory avoids some of the passivity problems associated with some other models. Although the SPFit model has positive feedback, there are many factors in the social environment of the individual that limit it. These can be as diverse as social feedback from friends, looking at oneself in the mirror, and frequent self-observation of obvious mistakes and lack of perfection. These factors that weigh SPFit down serve as negative feedback that is generally limiting and helps stabilize the model.

SPFit theory, like autoshaping, accounts well for many human-like addictive phenomena. In fact there are strong structural parallels between the two models, although this would not be obvious without control theoretic modeling.

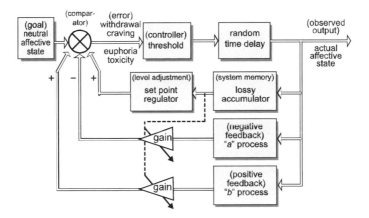

Fig. 3.14 Opponent process theory of chronic tolerance/sensitization. Technical implementation of opponent process theory. Note the addition, relative to Fig. 3.5, of random time delay (white noise), level adjustment (set point regulator), system memory (lossy accumulator), and gain blocks in the negative and positive feedback loops

3.7 Simulations of Two Addiction Systems

3.7.1 Simulation of Opponent Process Theory

To demonstrate implementation of a control system, we performed simulations on the opponent process theory model presented in Fig. 3.5 and more technical implementation as presented in Fig. 3.14. Since the respondent drug conditioning model is virtually identical in structural terms, Fig. 3.7 can be taken as an implementation of that theory as well. We chose opponent process theory to simulate for several reasons. First, opponent process theory is both historically important and in current use (e.g., Koob and Le Moal 2006). Second, it represents a potentially stable system because it is not based on positive feedback alone. Third, it is of particular interest in control theory terms because it has elements of both positive and negative feedback. Finally, it includes aspects that are characteristic of a number of the models of addiction presented here. For example, the negative feedback from the drug effect is similar to the heroin addiction model of Himmelsbach (1943). The positive feedback from the "b" process has similarities to instrumental conditioning and incentive-sensitization theory.

3.7.2 Opponent Process Theory Implementation

It is useful to consider the simplest model of an opponent process loop, as in Fig. 3.14. The error signal represents euphoria for negative swings, and withdrawal for positive swings; it drives a threshold comparator such that when the signal rises

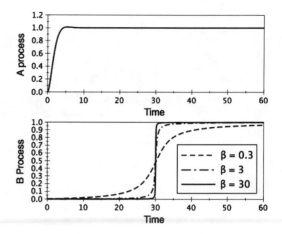

Fig. 3.15 Signals triggered by the A and B processes. Showing the signals triggered by the A and B processes. The slope of the B process is controlled by a parameter β

above a certain threshold, the "a" and "b" processes are triggered. These processes counteract each other in their influence on the error signal.

The "a" process that is triggered is taken as

$$A(t) = \frac{\exp(-\sigma t)\cos(\theta + \omega t)}{\sigma} + 1 \qquad (3.4)$$

using $\theta = 0.8\pi$, along with $\sigma = -\cos\theta$ and $\omega = \sin\theta$, giving the step response of a second-order lowpass filter. The "b" process is a slowly dropping function that lags behind the "a" process; here we consider a sigmoidal falling function of the form $B(t) = -[f(t) - f(0)]/[1 - f(0)]$ using

$$f(t) = \frac{1}{2} + \frac{\arctan(\beta(t - t_0))}{\pi}. \qquad (3.5)$$

The two functions are graphed in Fig. 3.15 using $t_0 = 30$ (arbitrary time units) which sets the lag time of the "b" process; the lower graph illustrates how the slope of the "b" process decreases with decreasing values of the parameter β. Subtracting the "b" process from the "a" process results in a square pulse with an initial overshoot (from the "a" process) and a return to zero with variable slope (from the "b" process).

3.7.3 "Lossy Accumulator"

To account for drug sensitization and tolerance, a lossy accumulator is included, which works akin to a leaky bucket: with each drug dose, "water" is added to the bucket but, due to "holes," the bucket is perpetually leaking some water away. The mathematical form obeys the differential equation

$$\frac{dw}{dt} = \alpha w(t) + g(t) \qquad (3.6)$$

Fig. 3.16 Gain and slope parameters. Indicating how the gain and slope parameters vary with the "water level" of the lossy accumulator

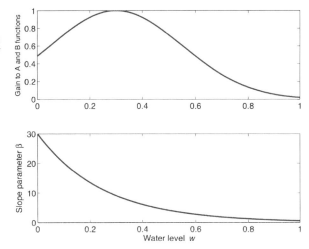

in which $w(t)$ plays the role of the "water level" in the leaky bucket, and the parameter α ($= -0.002$ in the simulations) controls the leakage rate. Note that stability occurs only when α is negative. The driving function $g(t)$ turns on during a drug dose that commences at time $t = t'$, and then returns to zero:

$$g(t) = \begin{cases} 0.02(1 - w(t')), & t' \leq t \leq t' + 5; \\ 0, & \text{otherwise.} \end{cases} \quad (3.7)$$

The "water level" w then determines the gain given to the opponent process functions, according to Fig. 3.16. It is seen that for initial doses (corresponding to a small "water level" in the bucket), the intensity of the "a" and "b" functions increases, to simulate sensitization, but as the "water level" rises further, the gain functions decrease again, to simulate tolerance. In addition, an offset equal to $w(t)$ is added to the error signal so that as the "water level" increases, the intensity of the withdrawal effect likewise increases. Finally, the slope parameter β of the b process is decreased as the "water level" increases, as shown in the lower graph of Fig. 3.16, to simulate faster counteraction of the drug dose that comes with tolerance. The lossy accumulator plays a role similar to the level adaptation mechanism of Peper (2004b), but the present scheme also accounts for sensitization by virtue of modifying the gains given to the "a" and "b" functions.

Figure 3.17 plots the error signal versus time of the closed-loop system, using a random (white noise) time delay between the trigger function and the "a" and "b" process activations; the trigger threshold is set to -0.01. The top graph shows the autonomous behavior of the system, in which successive positive-going pulses represent successive drug doses. The euphoria (indicated by positive swings of the error signal) initially increases in amplitude with successive doses, but eventually settles to smaller amplitudes, consistent with initial chronic sensitization followed by tolerance.

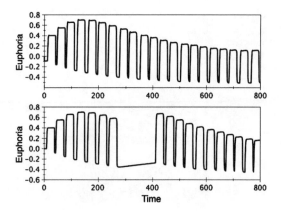

Fig. 3.17 Rise/fall characteristics. Each positive-going pulse represents a drug dose, which then wears off, causing the euphoria to drop below a set threshold, which triggers the next drug dose. (*Top*): The euphoria produced by successive drug doses initially increases, and then settles into a lower value. (*Bottom*): Illustrating the effects of withdrawal between time instants 250 and 400 (arbitrary time units); the negative value of euphoria slowly works its way closer to zero (a neutral condition) due to the lossy accumulator

3.7.4 Drug Withdrawal

The lower plot simulates a period of drug abstinence by removing the trigger function between time instants 250 and 400 (arbitrary time units). The dropping error signal indicates the intensity of the withdrawal symptoms, and is provoked by the offset included in the error signal. The intensity of the withdrawal effect slowly decreases due to the "leaky bucket" mechanism, until the next dose is administered after time 400.

These simulations illustrate how the interconnection of basic building blocks can simulate the "euphoria" and chronic tolerance characteristics of successive drug dosings, including conditioning. The "a" and "b" process functions are not precise, but rather are common mathematical functions that suffice for illustrating the interaction between mechanisms in the feedback loop. Replacing these with other functions displaying similar rise/fall characteristics would lead to interactions similar to those plotted in Fig. 3.17. Although the precise waveforms might change, the general rise/fall tendencies should still behave as plotted. Further tailoring of the model to specific habitual tendencies can be readily envisaged.

3.7.5 Simulation of Instrumental Conditioning Theory

We distinguish clearly between short- and long-term responses. Short-term responses refer to those occurring over a small number of successive self-administrations of the drug. Without external limitation, instrumental conditioning could lead to overdose and death because it is conceptualized as a simple positive feedback

3 Control Theory and Addictive Behavior

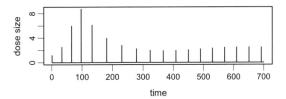

Fig. 3.18 Effects of instrumental conditioning. Illustration of the effects of instrumental conditioning on the pattern of subsequent self-administration with forced cessation. After the cessation the subject's substance use follows the same pattern of sensitization and habituation although the size of the sensitization is not as pronounced as in the naïve case due to residual long-term memory

loop. Internal response mechanisms that usually prevent overdose could be related to toxicity and eventually to an inability to administer the dose. Escalating drug self-administration can occur either as an increase in the size of the dose or as an increase in the frequency of dosing.

The short-term time scale is governed by a positive feedback loop as each drug self-administration is positively reinforced by the primary reward of the drug effect. In contrast, the longer time scale is associated with chronic use of the drug and slower processes that determine variations in dosage and timing—that is, sensitization and habituation (McSweeney et al. 2005). For the purposes of presentation, we focus on the longer time scale such that the short term processes are represented simply by a delay between doses; we ignore for the moment the complexities of the pharmacokinetics and dynamics of an individual drug self-administration, as did McSweeney et al. (2005).

To illustrate this concept, we begin with a simple model of drug self-administration in which the dose is administered following the principles of instrumental conditioning. This is expressed as a positive feedback loop (i.e., the size of the next dose is larger than the previous one by an arbitrary factor (1.1) in the model, and the length of time between doses is shortened by 5% (another arbitrary value)). In the absence of sensitization or habituation processes, the subject would increase the dose and the frequency of use and would eventually self-destruct. Sensitization and habituation are expressed in this model as variations in dosage and frequency necessary for stability. The dose is thus modulated in this model by two factors: the previous dose and the sensitization-habituation process that increases or decreases the dose depending on the lossy accumulation of drug experience.

Figure 3.18 shows instrumental conditioning for a situation of continuous use and Fig. 3.19 demonstrates the behavior of the model when use is forced to cease and then restarted with a small priming dose. We use the same lossy accumulator as in the previous opponent process theory model.

The gain function determines the shape of sensitization/habituation and has a shifted bell shape: Gain = $\exp(-(w(t) - 0.3)2/0.2)$. It reaches maximum at the value of $w(t) = 0.3$. As shown in Fig. 3.18, after a significant period of abstinence a small dose is able to trigger new consumption.

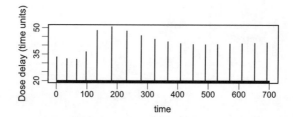

Fig. 3.19 Effect of self-adjustment. The effect of self-adjustment to the delay between the doses due to sensitization and habituation. During the sensitization phase the interval between the doses is shorter, which increases the speed at which the lossy accumulator fills. As habituation comes to predominate, the interval between self-administrations becomes longer and finally settles to a steady self-administration regime

This model has mathematical similarities to that for opponent process theory described above, although there are substantial differences between these models. In Instrumental conditioning, it is the dose and the frequency that are being increased, while for the opponent process theory model the increase was in the drug response and hedonics. In the current model, repetitive use is generated through positive feedback triggering the next dose after a delay from previous use. This repetitive mechanism could be generated even without an increase in dose or frequency and would result in repetitive consumption of a fixed dose and rate.

3.7.6 Sensitization and Habituation

The size and frequency of drug self-administrations are further controlled by sensitization/habituation processes that are mathematically expressed through a gain function, which is in turn a function of the amount in the lossy accumulator. As the "water level" increases in a lossy "bucket," gain first provides sensitization, and with further increases, habituation. The control theoretic model of instrumental conditioning does not have a preset error per se, but rather leads to a long term stationary process with a stable dose and regular consumption intervals.

Figure 3.20 illustrates this behavior of the model. While the definitions of sensitization and habituation here are different than the ones in the opponent process theory model, the mathematical forms of the lossy accumulator and gain functions remain the same. The lossy accumulator represents the accumulated history and experience of drug use and is an internal controlling factor for drug self-administration.

Positive and negative feedback operate such that as the dose increases, so does the value in the lossy accumulator. As this value becomes large, the gain function decreases it, which in turn decreases the value in the accumulator. When the value in the accumulator is low, the gain function increases the dose which in turn increases the accumulator (Fig. 3.20).

3   Control Theory and Addictive Behavior

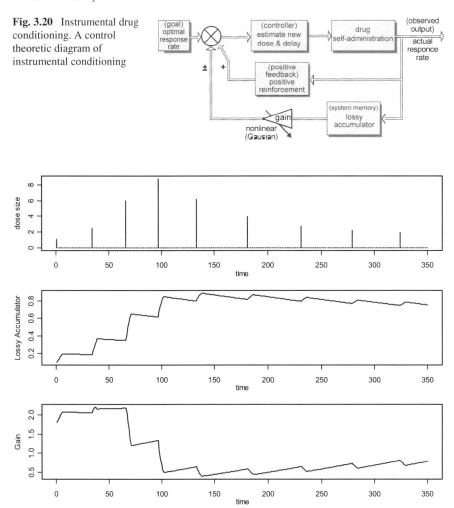

Fig. 3.20 Instrumental drug conditioning. A control theoretic diagram of instrumental conditioning

Fig. 3.21 Instrumental drug conditioning. Illustration of the relationship between the dose size, the amount in the lossy accumulator and the gain function. Note the delay in response of the dose compared to the lossy accumulator and gain function status. After the first two doses the value of the lossy accumulator has increased close to 0.3, which corresponds to the maximum of the gain function. This, in turn, resulted in the increase in the third dose. Although the gain reached the value of 0.6, the fourth dose still has positive gain compared to the previous dose resulting in the lossy accumulator reaching the value of 0.8. This dramatically decreases the gain function and thus the next dose. Gradually, the size of the lossy accumulator decreases such that the gain, and thus the dose size, is slowly increasing again

Similar processes control the timing of the subsequent dose (i.e., the gain function controls the timing of the next dose). Finally, both the dose size and its timing could be simultaneously controlled by sensitization/habituation processes. Figure 3.21 shows the delay between doses.

As noted above, this is not a theory of addiction. It is instead a theory of drug self-administration, which is a common component in models of addiction.

3.8 General Discussion

The seven models presented here span a vast range of different theories and proposed psychobiological mechanisms. Control theory highlights stark contrasts among theories. At the most fundamental level, models that employ only negative feedback are very different from those that use only positive feedback or both positive and negative. Even models derived from a common lineage (e.g., behavioral learning theory) can be remarkably different from each other in terms of their control theory structures (e.g., respondent conditioning versus instrumental conditioning). Yet there are some general similarities that are highlighted when they are modeled using control theory.

3.8.1 Linear

All of the seven models presented here are linear in conception, although implementation led to adding nonlinear components. In constructing these models, we used the approach suggested by Einstein that the model should be as simple as possible but not simpler. Adding more complexity and non-linearity might make the models better describe the complexities of actual drug use, but would make the representation of the theories more obscure. For example, both opponent process theory (Solomon 1980) and Siegel's (1975) respondent conditioning model assume that the two opposing processes ("a" and "b," or unconditioned and conditioned responses, respectively) combine additively (i.e., linearly). Clear and persistent evidence might be needed to abandon this (convenient) assumption.

3.8.2 Dynamical

All of the theories presented here (see Table 3.1) are intended to account for repetitive behavior that unfolds and changes over time. This is the nature of addictive behavior. For the addicted individual, there are a number of possible time scales that are relevant. For example, if the drug of choice is cocaine, then the time scale of the acute drug effect can be in minutes or tens of minutes, while bouts of cocaine use may last hours or days. The initial, intense subjective effect of cocaine is measured in seconds. Craving may last seconds or minutes, while the abstinence syndrome can be very protracted. In contrast, if an individual with opioid dependence receives regular methadone treatments, then these brief time scales are increased 10- to 100-fold because the acute methadone effect is so much longer lasting that of cocaine.

3 Control Theory and Addictive Behavior

Table 3.1 Proposed mechanisms for different stages of drug use and addiction

Model	Naïve use	Experimental drug use	Asymptotic drug use	Relapse
Heroin addiction			prevent withdrawal	
Opponent process		"a" process	"a" process & "b" process	"b" process
Respondent conditioning		unconditioned response	unconditioned & conditioned	conditioned response
Instrumental conditioning		learn contingency	obtain reward	obtain reward
Incentive—sensitization		drug "liking"	drug "wanting"	drug "wanting"
Autoshaping		learn contingency	sign-tracking	sign-tracking
Evolutionary psychology	increase SPFit	increase SPFit	increase SPFit	trigger SPFit

The length of time required for subchronic drug use to lead to addiction is an important time scale that is unclear at the present time. Finally, the career of the addicted individual can be often described in years or decades, with the potential for rapid (hours or days) readdiction after long (up to decades) periods of abstinence.

The extreme variation in times scales for the pharmacokinetics and pharmacodynamics of addictive behavior presents problems for control theory modeling (or any other mathematical modeling, for that matter). The linear dynamical model that describes well the acute drug response may be very different from that for chronic drug use over decades. Only an ambitious control theory model will attempt to apply the same structural model to functions over the full range of time scales.

We have been concerned mainly with system dynamics over brief to medium-length time scales. Drug use also develops over a very long time scale, from initial use in the drug-naïve individual, experimental use, addictive use, and relapse after periods of abstinence. Some of these models make different assumptions about the mechanisms that apply to different stages. This is illustrated in Table 3.1.

Note in Table 3.1 that SPFit theory is the only model that attempts to account for all stages of drug use, including the first use in a person who is naïve to drugs, with the same mechanism; this is the motivation to increase or to protect SPFit. Koob and Le Moal (1997) invoked a two-stage model with very different mechanisms in the early stages of drug use (incentive-sensitization) and the later addictive stage (opponent process theory). The early stage in Koob and Le Moal's (1997) model is characterized by positive feedback, while later stages involve both positive and negative feedback. In their later writings (Koob and Le Moal 2008a, 2008b), these authors have emphasized negative feedback (or negative reinforcement, which is a similar construct) in the addiction to drugs.

Implied in incentive-sensitization theory is that while drug "wanting" increases with persistent use, drug "liking" either remains constant or habituates; this is suggestive of a two-stage model because people in the early stages of drug use take the drug for a different reason ("liking" the drug effect) than in the later stages ("wanting" the drug effect). Traditional learning theories (respondent and instrumental conditioning, as well as autoshaping) imply an early learning phase in which the mechanism is primarily unconditioned, and a later stage in which conditioning has developed fully.

One might assume that two-stage models would account for more addictive phenomena than single-stage models, but inspection of Tables 3.1 and 3.3 suggests that this is not necessarily the case. Again, parsimony is an important consideration.

Most of these theories are nebulous about transitions between stages of drug use. This is striking considering that the transitions between stages, and the psychobiological mechanisms that determine those transitions, are considered critically important by many psychopathologists in the addictions field. For example, what determines when there is a transition from naïve use of a drug or drugs to subchronic use, and who specifically makes this transition and who does not? The structural modeling that we have performed sharply delineates these stages, but does not really speak to the mechanisms of transitions between stages. It does, however, encourage the theorist to specify the psychobiological mechanisms that control each stage, and these same mechanisms then become hypotheses for those mechanisms that control transitions between stages. We leave individual differences to future work, although it represents a substantial part of the literature on the etiology of substance use disorders.

3.8.3 Systems

A system is characterized by different processes or components that are functionally or statistically related to each other. In these systems, a perturbation in one component will reverberate throughout the system because they are mutually interconnected. There is a sense in which control theory allows one to predict how such a perturbation of one variable will affect others in the system. In many cases, these effects could not be predicted by knowing the isolated effect of changing one variable on one other.

The parameters of the model will likely differ as a function of the specific drug or drugs that the individual is taking, person variables (such as genotype, phenotype, and social or physical environment), and pharmacokinetic and pharmacodynamic factors. Most of these models (with the possible exception of the heroin addiction model) were either originally constructed to account for a wide range of addictive drugs, or were generalized to them. This generality is reflected in the Diagnostic and Statistical Manual of the American Psychiatric Association – IV (American Psychiatric Association 2000) as diagnostic criteria for abuse and dependence that are similar or identical for very different drugs of abuse. These drugs are a loose collection of substances that share "abuse liability," and have very little else in common. Modern conceptions of abuse liability go beyond self-administration in humans or nonhuman animals to common neurochemical functions, such as dopaminergic stimulation in the ventral striatum of rodents (Di Chiara and Imperato 1988).

In control theoretic analyses, the application to a specific drug or person becomes a matter of parameter estimation within the particular control theory model. Parameter estimation involves finding parameter values which are best compatible or best explain observed responses of phenomena, and is closely connected to sufficient statistics in statistical modeling (Kay 1993). Parameter estimation is beyond the scope of this discussion.

Noise Although these seven models of addiction are all deterministic systems, in real-world applications they have noise (stochastic) components. In constructing time-series simulations of these systems, the modeler must choose what types of noise (e.g., Gaussian or Poisson) components to add and where to add them.

Feedback All of these models employ some type of feedback. This can be negative feedback, positive feedback, or a combination of the two. Note in Table 3.2 that models with negative feedback tend to be more stable, and (in Table 3.3) to account well for drug withdrawal effects. On the other hand, models with only positive feedback are often unstable, and account well for acquisition of drug-taking and binging (rapid, repetitive use that leads to very high or toxic levels in the blood), but do not model the withdrawal syndrome well. Models with only one type of feedback (i.e., positive or negative but not both) have fewer parameters and are more parsimonious. In contrast, models with more than one type of feedback (i.e., opponent process theory, respondent conditioning, and instrumental conditioning) are inherently complex and can account for a wide range of empirical phenomena at the inevitable expense of parsimony. For example, Solomon's (1980) notion that both and the "a" and "b" processes are conditionable allows the model to account for almost any eventuality, but makes it more difficult to test the theory (i.e., it is difficult to falsify).

Note in Table 3.2 that all the models are linear in theory. However, in order to implement two of these models (opponent process theory and instrumental conditioning) it was necessary to add nonlinear components such as stochastic noise. The 'lossy accumulator' is a linear component (in these cases) that introduces memory into the system.

Set Points It may seem ironic that only one (heroin addiction model) of the seven models that we discuss has a simple set point. Even in this case, the set point is only at asymptotic levels of drug use; it must have had some (unspecified) mechanism by which it changed from a desired blood heroin level of zero (in the never user) to a nonzero positive level after progression to addictive drug use. Several of the models have "neighborhoods" that the system "seeks" or tracks rather than specific set points. For some limited purposes, the mean of this neighborhood could be taken as an estimated set point. These models predict oscillatory system behavior, a characteristic of addictive behavior, and these semi-periodic changes can be fairly regular, particularly when there are minimal social environmental constraints.

Table 3.2 Control theoretic aspects of models of addiction

Model	Feedback	Error signal	Stages	Set point	Stability	Linearity
Heroin addiction	neg (−)	mu-opioid?	1	yes	high	linear
Opponent process	neg (−) pos (+)	affective	2	equilibrium functions	high	linear nonlinear[a]
Respondent conditioning	neg (−) pos (+)	?	2	equilibrium functions	high	linear nonlinear[a]
Instrumental conditioning	neg (−) pos (+)	reinforcement	1	varying set points	moderate	linear nonlinear[a]
Incentive—sensitization	pos (+)	drug "wanting"	2	no	no	linear
Autoshaping	pos (+)	skeletal approach	1	no	no	linear
Evolutionary psychology	pos (+) neg (−)	SPFit	1	varying set points	moderate	linear

[a]The theoretical models are all linear, but implementation for simulations required the inclusion of some nonlinear components (see text)

At the same time, several of the models admitted no set points. This could be due to the fact that they are simple positive feedback systems that do not afford a set point (e.g., incentive-sensitization and autoshaping), or the system generally "strives" toward a higher level (e.g., SPFit theory) rather than a specific optimal level.

Structural Similarity Table 3.3 lists the addictive phenomena for which different models account. We noted (above) that control theory models of opponent process theory and respondent conditioning are so structurally similar that they are virtually identical in control theoretic terms (even though they were developed independently). Another example of structurally similar models is instrumental conditioning, incentive-sensitization, and autoshaping models, whose theoretical similarity is more subtle. All three models have positive feedback arms and the value of the drug "reward" becomes inflated by that positive feedback. These models account for the same addictive phenomena (see Table 3.3), except that McSweeney et al.'s (2005) instrumental model also has a negative feedback arm that accounts for drug tolerance, and their system does reach asymptote (limited by habituation).

Mutual Exclusivity Since some of the models discussed herein propose markedly different mechanisms and systems of addiction, it begs the question of whether these models are mutually exclusive for an individual person (or nonhuman animal), or for addictive processes in general. In fact, Koob and Le Moal's (1997) model incorporates two and perhaps three (including stress allostasis) into one (actually, a

Table 3.3 Addictive phenomena and theoretical models that may account for them

Theory	Acquisition	Craving	Withdrawal	Tolerance	Sensitization	Bingeing	Asymptote	Relapse
Homeostatic								
Heroin addiction	no	yes	yes	no	no	no	yes	no
Opponent process	no	yes	yes	yes	no/yes[a]	no	yes	yes
Respondent conditioning	no	yes	yes	yes	no/yes[a]	no	yes	yes
Nonhomeostatic								
Instrumental conditioning	yes	yes	no	yes	yes	yes	yes	yes
Incentive—sensitization	yes	yes	no	no	yes	yes	no	yes
Autoshaping	yes	yes	no	no	yes	yes	no	yes
Evolutionary psychology	yes	yes	yes	no	yes	yes	yes	yes

[a]The respondent conditioning model can account for sensitization if one assumes that the conditioned response is drug isodirectional (but it cannot then account for tolerance or withdrawal). The same is true of opponent process theory and the "b" process

two-stage model). This issue of mutual exclusivity is facilitated by control theoretic analyses. The combination of different addiction theories into an integrated control theoretic model is an interesting and important area for future research.

3.8.4 Limitations

Alternative Models This discussion focused on several homeostatic and non-homeostatic models of drug addiction that were selected primarily for didactic reasons. These models represent a number of alternative formulations that frequently contradict each other. Of course, there are many theories of addictive behavior that are beyond the scope of this presentation. Some of these other theories may be difficult to model in control theory terms, whether by linear or nonlinear dynamical systems (Khalil 2002). The potential usefulness of research on addictive behavior in the context of control theory should be apparent. Our attempt to make control theory explicit rather than implicit may facilitate research of this nature.

3.8.5 Surrogate Measures

These theoretical models are abstract representations or summaries of complex phenomena. In most cases they are oversimplifications for the purposes of better understanding, communicating, and testing hypotheses about the phenomena. In turn, a mathematical treatment—such as control theoretic modeling—is a further simplification. Nonetheless, modeling plays an important role in understanding the theories and the phenomena to which they relate. At some point, the theory (or mathematical treatment) must "touch down" to measurable variables that are considered key to testing the hypotheses derived from these theories. We noted above that most measures in these psychobiological systems are surrogate measures—that is, they relate, closely we hope, to the actual measures of interest, which may not be available for technical or ethical reasons.

Table 3.4 lists common surrogate measures associated with the seven theories that we have discussed, as well as the target measures that we might prefer to measure but are generally unavailable at this time. In the case of instrumental conditioning, the rate of operant responding is the theoretical *sine qua non* of the model, so the surrogate and the target measures are the same. Again, the variables in Table 3.4 represent a remarkably wide range of measures, from rate of pressing a lever or nose-pokes (rate of drug self-administration) to a questionnaire measure of power and sexual functioning (SPFit). These may or may not have been the measures that we would have chosen in a perfect world, but they may suffice to test theoretical hypotheses. The degree of discrepancy between the surrogate and target measures is of great importance to testing the theories. We hope that control theoretic analyses sharpen the focus on these surrogate measures and issues concerning the congruence between them and the variables we might prefer to measure.

Table 3.4 Common surrogate measures and the target measures often associated with different addiction theories

Model	Surrogate measure	Target measure
Heroin addiction		ventral striatal β-endorphin
Opponent process	intense affect	brain substrates
Respondent conditioning	panorama of the drug effect and its opposite	brain substrates of these two responses
Instrumental conditioning	rate of drug self-administration	rate of drug self-administration
Incentive—sensitization	drug "wanting"	neural plasticity of mesolimbic dopamine circuitry
Autoshaping	physical approach	psychological approach (sign-tracking)
Evolutionary psychology	self-perceived fitness (SPFit)	survival and reproductive fitness motivation system

A further consideration is that most theories do not depend on a single measure, whether surrogate or target. There is typically a vector of measures rather than a single one in the theory and the control theoretic model; this vector may be of surrogate or target variables. For example, SPFit theory uses the variable "SPFit" as the surrogate measure of oversight of the survival and reproductive fitness motivation system. The actual vector of variables might include SPFit-power motivation, self-assessment of social rank and influence, SPFit-reproductive fitness motivation, prefrontal dopaminergic stimulation, ventral striatal dopamine and its modulating influences, etc. 'SPFit' is shorthand for this vector. Our denoting a single variable is a further simplification to already concise models of addiction. Modern control theoretic modeling easily embraces this complexity if the vector of variables is measurable.

3.8.6 Issues for Further Theoretical Development

Our review of theories did not touch on differences among drugs of abuse, among different dosages of the same drug, or even between groups of people or animals that respond differently to the same drug (or to different dosages of that drug), although it is conceivable that these merely involve different sets of parameters in essentially a simple structural model. Moreover, we reviewed mainly models of *chronic* drug use. We discussed aspects of the *acute* response to drugs (e.g., acute tolerance/sensitization, biphasic or multiphasic pharmacodynamic responses, nonlinearities in dose–effect curves, etc.) only as they might relate to models of chronic, relapsing addiction. Whether these many factors can be modeled by changes in parameter values or require entirely different structural models is an open question. However, we do suggest that control theory carries with it powerful tools for addressing these issues, as well as raising questions for study that might not have been asked without this explicit approach.

3.9 Conclusions

We have shown that many (seven) theories of addiction and their variations are amenable to control theoretic modeling. This is only the tip of the iceberg. The control theory literature is immense, and the tools are powerful and fully developed for analyzing feedback systems. We conclude with a set of recommendations for further applications of control theory to addictive behavior that summarize some of the points that we have made:

- Control theoretic diagrams are simple and useful means to summarize and communicate many (but not all) theoretical models of addiction. These diagrams make explicit the feedback loops that are implicit in many theories and highlight limitations of specific models. Moreover, they encourage the theorist to explicitly designate surrogate measures for (currently) immeasurable variables that are central to the operation of the controlled system. Most importantly, these diagrams lead rather directly to mathematical implementations of the models.
- Control theory has been thoroughly developed because of its many applications in engineering, computer science, and biology. It has proven highly flexible in solving theoretical and practical problems. The addictions theorist is more likely to be limited by gaps in the addictions literature than they are of control theoretic implementations. Addiction theories that do not involve feedback—open-loop models—may not benefit by control theoretic modeling, except when the theorist proposes that the model is "normally" closed-loop, but the feedback loop is "broken" in this particular psychopathology.
- The addictions theorist interested in control theory modeling is certainly not limited to the types of control theoretic models presented herein (e.g., hybrid and adaptive models [Goebel et al. 2009; Astrom and Wittenmark 1989; Dewilde and van der Veen 1998]).
- Researchers might consider carefully whether linear control models are sufficient to account for available evidence. Not unlike the "general linear model" in classical statistics, linear dynamical models with some degree of negative feedback are robust and stable. Nonlinear dynamical models and chaotic systems may be necessary to account for some phenomena that are not encompassed by linear models.
- Control theoretic modeling may facilitate the determination of congruence between empirical evidence and theoretical models. Although an in-depth discussion of parameter estimation is beyond the scope of this discussion, we note a general approach. Once one has fixed on a particular structure for a proposed

model, there will typically be several parameters involved, such as time scales, relative units, interaction coefficients, etc. If this structure is really appropriate, we expect that good agreement with experimental data occurs for some correct specifications of parameter values. Without going into details of the theory of parameter estimation, we note that this procedure typically involves an optimization of the choice of values as matching the results/observations of a number of experiments. Depending on the amount of measurement noise and the sensitivity of the measurements to variation of the parameter choices, this procedure can be quite effective in providing a satisfactory match—subject, of course, to the correctness of our assumption that the right general structure has been chosen for analysis.

- Multiple stages of addiction are amenable to control theoretic modeling. Different stages of addiction need not have the same feedback structures (e.g., Koob and Le Moal 1997). Addiction staging highlights the critical importance of factors or psychobiological mechanisms (which may or may not be modeled in control theoretic terms) that determine the transitions between stages.
- Theoretical integration of models in control theoretic terms that might appear irreconcilable or unrelated is a priority for future research. Different theories of addiction may simply emphasize different feedback arms in a controlled system (i.e., control theory may aid reconciliation of apparently unrelated theories). Recognition of the vastly different time scales for different addiction phenomena may lead to controlled systems embedded within other regulated systems, each with different time scales and potentially different feedback mechanisms. This recognizes the multi-dimensional nature of addiction phenomena;

3.9.1 Final Comment

In 1978, Meehl argued that psychological science must go beyond null hypothesis testing and "significant differences between means," in the direction of estimating point values for mathematical functions. Control theoretic modeling is one possible approach to attaining this goal. We have taken a first step in terms of specifying several theoretical models of addiction in control theoretic terms. These specifications lead rather directly to constructing simulations that can be tested against empirical realities. The "goodness of fit" of these simulations has important implications for testing, rejecting, modifying, or developing new theories that can best mimic empirical data. While this is not the only contribution that control theoretic modeling can bring to drug abuse studies, it is arguably the most compelling. We hope that control theory gains traction in addictions research.

Acknowledgements The authors are particularly indebted to Thomas Piasecki for invaluable comments on two separate drafts of this manuscript. We thank Daniel Shapiro, Kevin Strubler and Rachael Renton for editorial assistance and Diane Philyaw for extensive graphics work. The authors have no conflicts of interest in this research. This research was supported in part by NIAAA grant 1R21AA015704 and NIDA grant 1R21DA020592 to DBN and by RTI International.

References

Aarmodt S (2006) Putting the brain back together. Nat Neurosci 9(4):457
Ahmed SH, Bobashev G, Gutkin BS (2007) The simulation of addiction: Pharmacological and neurocomputational models of drug self-administration. Drug Alcohol Depend 90(2–3):304–311
Airy GB (1840) On the regulator of the clock-work for effecting uniform movement of equatorials. Mem R Astron Soc ll:249–267
American Psychiatric Association (2000) Diagnostic and statistical manual of mental disorders, 4th edn. Author, Washington, DC
Anderson BDO, Moore JB (1979) Optimal filtering. Prentice Hall, Upper Saddle River
Astrom KJ, Wittenmark B (1989) Adaptive control. Addison-Wesley, Reading
Baker TB, Tiffany ST (1985) Morphine tolerance as habituation. Psychol Rev 92(1):78–108
Baker TB, Piper ME, McCarthy DE, Majeskie MR, Fiore MC (2004) Addiction motivation reformulated: An affective processing model of negative reinforcement. Psychol Rev 111:33–51
Bobashev G, Costenbader E, Gutkin B (2007) Comprehensive mathematical modeling in drug addiction sciences. Drug Alcohol Depend 89(1):102–106
Bolles RC (1972) Reinforcement, expectancy, and learning. Psychol Rev 79(5):394–409
Brown JS, Jenkins HM (1968) Auto-shaping of the pigeon's key peck. J Exp Anal Behav 11:1–8
Cannon WB (1929) Organization for physiological homeostasis. Physiol Rev 9(3):399–431
Carroll M, Lac ST (1993) Autoshaping i.v. cocaine self-administration in rats: Effects of nondrug alternative reinforcers on acquisition. Psychopharmacology 110:5–12
Carver CS, Scheier MF (1990) Origins and functions of positive and negative affect: A control-process view. Psychol Rev 97(1):19–35
Chen C (1999) Linear system theory and design, 3rd edn. Oxford University Press, Oxford
Cox WM, Klinger E (1988) A motivational model of alcohol use. J Abnorm Psychol 97(2):168–180
Csete ME, Doyle JC (2002) Reverse engineering of biological complexity. Science 295:1664–1669
Darwin C (1871) The descent of man, and selection in relation to sex. Murray, London
Day TA (2005) Defining stress as a prelude to mapping its neurocircuitry: No help from allostasis. Prog Neuro-Psychopharmacol Biol Psychiatry 29(8):1195–1200
DePetrillo PB, White KV, Liu M, Hommer D, Goldman D (1999) Effects of alcohol use and gender on the dynamics of EKG time-series data. Alcohol Clin Exp Res 23(4):745–750
Deroche-Gamonet V, Belin D, Piazza PV (2004) Evidence for addiction-like behavior in the rat. Science 305:1014–1017
Dewilde PM, van der Veen A-J (1998) Time varying systems and computations
Di Chiara G, Imperato A (1988) Drugs abused by humans preferentially increase synaptic dopamine concentrations in the mesolimbic system of freely moving rats. Proc Natl Acad Sci USA 85(14):5274–5278
Doyle JC, Glover K, Khargonekar PP, Francis BA (1989) State-space solutions to standard H-2 and H-infinity control problems. IEEE Trans Autom Control 34(8):831–847
Ehlers CL (1995) Chaos and complexity: Can it help us to understand mood and behavior? Arch Gen Psychiatry 52(11):960–964
Ehlers CL, Havstad J, Prichard D, Theiler J (1998) Low doses of ethanol reduce evidence for nonlinear structure in brain activity. J Neurosci 18(18):7474–7486
Eikelboom R, Stewart J (1982) Conditioning of drug-induced physiological responses. Psychol Rev 89(5):507–528
Eisenberg RS (2007) Look at biological systems through an engineer's eyes. Nature 447:376
Francis BA (1988) A course in H-infinity control. Springer, New York
Goebel R, Sanfelice RG, Teel AR (2009) Hybrid dynamical systems. IEEE Control Syst Mag 29(2):28–93
Goldstein DS, McEwen B (2002) Allostasis, homeostats, and the nature of stress. Stress 5(1):55–58

Gutkin BS, Dehaene S, Changeux J (2006) A neurocomputational hypothesis for nicotine addiction. Proc Natl Acad Sci USA 103(4):1106–1111

Hamming RW (1962) Numerical methods for scientists and engineers. McGraw-Hill, New York

Hearst E, Jenkins HM (1974) Sign-tracking: The stimulant-reinforcer relation and directed action. Monograph of the Psychonomic Society. Austin

Himmelsbach CK (1943) Morphine, with reference to physical dependence. Fed Proc 2:201–203

Hufford MR, Witkiewitz K, Shields AL, Kodya S, Caruso JC (2003) Relapse as a nonlinear dynamic system: Application to patients with alcohol use disorders. J Abnorm Psychol 112(2):219–227

Isidori A (1989) Nonlinear control systems. Springer, New York

Johnson RE, Chang C, Lord RG (2006) Moving from cognition to behavior: What the research says. Psychol Bull 132(3):381–415

Kailath T (1980) Linear systems. Prentice Hall, Upper Saddle River

Kalman RE (1963) Lyapunov functions for the problem of Lur'e in automatic control. Proc Natl Acad Sci USA 49(2):201–205

Kay SM (1993) Fundamentals of statistical signal processing, vol. I—Estimation theory. Prentice Hall, Upper Saddle River

Khalil HK (2002) Nonlinear systems, 3rd edn. Prentice Hall, Upper Saddle River

Koob GF, Le Moal M (1997) Drug abuse: Hedonic homeostatic dysregulation. Science 278(5335):52–58

Koob GF, Le Moal M (2006) Neurobiology of addiction. Academic Press, London

Koob GF, Le Moal M (2008a) Addiction and the brain antireward system. Annu Rev Psychol 59:29–53

Koob GF, Le Moal M (2008b) Review. Neurobiological mechanisms for opponent motivational processes in addiction. Philos Trans R Soc Lond B, Biol Sci 363(1507):3113–3123

Lurie BJ, Enright PJ (2000) Classical feedback control with MatLab. CRC Press, New York

Maxwell JC (1868). On governors. Proceedings of the Royal Society, 100

McSweeney FK, Murphy ES, Kowal BP (2005) Regulation of drug taking by sensitization and habituation. Exp Clin Psychopharmacol 13(3):163–184

Meehl P (1978) Theoretical risks and tabular asterisks: Sir Karl, Sir Ronald, and the slow progress of soft psychology. J Consult Clin Psychol 46:806–834

Nesse RM, Berridge KC (1997) Psychoactive drug use in evolutionary perspective. Science 278(5335):63–66

Newlin DB (1992) A comparison of drug conditioning and craving for alcohol and cocaine. Recent Dev Alcohol 10:147–164

Newlin DB (2002) The self-perceived survival ability and reproductive fitness (SPFit) theory of substance use disorders. Addiction 97(4):427–445

Newlin DB (2007) Self-perceived survival and reproductive fitness theory: Substance use disorders, evolutionary game theory, and the brain. In: Platek SM, Keenan JP, Shackelford TK (eds) Evolutionary cognitive neuroscience. MIT Press, London, pp 285–326

Nyquist H (1932) Regeneration theory. Bell Syst Tech J 11:126–147

Pavlov IP (1927) Conditional reflexes: An investigation of the physiological activity of the cerebral cortex. Oxford University Press, Oxford

Peper A (2004a) A theory of drug tolerance and dependence I: A conceptual analysis. J Theor Biol 229(4):477–490

Peper A (2004b) A theory of drug tolerance and dependence II: The mathematical model. J Theor Biol 229(4):491–500

Petraitis J, Flay BR, Miller TQ (1995) Reviewing theories of adolescent substance use: Organizing pieces in the puzzle. Psychol Bull 117(1):67–86

Pickens RW, Thompson T (1968) Cocaine-reinforced behavior in rats: Effects of reinforcement magnitude and fixed ratio size. J Pharmacol Exp Ther, 161(1):122–129

Poulos CX, Cappell H (1991) Homeostatic theory of drug tolerance: A general model of physiological adaptation. Psychol Rev 98(3):390–408

Redish AD, Jensen S, Johnson A (2008) A unified framework for addiction: Vulnerabilities in the decision process. Behav Brain Sci 31(4):415–437; discussion 437–487. Review

Robinson TE, Berridge KC (1993) The neural basis of drug craving: An incentive-sensitization theory of addiction. Brain Res Brain Res Review 18(3):247–291

Robinson TE, Berridge KC (2003) Addiction. Annu Rev Psychol 54(1):25–53

Siegel S (1975) Evidence from rats that morphine tolerance is a learned response. J Comp Physiol Psychol 89(5):498–506

Siegel S (1983) Classical conditioning, drug tolerance, and drug dependence. Res Adv Alcohol Drug Probl 7:207–246

Skinner BF (1938) The behavior of organisms. Appleton-Century-Crofts, New York

Solomon RL (1980) The opponent-process theory of acquired motivation: The costs of pleasure and the benefits of pain. Am Psychol 35(8):691–712

Solomon RL, Corbit JD (1973) An opponent-process theory of motivation. II. Cigarette addiction. J Abnorm Psychol 81(2):158–171

Stewart J, de Wit H, Eikelboom R (1984) Role of unconditioned and conditioned drug effects in the self-administration of opiates and stimulants. Psychol Rev 91:251–268

Tiffany ST (1990) A cognitive model of drug urges and drug-use behavior: Role of automatic and nonautomatic processes. Psychol Rev 97(2):147–168

Umulis DM, Gürmen NM, Singh P, Fogler HS (2005) A physiologically based model for ethanol and acetaldehyde metabolism in human beings. Alcohol 35(1):3–12

Warren WH (2006) The dynamics of perception and action. Psychol Rev 113(2):358–389

Wiener N (1948) Cybernetics or control and communication in the animal and the machine. Wiley, New York

Wikler A (1973) Dynamics of drug dependence: Implications of a conditioning theory for research and treatment. Arch Gen Psychiatry 28(5):611–616

Part II
Neurocomputational Models of Addiction

Chapter 4
Modelling Local Circuit Mechanisms for Nicotine Control of Dopamine Activity

Michael Graupner and Boris Gutkin

Abstract Nicotine exerts its reinforcing action by boosting dopamine output from the ventral tegmental area (VTA). This increase results from stimulation of nicotinic acetylcholine receptors (nAChRs). However while much is known about the receptor mechanisms of nicotine actions several issues remain to be clarified. One is how the receptor-level action results in acquisition and maintenance of nicotine addiction. Another is what are the specific circuit level neural pathways of nicotine acute action on the dopamine system. In fact in vivo and in vitro experiments reach contradictory conclusions about the key target of nicotine action: direct DA cell stimulation or indirect effects mediated through GABAergic interneurons. We address these issues through computational modeling first through a global framework taking into account multiple-time scales of nicotine effect and second modeling the VTA circuitry and nAChR function which allows to pinpoint the specific contributions of various nAChRs to the DA signal. We show that the GABA interneurons play a central role in mediating nicotine action. Our results propose mechanisms by which the VTA mediates the rewarding properties of nicotine.

4.1 Introduction

The ventral tegmental area (VTA) is a key dopaminergic structure that is involved in signaling of reward and motivation as well as in the acquisition of drug reinforced behavior (Hornykiewicz 1966; Nestler and Aghajanian 1997; Chiara 2000). Nicotine (Nic) stimulates nicotine acetylcholine receptors (nAChRs) in the VTA (Mereu et al. 1987) thereby boosting dopamine levels in its target brain structures, such as the nucleus accumbens (NAcc) (Clarke 1991). Several lines of evidence suggest that the nACh receptors in the mesolimbic dopamine (DA) system dominantely mediate the motivational properties of nicotine (Chiara and Imperato 1988; Corrigall and Coen 1991; Corrigall et al. 1994; Nisell et al. 1994; Picciotto et al. 1998). Despite ample data on the outcome of nicotine action, the exact mechanisms

---

M. Graupner (✉) · B. Gutkin
Group for Neural Theory, Laboratoire de Neurosciences Cognitives, INSERM Unité 969, Départment d'Etudes Cognitives, École Normale Supérieure, 29, rue d'Ulm, 75005 Paris, France
e-mail: michael.graupner@ens.fr

how nicotine usurps dopamine signaling in the VTA have not been conclusively resolved. A significant open question is how the local effects of nicotine lead to nicotine addiction. Another major outstanding question is whether nicotine acts directly on the dopamine neurons or changes the interplay between the dopamine and the local GABA neurons. In other words, is the nicotine boost of DA signaling a single cell—or a local circuit phenomenon? In this chapter we go over our modelling efforts toward answering these open questions.

4.2 The Ventral Tegmental Area as a Local Circuit: A Brief Overview

The VTA is a neuronal microcircuit, containing approximately 80% DA neurons targeted locally by the GABAergic cells (∼20%) (Lacey et al. 1989; Johnson and North 1992b; Sugita et al. 1992; Ikemoto et al. 1997). This local circuit receives glutamatergic (Glu) afferents from the prefrontal cortex (PFC) (Christie et al. 1985; Sesack and Pickel 1992; Tong et al. 1996; Steffensen et al. 1998) and is furthermore innervated with glutamatergic (Clements and Grant 1990; Cornwall et al. 1990; Forster and Blaha 2000) and cholinergic projections (Oakman et al. 1995) from the brainstem. In turn DA and GABAergic projections from the VTA target numerous areas of the brain including the PFC (Thierry et al. 1973; Berger et al. 1976; Swanson 1982; Steffensen et al. 1998; Carr and Sesack 2000a) and limbic/striatal structures (Andén et al. 1966; Ungerstedt 1971; Oades and Halliday 1987; Bockstaele and Pickel 1995). Thus the VTA DA and GABAergic neurons send projections throughout the brain generating DA and GABAergic signals in response to cortical (Taber and Fibiger 1995; Tong et al. 1998) and subcortical inputs (Floresco et al. 2003; Lodge and Grace 2006) as well as to nicotine.

Release of the endogenous ligand ACh into the VTA causes nearly synchronous activation of nAChRs (Dani et al. 2001). The rapid delivery and breakdown of ACh by acetylcholinesterase precludes significant nAChR desensitization (Feldberg 1945). Nicotine (Nic) activates and then desensitizes nAChRs within seconds to minutes (Katz and Thesleff 1957; Pidoplichko et al. 1997; Fenster et al. 1997) since it remains elevated in the blood of smokers during and after smoking (∼500 nM for ∼10 min; Henningfield et al. 1993). Importantly, the various subtypes of nAChRs have distinct affinities for acetylcholine as well as nicotine, exhibit markedly different activation/desensitization kinetics (Changeux et al. 1998), and have different expression targets: (i) low affinity $\alpha 7$ containing nAChRs desensitize rapidly (∼ms) (Picciotto et al. 1998) and are found on glutamatergic terminals; (ii) high affinity, slowly desensitizing $\alpha 4 \beta 2$ containing nAChR on GABAergic cells; and (iii) $\alpha 4$- and $\alpha 6$-containing nAChRs on DA neurons (Calabresi et al. 1989; Klink et al. 2001). Most of the nAChR-mediated currents and the reinforcing properties in response to nicotine are mediated by the $\alpha 4 \beta 2$ nAChR subtype (Picciotto et al. 1998), whereas $\alpha 7$ nAChRs have been suggested to contribute to the fine-tuning of the DA response to nicotine.

On a longer time scale nicotine is known to lead to upregulation of the α4β2 nAChRs. This increases the density of receptors, boosting their number, while the receptors remain fully functional (Picciotto et al. 1998; Champtiaux et al. 2003). A further longer time scale, homeostatic re-dress of the receptor function—the downregulation—has also been hypothesized. In contrast, the α7 nAChRs have therefore been suggested to contribute to long-term potentiation of glutamatergic afferents onto DA neurons (Bonci and Malenka 1999; Mansvelder and McGehee 2000).

4.3 Global Neurocomputational Framework Shows How Receptor-Level Effects of Nicotine Result in Self-administration

We tackled the first question posed above by constructing a neuro-computational framework for nicotine addiction that integrates nicotine effects on the dopaminergic (DAergic) neuron population at the receptor level (signaling the reward-related information), together with a simple model of action-selection. This model also incorporates a novel dopamine-dependent learning rule that gives distinct roles to the phasic and tonic dopamine (DA) neurotransmission. We strived to clarify the relative roles of the positive (rewarding) and opponent processes in the acquisition and maintenance of drug taking behavior, and the development of such behavior into a rigid habit.

The major hypothesis for the approach is that the nicotine effects on dopamine signaling in the ventral tegmental area initiate a cascade of molecular changes that in turn bias glutamatergic (Glu) learning processes in the dorsal striatum-related structures that are responsible for behavioral choice, leading to the onset of stable self-administration. Gutkin et al. (2006) specifically hypothesized that nicotine, through activation and up-regulation of nicotinic acetylcholine receptors (nAChRs) in the VTA (e.g., Picciotto et al. 1998; Champtiaux et al. 2003), dynamically changes the gain of the dopaminergic signaling. Hence, nicotine both potentiates the phasic DA response to rewarding stimuli and evokes such signal by itself (Picciotto et al. 1998; Changeux et al. 1998; Dani and Heinemann 1996). Note that this is rather different than in the reinforcement learning models of addiction (e.g. Redish 2004) where the pharmacological and reward signals are independent. In the neurodynamical framework, the reward signal is in fact modulated by the pharmacological effect of the drug. The phasic DA in turn instructs the learning and plasticity in the action-selection neural machinery that is modeled as a stochastic winner-take-all network (Usher and McClelland 2001; Cho et al. 2002) and reflecting activity in the dorsal nigro-striatal-cortical loops (Beiser et al. 1997; Dehaene and Changeux 2000). Since both DA and nicotine potentiate Glu plasticity in the dorsal striatum (Reynolds and Wickens 2002), the authors proposed a specific Hebbian learning rule for the excitatory (cortico-striatal-cortical) synapses gated by the tonic DA. Persistent nicotine-dependent depression in tonic DA then causes the learned behavioral bias to become rigid. Here Gutkin et al. (2006) hypothesize a slow-onset opponent process that is recruited and that in turn disrupts DA neurotransmission to the point that extinction learning or response unlearning is impaired; hence, progressively, nicotine self-administration escapes from the control of the DA signal. This effectively models the ventral–dorsal progression of long-term addiction hypothesized by DiChiara (1999). Further supporting data for the framework was discussed in Gutkin et al. (2006).

The general framework is applied to simulating self-administration of nicotine. In the computational framework, nicotine affects the DA response through a three-time scale model of drug action on the dopaminergic neuron population; the phasic nicotine dependent activation of nicotinic ACh receptors, slower nicotine dependent upregulation or increase in number of receptors (modeled as a multiplicative term in the model) and subsequent upregulation-evoked opponent homeostatic down-regulation of nAChRs (and hence their responses to nicotine).

Injections of nicotine in sufficient doses potentiate the DA signal so as to gate plasticity in the action-selection machinery. Since nicotine is contingent on a specific action choice (encoded in the model as activity of a specific neuronal population), the excitatory synaptic weights of the corresponding neural population increase and bias the action-selection towards the self-administration of nicotine. With prolonged self-administration, the influence of the DA signal diminishes due to the opponent process (consequence of the receptor down-regulation)—the behavioral bias for the action leading to nicotine becomes "stamped in". Drug seeking behavior becomes routinized, and inelastic the motivational value of nicotine or the cost and is associated with hypodopaminergic withdrawal (Rahman et al. 2003).

Simulations of the above framework, showed that positing drug induced neuroadaptations in the ventral dopaminergic circuitry and drug-modulated learning in the dorsal cortico-striatal action-selection system is sufficient to account for the development and maintenance of self-administration. Importantly, the positive rewarding effect of the drug is translated into biased action selection and choice making, whereas the slow opponent process plays a key role in cementing the drug-associated behavior by removing the DA signal from the range where learning (and unlearning) can take place. Hence, the model predicts that in the long-term the self-administration behavior would tend to become progressively more difficult to extinguish. The model speculates that this effect on action-selection learning may be the reason why nicotine has reportedly high addictive liability despite its limited hedonic impact.

The major strength of the model framework is that it neatly integrates the various processes involved in nicotine self-administration identifying the various functional effects with biological mechanisms and brain structures. This framework can be viewed as a "knowledge repository model" (Bobashev et al. 2007) synthesizing a host of known effects at multiple levels of organization. For example, it links receptor level effects to behavior. The model further makes a number of interesting predictions. An important prediction of the model is that plasticity in the dorsal striatum of animals that are chronically exposed to nicotine should be reduced. These animals should show deficits in re-adjusting their behavior under new conditions (see Granon et al. 2003 for possible experimental equivalent). The above framework implies a hierarchy of thresholds for the progressive stages of addiction. This is an outcome of the distinct roles of the direct motivational (rewarding), and opponent processes in drug addiction such that the dose and duration of the exposure to nicotine for the initial sensitization by the drug is below that for the acquisition of the self-administration, followed by higher thresholds for the stabilization of the self-administration and for the transfer to habit-like rigidity. The computational framework implies that the sensitization of behavior by nicotine through DA-dependent processes may be disassociated from the acquisition of self-administration. At low doses/short duration, nicotine may lead to apparent behavioral sensitization, but not self-administration. Drug-related behaviors may be acquired due to the action of the positive "reinforcement" or "reward" DA-related process. Hence, the acquisition of self-administration would be under motivational control. The behavioral choices will be selected probabilistically in agreement with their relative value. The development of rigidity in actions is a major point of the neurocomputational framework proposed by Gutkin et al. (2006). The model suggests how, in the long run, processes that oppose the primary reward ingrain the drug-related behavior making it independent of the motivation state and value of various action choices and difficult to modify in the face of changing contingencies. This further implies that drug-related behaviors would be extremely difficult to unlearn, even when the environment is enriched by new rewarding stimuli.

Like other models, the neurodynamical approach rests on a number of assumptions to be confirmed and leaves questions that are not directly addressed. For example, explaining why nicotine self-administration can be difficult to acquire remains a challenge. One clue may come from the hypothesized multiplicative role of nicotine on dopaminergic signaling: at low doses nicotine may not boost the phasic dopamine signal (burst) sufficiently to lead to learned self-administration, yet when the dopamine burst is evoked by another rewarding stimulus, the multiplicative effect of nicotine would boost such DA response nonlinearly, subsequently leading to a preference for drug-related behavior. Finally, the global model does not pin-point the specific local mechanisms by which nicotine may bias the DA signaling. We now turn to reviewing a novel hybrid pharamcodyamics- neural model that focuses exactly on the question of local circuitry and nicotinic receptor mechanisms.

4.4 Local Circuit Model of the VTA Shows the Mechanisms for Nicotine Evoked Dopamine Responses

Despite the accumulated knowledge, the mechanisms that translate the nAChR activation by nicotine into the observed DA response in the VTA remains controversial. In particular, in vitro studies attribute a crucial role to nicotine-mediated activation of $\alpha 4\beta 2$ nAChRs on GABA cells, whereas in vivo data point to DA cells, leaving the dominant site of nicotine action elusive. In vitro data show that nicotine transiently increases afferent glutamatergic and local GABA input to VTA DA cells (Mansvelder and McGehee 2000; Mansvelder et al. 2002). Mansvelder et al. (2002) show that the change in inhibitory postsynaptic current (IPSC) frequency in response to nicotine is biphasic: a robust increase during the presence of nicotine is followed by a drop below baseline after the removal of nicotine. Blocking $\alpha 7$-containing nAChRs with methyllycaconitine (MLA), a specific $\alpha 7$ nAChR antagonist, reduces the relative IPSC frequency increase from 320% to 280%, whereas mecamylamine (MEC), an antagonist of non-$\alpha 7$-containing nAChRs, completely abolishes the change in IPSC frequency. Together with the fact that GABAergic neurons mainly express $\alpha 4\beta 2$ nAChRs (Klink et al. 2001), Mansvelder et al. (2002) conclude that the increase in IPSC frequency is due to activation of $\alpha 4\beta 2$ nAChRs on those cells. Based on in vitro recordings, Mansvelder et al. (2002) reason that the increase in DA is due to disinhibition. The idea being that nicotine transiently boosts GABA transmission to DA cells, followed by $\alpha 4\beta 2$ nAChR desensitization which in turn removes the tonic cholinergic drive to GABAergic neurons, i.e. disinhibits DA cells (Mansvelder and McGehee 2000; Mansvelder et al. 2002). In a stark contrast, in vivo studies of wild-type as well as $\alpha 7$- and $\beta 2$ knockout mice emphasize the importance of $\beta 2$-containing nAChRs on DA neurons (Mameli-Engvall et al. 2006). Mameli-Engvall et al. (2006) conclude that $\beta 2$-nAChRs on DA cells act like a nicotine-triggered switch between basal and excited state of these cells.

To pin-point the mechanisms by which nicotine acutely controls the DA neuron activity in the VTA we apply computational methods. We build a simple neuronal network model that accounts for the afferent inputs to the VTA, the local VTA connectivity as well as the location and activation/desensitization properties of the different nAChR subtypes. We examine to which extent nicotine influences DA signaling through $\alpha 4\beta 2$ nAChRs on GABAergic or DA cells, revealing that disinhibition and direct stimulation of DA cells, respectively, may in principle account for experimental data. However, and crucially, the general conditions required suggest the disinhibition scenario to be more likely, i.e. $\alpha 4\beta 2$ nAChRs on GABA cells predominantly mediate the reinforcing role of nicotine. Lastly, we develop a series of experimental predictions and protocols that should disambiguate the predominance of the local circuit (disinhibition via GABA) vs. single cell (direct DA stimulation) pathways.

4.5 VTA and nAChR Model

In order to examine the mechanisms of nicotine action, we build a neural population model of the ventral tegmental area microcircuit. The temporal dynamics of the dopaminergic and GABAergic neurons in the VTA are modeled using a mean-field description, that is we model the dynamics of the firing rates (see Wilson and Cowan 1972 for the derivation of mean-field equations, and Hansel and Sompolinsky 1998 for an example). The average activities of both neuronal populations are accounted for with respect to afferent inputs to the VTA, local circuitry and the location as well as activation/desensitization properties of nicotinic acetylcholine receptors. Specifically, we take into account the two main classes of nAChRs responsible for nicotine evoked responses in the ventral tegmental area, i.e. high affinity slowly desensitizing ($\alpha 4\beta 2$-type) and low affinity rapidly desensitizing nAChRs ($\alpha 7$-containing) (Mansvelder and McGehee 2000; Champtiaux et al. 2003; Gotti et al. 2007).

4.5.1 Mean-Field Description of Dopaminergic and GABAergic VTA Neurons

The temporal dynamics of the average activities of dopaminergic and GABAergic neuron populations in the VTA is characterized by

$$\tau_D \dot{v}_D = -v_D + \Phi(I_0 - I_G + I_{\text{Glu}} + rI_{\alpha 6\beta 2}), \quad (4.1)$$

$$\tau_G \dot{v}_G = -v_G + \Phi(I_{\text{Glu}} + (1-r)I_{\alpha 4\beta 2}). \quad (4.2)$$

v_D and v_G are the firing rates of the DA and GABAergic neuron populations, respectively. τ_D and τ_G are membrane time constants of the neurons specifying how quickly the neuron integrates input changes, i.e. $\tau_D = \tau_G = 20$ ms. I_{Glu} and $I_{\alpha 4\beta 2}$ characterize excitatory inputs to both neuron populations mediated by glutamate receptors and $\alpha 4\beta 2$-containing nAChRs, respectively, expressed by the neurons. I_G is the local inhibitory input to DA neurons emanating from VTA GABAergic neurons. I_0 is an intrinsic current of DA cells giving rise to intrinsic activity in the absence of external inputs (Grace and Onn 1989). We assume furthermore that I_0 accounts for other input sources with are not affected by nicotine exposures and therefore provide a constant background input (e.g. $\alpha 6$-containing nAChR mediated cholinergic input to DA cells, Champtiaux et al. 2003; inhibitory input originating in other brain regions; etc.). $\Phi(I)$ is the steady-state current-to-rate transfer function. For simplicity, we assume that $\Phi(I)$ is threshold-linear, i.e. $\Phi(I) = I$ if $0 \leq I$ and $\Phi(I) = 0$ otherwise. The parameter r controls the balance of $\alpha 4\beta 2$ nAChR action through GABAergic or dopaminergic cells in the VTA. For $r = 0$: $\alpha 4\beta 2$ containing nAChRs act through GABAergic neurons only, whereas for $r = 1$: $\alpha 4\beta 2$ receptor activation influences DA neurons only. Both neuron populations are influenced by $\alpha 4\beta 2$ nAChR activation for intermediate values of r. In practice, this balance is determined by the expression level of $\alpha 4\beta 2$ nAChRs, the overall impact of local GABAergic inputs on DA activity and by the location of $\alpha 4\beta 2$ nAChRs on the somatodendritic tree of DA and GABAergic cells. We vary r in order to investigate the implications of $\alpha 4\beta 2$ nAChR action through GABAergic or dopaminergic cells. See Fig. 4.1A for a schematic depiction of the VTA as accounted for by the model.

The input currents in Eqs. (4.1) and (4.2) are given by

$$I_G = w_G v_G, \quad (4.3)$$

$$I_{\text{Glu}} = w_{\text{Glu}}[v_{\text{Glu}} + v_{\alpha 7}]_1, \quad (4.4)$$

$$I_{\alpha 4\beta 2} = w_{\alpha 4\beta 2} v_{\alpha 4\beta 2}, \quad (4.5)$$

where the w_x's (with $x = G$, Glu, ACh) specify the total strength of the respective input since the activation variables (v_G, $v_{\alpha 7}$, $v_{\alpha 4\beta 2}$) are normalized to vary between 0 and 1. For our qualitative investigations, we use $w_x = 1$ (with $x = $ Glu, ACh) without loss of generality. Inhibitory input to DA cells, I_G, depends on the GABAergic neuron population activity, ν_G. Glutamatergic input is activated either by upstream glutamatergic activity, ν_{Glu}, or by activation of $\nu_{\alpha 7}$ nAChRs on presynaptic glutamatergic terminals, $\nu_{\alpha 7}$ (see next section). Nicotine-evoked glutamatergic transmission is independent of action potential activation in presynaptic fibers (Mansvelder and McGehee 2000). Hence, either of both inputs can fully activate glutamatergic transmission

$$[\nu_{Glu} + \nu_{\alpha 7}]_1 = \begin{cases} \nu_{Glu} + \nu_{\alpha 7} & \text{if } \nu_{Glu} + \nu_{\alpha 7} \leq 1, \\ 1 & \text{if } \nu_{Glu} + \nu_{\alpha 7} > 1. \end{cases} \qquad (4.6)$$

The activation of $\alpha 4\beta 2$ nAChRs, $\nu_{\alpha 4\beta 4}$ (see next section), determines the level of direct excitatory input, $I_{\alpha 4\beta 2}$, evoked by nicotine or acetylcholine (Champtiaux et al. 2003).

4.5.2 Modeling the nAChR Activation and Desensitization Driven by Nic and ACh

The activation and desensitization of nicotinic acetylcholine receptors (nAChRs) is controlled by a number of endogenous and exogenous ligands including acetylcholine (ACh) and nicotine, respectively (Gotti et al. 2006). We included in our circuit model of the VTA a minimal description for the receptors that resolves the subtype specific activation and desensitization properties of the two considered nAChR classes, i.e. the high affinity slowly desensitizing $\alpha 4\beta 2$ nAChR, and the low affinity fast desensitizing $\alpha 7$ nAChR.

We implement nAChR activation and desensitization as transitions between two independent state variables. This yields four different states of the nicotinic acetylcholine receptor: inactivated/sensitized (also resting or responsive state), activated/sensitized, activated/desensitized and inactivated/desensitized state (see Fig. 4.1B). Of those states, three are closed and the activated/sensitized state is the only open state of the receptor in which it mediates an excitatory current. Note that compared to other model suggestions of allosteric transitions of the nAChR, we choose to leave aside the rapidly and slowly desensitized states (Changeux et al. 1984), deeper-level desensitized state or inactivated states (Dani and Heinemann 1996). Such states are collapsed in the desensitized state here. Our model is modified from Katz and Thesleff (1957) where "effective" and "refractory" in their model refer to sensitized and densensitized here, respectively. Assuming independent transitions of the activation and the desensitization variables entails another simplification compared to cyclic allosteric transition schemes. In our model, the reaction rates are the same on opposite sides of the reaction cycle (Fig. 4.1B), i.e. the rate from inactivated/sensitized to activated/sensitized is the same as the transition rate from inactivated/desensitized to activated/desensitized.

The model accounts for the opening of the channel (transition from inactivated/sensitive to activated/sensitive, Fig. 4.1B and C) in response to both, Nic and ACh; while desensitization is driven solely by nicotine (transition into the activated/desensitized state, Fig. 4.1B and C). Although acetylcholine may also desensitize nAChRs (Katz and Thesleff 1957; Quick and Lester 2002), its rapid removal from the synapse by hydrolization through acetylcholinesterase (Feldberg 1945) allows us to leave aside the ACh driven desensitization. The inverse transitions, i.e. from activated to inactivated and from desensitized to sensitized, occur after the removal of Nic and ACh.

The total activation level of nAChRs ($v_{\alpha4\beta2}, v_{\alpha7}$) is modeled as the product of the fraction of receptors in the activated state, a, and the fraction of receptors in the sensitized state, s. The total normalized nAChR activation is therefore with $v_x = a_x s_x$ with $x = \alpha4\beta2$ or $\alpha7$. The time course of the activation and the sensitization variables is given by

$$\frac{dy}{dt} = (y_\infty(Nic, ACh) - y)/\tau_y(Nic, ACh), \quad (4.7)$$

where τ_y (Nic, ACh) refers to the Nic/ACh concentration dependent time constant at which the maximal achievable state y_∞ (Nic, ACh) is exponentially attained. The maximal achievable activation or sensitization, for a given Nic/ACh concentration, a_∞ (Nic, ACh) or s_∞ (Nic, ACh) respectively, is given by Hill equations of the form

$$a_\infty(Nic, ACh) = \frac{(ACh + \alpha Nic)^{n_a}}{EC_{50}^{n_a} + (Ach + \alpha Nic)^{n_a}}, \quad (4.8)$$

$$s_\infty (Nic) = \frac{IC_{50}^{n_s}}{IC_{50}^{n_s} + Nic^{n_s}}. \quad (4.9)$$

EC_{50} and IC_{50} are the half maximal concentrations of nACh receptor activation and sensitization, respectively. The factor $\alpha > 1$ accounts for the higher potency of nicotine to evoke a response as compared to acetylcholine (Peng et al. 1994; Gerzanich et al. 1995; Buisson and Bertrand 2001; Eaton et al. 2003; see Table 4.2). n_a and n_s are the Hill coefficients of activation and sensitization.

The transition from the inactivated to the activated state is fast (~µs, ms) (Changeux et al. 1984) compared the time scales investigated here that are of the order seconds to minutes. We therefore simplify the activation time constant, τ_a, to be independent of the acetylcholine/nicotine concentration, i.e. $\tau_a(Nic, ACh) = \tau_a = $ const. The time course of nicotine driven desensitization is characterized by a nicotine concentration dependent time constant

$$\tau_d(Nic) = \tau_0 + \tau_{max} \frac{K_\tau^{n_\tau}}{K_\tau^{n_\tau} + (Nic)^{n_\tau}} \quad (4.10)$$

τ_{max} refers to the recovery time constant from desensitization in the absence of nicotine ($\tau_{max} \gg \tau_0$). τ_0 is the minimal time constant at which the receptor is driven in the desensitized state at high Nic concentrations. K_τ is the Nic concentration at which the desensitization time constant attains half of its minimum ($\tau_{max} \gg \tau_0$). The parameters describing activation and desensitization of the two nicotinic acetylcholine receptor subtypes are taken from a number of studies on human nAChRs and are listed in Table 4.2. Note that we use $\tau_{max} = 10$ min for $\alpha4\beta2$ nAChR in order to match the time course of DA activity recorded in vivo (see below), while Fenster et al. (1997) recorded a value of $\tau_{max} = 86.9$ min during experiments at room temperature (see Sect. 4.10). τ_0 is adjusted such that the qualitative time course of nAChR mediated current recordings is reproduced (see Sect. 4.6).

In our simulations, we assume that both the nicotine bath application and the intravenous injection implies a slow build up of the nicotine concentration at the site of the receptor. That is, the applied nicotine concentration is not immediately available but increases/decays exponentially with a time constant of 1 min. The fast activation of nAChRs, a (transition from inactivated/sensitive to activated/sensitive, Fig. 4.1B), is therefore taken to be in steady-state with the nicotine concentration at all times.

Clearly, the above presented simple model of nAChR activation and desensitization does not resolve all the details of nAChR kinetics. For example, it is assumed that ACh and Nic evoked responses reach the same maximal amplitude and that despite different potencies, Nic and ACh dose–response curves can be characterized by the same Hill coefficient. These assumptions are approximately met for $\alpha 7$-containing nAChRs (Chavez-Noriega et al. 1997). ACh evokes however twice the response of Nic with human $\alpha 4 \beta 2$ nAChRs in a study by Chavez-Noriega et al. (1997), but the same response according to Buisson and Bertrand (2001). We simplify the dose–response curve using a single Hill equation, rather than using a sum of two Hill equations as suggested by Buisson and Bertrand (2001). Nevertheless the simple model presented here captures the qualitative time course of nAChR currents evoked in response to Nic and ACh exposures (see below).

4.6 VTA Model Results

To address the question of the specific mechanisms of nicotine action we propose a minimal local circuit model of the VTA that reflects the glutamatergic and cholinergic afferent inputs to DA and GABA cells in the VTA, as well as local inhibition of DA cells by GABA neurons (see Fig. 4.1A). The $\alpha 4 \beta 2$- and the $\alpha 7$ nAChRs mediate nicotine effects in VTA. We describe their activation and desensitization in response to nicotine and ACh by a simple 4-state model adapted from Katz1957 (see Fig. 4.1B and C). We model $\alpha 7$ nAChRs at presynaptic terminals to affect Glu input strength. We introduce a parameter r which allows to shift continuously the balance of $\alpha 4 \beta 2$ nAChR action on DA cell activity from mediated purely through GABA cells ($r = 0$, disinhibition) to being exerted directly on DA cells ($r = 1$, direct stimulation; see Fig. 4.1A). We adjust the model to reproduce in vitro and in vivo nicotine application experiments using the same implementation of the VTA model and only changing the external input strength. The model combines a mean-field dynamic firing-rate approach with subtype specific receptor kinetics to study neuronal activity in response to endogenous (Nic) and exogenous (ACh) ligands acting on nAChRs (please refer to Sect. 4.5 for more details on the model).

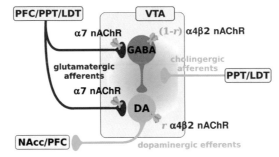

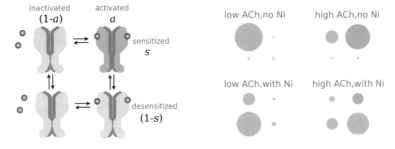

Fig. 4.1 Scheme of the ventral tegmental area and the states of nicotinic acetylcholine receptors. **A**, Afferent inputs and circuitry of the ventral tegmental area. GABAergic neuron population (*red*) and dopaminergic neuron population (*green*) receive excitatory glutamatergic input (*blue lines*) from the PFC, the LDT and the PPT. The LDT and the PPT furthermore furnish cholinergic projections (*cyan lines*) to the VTA. nAChRs are found at presynaptic terminals of glutamatergic projections (α7-containing receptors), on GABAergic neurons ($\alpha 4\beta 2$ nAChRs) and DA neurons ($\alpha 4\beta 2$ nAChRs). r is a parameter introduced in the model to change continuously the dominant site of $\alpha 4\beta 2$ nAChR action. Dopaminergic efferents (*green*) project, amongst others, to the NAcc and the PFC (see text for more details). **B**, State model of nicotinic acetylcholine receptors. Activation (*horizontal*) and desensitization (*vertical*) of nAChRs are two independent transitions in the model, i.e. the receptor can exist in four different states: (i) inactivated/sensitized (*up-left*), (ii) activated/sensitized (*up-right*), (iii) inactivated/desensitized (*down-left*), and (iv) activated/desensitized (*down-right*). Activation is driven by Nic and ACh and induces a transition from the inactivated/sensitized to the activated/sensitized state (*green*), the only open state in which the receptor mediates an excitatory current. Desensitization is driven by Nic only. a and s characterize the fraction of nAChRs in the activated and the sensitized state, respectively (modified from Katz and Thesleff 1957). **C**, $\alpha 4\beta 2$ nAChR state occupation as described by the model for different Nic and ACh concentrations. The area of the circle represents the fraction of receptors in one of the four states (alignment as in **B**). The occupation of receptor states is shown for long term exposures to low (0.1 μM) and high (100 μM) ACh, without and with 1 μM nicotine. A star means that the respective state is not occupied

4.7 Kinetics of the Subunit Specific nAChR Model Stimulated by Nic and/or ACh

In order to develop our hybrid neurodynamical—pharmacodynamical network we must first come up with a simple, yet sufficiently accurate model of the nACh receptor. Hence we define here the $\alpha 4\beta 2$- and $\alpha 7$ nACh receptor model by replicating the receptor responses to nicotine and acetycholine recorded with human nAChR subtypes expressed in *Xenopus* oocytes. We then use the receptor model to investigate to which extent the current evoked by ACh is affected in the presence of a constant, realistic level of nicotine. In particular, we emphasize the differences in response properties for the two nAChR subtypes responsible for nicotine action in the VTA: the $\alpha 4\beta 2$- and the $\alpha 7$.

Both, acetylcholine and nicotine induce the transition from the inactivated/sensitized to the activated/sensitized state of the receptor allowing an excitatory ionic current to pass through the receptor pore (see Fig. 4.1B). Sustained presence of the agonist drives the receptor in the desensitized state in which the receptor is in a closed conformation. The receptor recovers on the time scale of seconds to minutes back to the inactivated/sensitized state in the absence of the agonist (Changeux et al. 1984). Figure 4.2A and B show how simultaneous receptor activation and desensitization shape the current mediated by the $\alpha 4\beta 2$- and the $\alpha 7$-containing nAChR in our model in response to nicotine. Fast activation gives rise to an initial peak current and the concurrent slow desensitization reduces the current during the 200 ms of the agonist presence (blue lines in Fig. 4.2A and B). Figure 4.2C and D show the dose–response curves of the peak current and the evoked net charge (illustrated in Fig. 4.2A and B) with respect to agonist concentration for the $\alpha 4\beta 2$- and the $\alpha 7$-containing nAChR, respectively.

Fig. 4.2 Nicotinic acetylcholine receptor responses to nicotine and acetylcholine. Response properties of $\alpha 4\beta 2$- (panels **A**, **C**, **E**) and $\alpha 7$ nAChRs (panels **B**, **D**, **F**). **A** & **B**, Dynamics of $\alpha 4\beta 2$- (**A**) and $\alpha 7$-containing nAChRs (**B**) in response to nicotine. Activation, a, (*purple lines*) and sensitization, s, (*orange lines*) variables are shown during and after the exposure to a constant Nic concentration of 10 μM (**A**) and 100 μM (**B**) for 200 ms starting at $t = 50$ ms. The normalized receptor activation, $v = a \cdot s$, is shown in *blue*. Peak current and net charge mediated during the exposure are illustrated in the panels. The *inset* shows the dynamics of the same variables on a longer time scale. **C** & **D**, Dose–response curves of $\alpha 4\beta 2$- (**C**) and $\alpha 7$-containing nAChRs (**D**) in response to Nic and ACh. *Full lines* show the peak current (illustrated in **A** & **B**) and the *dashed lines* show the normalized net charge (illustrated in **A** & **B**) mediated by the receptor during a 200 ms exposure to the respective agonist concentration. The responses to nicotine (acetylcholine) are depicted in *blue* (*red*). Realistic nicotine concentrations are indicated by the *arrow*. Example currents evoked by Nic (*blue lines*) and ACh (*red lines*) are shown on the *top of the panel* for different agonist concentrations (in μM on top of the traces). **E** & **F**, Dose–response curves of $\alpha 4\beta 2$- (**E**) and $\alpha 7$-containing nAChRs (**F**) in response to ACh in the presence of constant Nic. The normalized peak current (*full lines*) and the normalized net charge (*dashes lines*) evoked by ACh (*green lines*) are shown. A constant concentration of $Nic = 0.5$ μM is present during the 200 ms exposures to ACh. *Red lines* depict the responses evoked by the respective ACh concentration in the absence of Nic (as depicted in **C** and **D**). The net charge is normalized to 163 unit current times ms for $\alpha 4\beta 2$ nAChRs (panel **C** and **E**) and to 73 unit current times ms for $\alpha 7$ nAChRs (panels **D** and **F**). See Table 4.2 for parameters

4 Modelling Local Circuit Mechanisms for Nicotine Control of Dopamine Activity   123

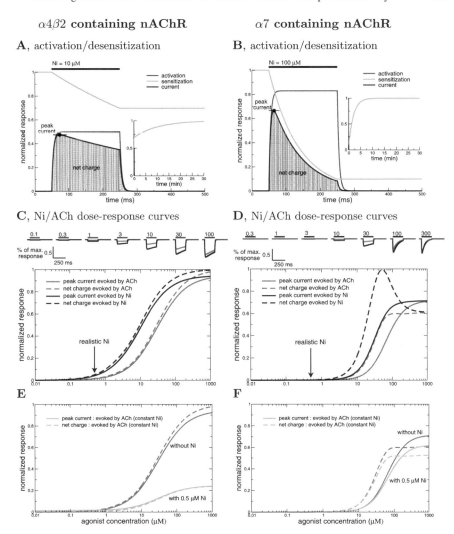

Inward current traces in response to different concentrations of ACh (red lines) and Nic (blue lines) are shown on top of Fig. 4.2C and D. The two nAChR subtypes exhibit a fast current increase with the onset of the agonist exposure and the current reaches a stable plateau level for sustained applications of low agonist concentrations (e.g. at $ACh = Nic = 3$ μM for $\alpha 4\beta 2$ nAChRs, or at $ACh = Nic = 10$ μM for $\alpha 7$ nAChRs). However, the currents after the initial peak mediated by the two nAChR subtypes exhibit different kinetics at higher agonist concentrations. The $\alpha 4\beta 2$ nAChR mediated currents decay slowly over the course of agonist exposure due to the slow transition in the desensitized state (Fig. 4.2C, for ACh and $Nic \geq 10$ μM). On the contrary, $\alpha 7$ nAChR mediated currents decay completely after the fast initial peak current in the presence of high agonist concentrations (Fig. 4.2C, for ACh and $Nic \geq 100$ μM). Here the fast desensitization of $\alpha 7$-containing receptors suppresses completely the inward current on the time scale of ~ 100 ms. In practice, we adjust the minimal time constant by which the receptors can be driven in the desensitized state, τ_0, such that the model currents capture qualitatively the experimentally observed behavior (see above). This fit yields a faster minimal desensitization constant for the $\alpha 7$ nAChR, i.e. $10\tau_0^{\alpha 7} = \tau_0^{\alpha 4\beta 2}$ (see Sect. 4.5 and Table 4.2, compare top panels of Fig. 4.2C and D with Briggs et al. (1995), Chavez-Noriega et al. (1997), Buisson and Bertrand (2001), Papke (2006)).

We furthermore investigate the model response of $\alpha 4\beta 2$- and $\alpha 7$-containing nAChRs, respectively, to acetylcholine applications in the presence of a constant level of 0.5 µM nicotine (green lines in Fig. 4.2E and F). While the presence of Nic does not affect the half-maximum concentrations of peak amplitude and transmitted net charge, it reduces the maximum amplitudes of both attained at high agonist concentrations. The maximal response of $\alpha 4\beta 2$ nAChRs is reduced to $\sim 25\%$ of the response in the absence of the drug (see Fig. 4.2E), whereas the maximum peak amplitude and net charge only drops to $\sim 85\%$ of the response at 1000 µM ACh for $\alpha 7$ nAChR in the absence of nicotine (Fig. 4.2F). This is because a smaller fraction of $\alpha 7$ nACh receptors is driven in the desensitized state due to the lower affinity of $\alpha 7$ nAChRs for desensitization as compared to $\alpha 4\beta 2$ nAChR, i.e. $IC_{50}^{\alpha 7} \gg IC_{50}^{\alpha 4\beta 2}$ (see Sect. 4.5 and Table 4.2). Nicotine not only triggers desensitization but also evokes activation that results in a small residual activation (see elevated green lines at very low agonist concentrations in Fig. 4.2E and F). This baseline current is more apparent with $\alpha 4\beta 2$ nAChR due to the higher affinity and has been seen experimentally (Pidoplichko et al. 1997). Overall, realistic nicotine concentrations reduce the response of both receptor subtypes to ACh, the response is however not completely abolished due to partial desensitization by 0.5 µM nicotine only.

The experimentally observed differences in response properties of the two receptor subtypes are reproduced by our model. We show below that these give rise to functionally distinct roles of the two considered nAChR subtypes in the VTA. Most importantly, the affinity of the $\alpha 4\beta 2$ nAChR for nicotine and acetylcholine is much higher than that of the $\alpha 7$ nAChR. This fact is crucial since the nicotine concentration remains relatively low (0.5 µM) in the blood of smokers (Henningfield et al. 1993; indicated in Fig. 4.2C and D). Moreover, nicotine is a more potent agonist than ACh since the half-maximum nicotine concentration for the peak current and the net charge is lower than the same measure for ACh for both receptor subtypes (the higher potency of Nic compared to ACh is reflected by the factor α in the model, i.e. $\alpha_{\alpha 4\beta 2} = 2$, $\alpha_{\alpha 7} = 3$, see Sect. 4.5 and Table 4.2). Another important difference is the slow desensitization of $\alpha 4\beta 2$ nAChRs in contrast to the fast desensitization dynamics of $\alpha 7$ nAChRs. The fast desensitization of $\alpha 7$ nAChRs yields a significantly lower half-maximum concentration of the transmitted net charge than that of the peak current (see Fig. 4.2D). High agonist concentrations drive the receptor rapidly in the desensitized state leading to an early saturation (and even reduction) of the transmitted charge, while the initial peak remains unaffected by desensitization ($\tau_a \ll \tau_0$) and keeps rising with agonist concentration. A difference in half-maximum concentrations of the same order of magnitude is observed experimentally for $\alpha 7$ nAChRs (Papke and Papke 2002; Papke 2006). This detail further validates our reduced receptor model.

In summary we have constrained our reduced nAChR model to capture the key properties of the subtype specific responses to nicotine and acetylcholine. Importantly, experimental whole-cell current recordings (e.g. from oocytes, human embryonic kidney 293 cells, neurons) expressing the respective nAChR subtype are typically done in the absence of acetylcholinesterase (Peng et al. 1994; Gerzanich et al. 1995; Fenster et al. 1997; Buisson and Bertrand 2001; Eaton et al. 2003; Papke 2006). We therefore assume in this section that induced inward currents rapidly activate *and* desensitize in response to nicotine and acetylcholine. Note however that desensitization is assumed to be driven by nicotine only in the rest of this study. In other words, acetylcholine drives the transition from inactivated to activated only and is hydrolyzed before it evokes significant desensitization (see Fig. 4.1B and C).

4.8 Modeling the VTA Response to Nicotine in vitro and in vivo

We now turn to the question whether the in vitro and in vivo data on nicotine action in the VTA can be reconciled? What are the specific pathways for nicotine action? We start by showing that in vitro excitatory and inhibitory input changes to DA neurons in response to bath perfusion of nicotine are captured by the model by adjusting afferent input strengths only (Mansvelder and McGehee 2000; Mansvelder et al. 2002). We then show that the DA recordings in vivo from wild type and nAChR knockout mice are reproduced by the same model of the VTA again changing only the afferent input strength (Mameli-Engvall et al. 2006). Hence we show that a minimal circuit model of the VTA model can elegantly account for both in vitro and in vivo for electrophysiological recordings from VTA DA neurons during nicotine applications. Finally we go on to show that two alternative mechanisms, direct stimulation and disinhibition of DA cells, may potentially account for the increase in DA activity recorded in vivo in response to intravenous nicotine injections. We then delineate the conditions for each mechanism.

4.8.1 Excitatory and Inhibitory Input Increases to DA Cells in vitro Reproduced by the Model

Data show that bath application of nicotine initially increases the frequency of IPSCs followed by a drop below baseline of the IPSC frequency after nicotine perfusion (Mansvelder et al. 2002). It has furthermore been shown that nicotine causes a robust enhancement of the spontaneous EPSC frequency onto DA neurons in slices of the VTA (Mansvelder and McGehee 2000). We show here that the qualitative behavior of VTA DA cell inputs is reproduced by the model if we assume weak afferent input strength, which characterizes the in vitro situation in which those data were recorded.

The model reproduces GABAergic input changes to DA neurons in response to nicotine (1 μM nicotine for 2 min). Note that Glu transmission is blocked during those experiments. In the model, this corresponds to setting $w_{Glu} = 0$ (see Sect. 4.5). The GABAergic input (I_G) to VTA DA cells increases initially and drops below baseline after nicotine application (Fig. 4.3C). This change is mediated through nicotinic action on $\alpha 4\beta 2$ nAChRs on GABA cells. The maximal relative increase in I_G during nicotine perfusion (green line in Fig. 4.3E) and the maximal relative decrease after nicotine washout (magenta line in Fig. 4.3E) as observed in the model (lines in Fig. 4.3E) matches the experiments (squares in Fig. 4.3E, Mansvelder et al. 2002). Note that the GABAergic input to the DA cells (I_G) corresponds to the activity of the VTA GABA cells (ν_G) since they furnish local inhibitory connections.

What are the receptor level mechanisms for this effect? The $\alpha 4\beta 2$ nAChR expressed by GABAergic cells are activated by nicotine increasing GABAergic population activity and in turn leading to an increase in GABAergic input (I_G) to DA cells (see green lines in Fig. 4.3C and E). The biphasic nature of this increase (Fig. 4.3C during Nic presence) stems from the single-exponential build-up of Nic concentration in the model (time constant of 1 min). The initial peak arises from fast activation counterbalanced by slower desensitization. The subsequent steady and smaller increase follows the time course of Nic concentration build-up in the model, i.e. activation and desensitization can be considered in quasi steady-state during that phase. A fraction of nAChRs remains desensitized after the washout of the drug and recovers slowly back to the sensitized state. This desensitized fraction of nAChR reduces the constant excitatory drive from cholinergic input to GABAergic cells, i.e. I_G falls below baseline levels after nicotine is removed (illustrated in magenta in

Fig. 4.3 VTA response to nicotine in vitro. *Left hand panels* (**A**, **C**, **E** and the left-hand side of panel **G**) show the results on GABAergic input changes to VTA DA cells, while the panels on the *right-hand side* (**B**, **D**, **F** and the *right-hand side* of panel **G**) depict results on glutamatergic input increases to VTA DA cells in response to 1 μM nicotine for 2 min starting at $t = 1$ min. **A** & **B**, In vitro input changes to VTA DA cells in response to Nic. *Grey shaded parts* and *black crosses* show blocked transmission pathways, and the scissors illustrate the truncation of the input pathways in vitro (compare with Fig. 4.1A). **C** & **D**, Time course of GABAergic, I_G, (**C**) and glutamatergic input, I_{Glu}, (**D**) changes to VTA DA neurons during and after Nic exposures (*black bar on top of the panels*). The increase (*green*) and the decrease (*magenta* in **C**) of the input currents with respect to baseline are illustrated in both panels. **E** & **F**, Maximal change of GABAergic (**E**) and glutamatergic input currents (**F**) as a function of the nicotine concentration applied. The *lines* show the results of the model for control conditions (in *green* in both panels and *magenta* for decrease in panel **E**), with $\alpha 4\beta 2$ nAChRs blocked (*green* in panel **E** and **F**, and *cyan* in panel **E**), and with $\alpha 7$ nAChRs blocked (*green* and *magenta* in panel **E**, *orange* in panel **F**). The *squares* show experimental results adapted from Mansvelder et al. (2002) (panel **E**) and Mansvelder and McGehee (2000) (panel **F**) for different experimental situations: control conditions—*green squares*; with $\alpha 7$ nAChR specific antagonist—*orange squares*; and with antagonist for non-$\alpha 7$ nAChRs—*cyan squares*. **G**, Comparison of relative input changes between model and experiment for the case of 1 μM nicotine for 2 min. Model results are shown with shaded and experimental results with *filled bars*. Both, GABAergic- and glutamatergic input changes are shown for the three discussed cases: control conditions—*green and magenta*, $\alpha 4\beta 2$ nAChR blocked—*cyan*, and $\alpha 7$ nAChR blocked—*orange* and *magenta* (experimental data adapted from Mansvelder and McGehee 2000; Mansvelder et al. 2002; $ACh = 0.384$ μM and $\nu_{Glu} = 5.68 \cdot 10^{-14}$ in all panels, see Table 4.2 and Sect. 4.5 for other parameters)

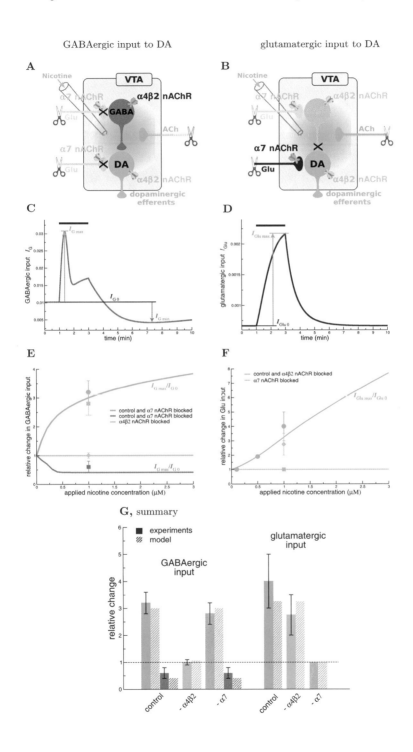

Fig. 4.3C and E). The recovery time course is governed by the maximal desensitization time constant τ_{max} of $\alpha 4\beta 2$ nAChRs. This time constant in the model (10 min, see Table 4.2) is on the same order of magnitude as observed experimentally, i.e. Mansvelder et al. (2002) fit the recovery of the IPSC frequency back to baseline at room temperature with a time constant of 20 min.

We find that in line with the experiments, the increase in IPSC frequency in the model is exclusively mediated by $\alpha 4\beta 2$ nAChRs. Hence, when these receptors are blocked in the model, the input changes are abolished (see cyan line in Fig. 4.3E). In contrast, $\alpha 7$ nAChRs have little or no impact on the response (green line and orange square in Fig. 4.3E). In the model, we account for the experimentally measured relative increases of $I_{G\,max}/I_{G0}$ under control conditions and with $\alpha 7$ nAChRs blocked by adjusting the average cholinergic drive to GABAergic cells such that the relative increase of GABAergic input attains 300% of baseline ($ACh = 0.384$ μM). Interestingly, the drop below baseline matches experimental data without further fitting of the model (see Fig. 4.3E). Importantly, the constant ACh tone signifies ongoing cholinergic afferent activity in the model. We therefore do not assume that ACh was present in the external bath during the in vitro experiments considered.

Our model furthermore accounts for the increase of glutamatergic (I_{Glu}) input to VTA DA cells in response to the same stimulation protocol as above (1 μM Nic for 2 min, see Fig. 4.3D). This increase stems from the activation of presynaptic $\alpha 7$ nAChRs by nicotine. The maximal relative increase of I_{Glu} is depicted with respect to the applied nicotine concentration in Fig. 4.3F. We can see that the model accounts for experimental data recorded in control conditions (green line and squares); in the presence of an $\alpha 7$ nAChR specific antagonist (orange line and square); and in the presence of an antagonist for non-$\alpha 7$ subunit-containing nAChRs (green line and cyan square).

In order to account quantitatively for the observed nicotine-induced increase in Glu input to DA cells ($I_{Glu\,max}/I_{Glu0}$) we choose v_{Glu} such that the relative increase attains 325% of baseline level. This satisfies both the control data (400% increase) and the increase measured when $\alpha 4\beta 2$ nAChRs are blocked (275% increase; and $ACh = 0.384$ μM; see Fig. 4.3F). Note that increasing basal Glu input in the model reduces the relative increase in I_{Glu}, for example, since nicotine induced changes are added to the basal Glu upstream activity (see Sect. 4.5; Eq. (4.4)). Blocking the $\alpha 4\beta 2$ nAChRs does not affect the glutamatergic input increase in the model since the change stems from $\alpha 7$ nAChR activation only (see Fig. 4.3B). Note that the constant cholinergic tone in the model (ACh), fixed through the experiments described above, results in a weak activation of $\alpha 7$ nAChRs on average in the absence of nicotine.

In summary, we show that the model reproduces qualitatively the relative change in IPSC and EPSC frequency to VTA DA neurons during in vitro nicotine perfusions (see comparison of model and experiments in Fig. 4.3G). The model reproduces furthermore the supralinear increase in EPSC frequency in the nicotine range from 0.1 to 1 μM Nic (green line and squares in Fig. 4.3F). Note that the increase in IPSC frequency is predicted to show a sublinear increase in the same range of nicotine (4.3E). This difference stems from the difference in affinity of the $\alpha 7$ and the $\alpha 4\beta 2$

receptor, i.e. the response of the high affinity $\alpha 4\beta 2$ receptor starts to saturate in this range. We argue that the afferent input strength, reflected by the constant cholinergic tone (*ACh*) and the relative upstream glutamatergic activity (ν_{Glu}), is weak in vitro (compare with in vivo conditions below). The low input tones reflect the disruption of the afferent pathways in this experimental preparation (illustrated by the scissors in Figs. 4.3A and B).

4.8.2 Direct Stimulation (Intrinsic Cellular) vs. Disinhibition (Local Circuit) Mechanisms for Nicotine-Evoked Increase of DA Cell Activity in vivo

We now show that the minimal VTA-circuit model accounts for the in vivo experiments. To do so we leave the VTA circuit unchanged and alter only the afferent input strength. We require the model to account for the following data: (i) An intravenous injection of nicotine in anesthetized wild type mice increases the firing rate of DA cells in vivo. (ii) This increase is diminished in $\alpha 7$ knockout mice and is completely abolished in $\alpha 4\beta 2$ knockout mice (Mameli-Engvall et al. 2006).

We identify two different instances of the model which both could account for the experimentally observed behavior in wild type, $\alpha 7$- and $\beta 2$ nAChR knockout mice. We refer to the scenario in which $\alpha 4\beta 2$ nAChR-mediated action is predominantly exerted onto the DA cells as to "direct stimulation" (see Fig. 4.4A). On the

Fig. 4.4 VTA response to nicotine in vivo. Panels on the *left-hand side* (**A**, **C** and **E**) show results of the direct stimulation scenario ($I_0 = 0.0202$, $ACh = 0.1$ μM, $\nu_{Glu} = 0.1$) and panels on the *right-hand side* (**B**, **D** and **F**) depict results for disinhibition ($I_0 = 0.1$, $ACh = 1.77$ μM, $\nu_{Glu} = 0.1$). The increase in DA activity stems from activation of $\alpha 4\beta 2$ nAChRs on DA cells in case of direct stimulation, whereas $\alpha 4\beta 2$ nAChR desensitization on GABAergic cells boosts DA activity for disinhibition (see text for more details). **A** & **B**, Illustration of the experimental situation in vivo. Note the difference in $\alpha 4\beta 2$ nAChR distribution between the direct stimulation (**A**, $r = 0.8$) and the disinhibition case (**B**, $r = 0$). **C** & **E**, Normalized GABAergic (**C**) and DA neuron activity (**E**) in response to the application of 1 μM nicotine in case of direct stimulation. The *full lines* show the time course of the normalized ν_G (**C**) and ν_D (**E**) for three different durations of nicotine exposure, T_{Nic} (as indicated, and illustrated by the bar on top of panel **C**). The *full* and the *dashed blue lines* depict the responses for low ($ACh = 0.1$ μM) and high cholinergic input levels ($ACh = 1.77$ μM), respectively ($T_{Nic} = 10$ min). **D** & **F**, Normalized GABAergic (**D**) and DA neuron activity (**F**) in response to 1 μM nicotine for 2 min in case of disinhibition. The *full lines* show the time course of the normalized ν_G (**D**) and ν_D (**F**) for three different maximal desensitization time constants of $\alpha 4\beta 2$ nAChRs, τ_{max} (as indicated in the panel). The *full* and the *dashed blue lines* depict the responses for high ($ACh = 1.77$ μM) and low cholinergic input levels ($ACh = 0.1$ μM), respectively ($\tau_{max} = 10$ min). **G**, Comparison of model results (*shaded bars*) and experimental data (*full bars*) on relative DA neuron activity changes in response to 1 μM nicotine. The maximal relative increase of DA activity in wild type ($T_{Nic} = 10$ min for direct stimulation; and $\tau_{max} = 10$ min for disinhibition) and mutant mice is shown (experimental data adapted from Mameli-Engvall et al. 2006). **H**, Comparison of the total duration of elevated DA neuron activity with respect to the duration of Nic application, T_{Nic}. The duration of elevated activity is taken to be the time between the two points where ν_D attains half-of-maximum activity (as illustrated in **E** and **F** and depicted by *square* and *circle*, respectively). This duration is plotted for direct stimulation (*green*) and disinhibition (*red*)

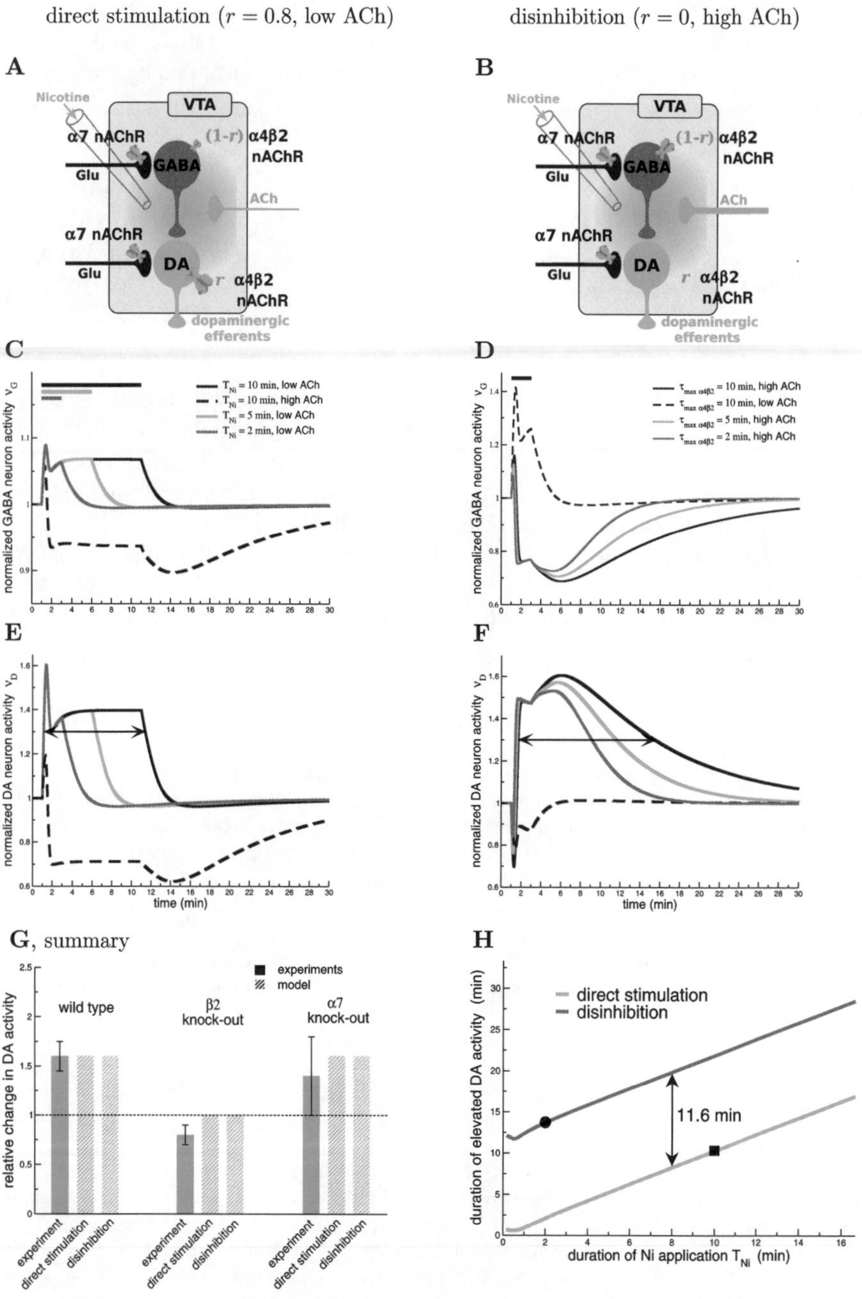
direct stimulation ($r = 0.8$, low ACh)
disinhibition ($r = 0$, high ACh)
A
B
Nicotine
VTA
α7 nAChR
(1-r)
α4β2 nAChR
Glu
GABA
ACh
DA
r
dopaminergic efferents
C
D
normalized GABA neuron activity v_G
T_{Ni} = 10 min, low ACh
T_{Ni} = 10 min, high ACh
T_{Ni} = 5 min, low ACh
T_{Ni} = 2 min, low ACh
$\tau_{max\ \alpha4\beta2}$ = 10 min, high ACh
$\tau_{max\ \alpha4\beta2}$ = 10 min, low ACh
$\tau_{max\ \alpha4\beta2}$ = 5 min, high ACh
$\tau_{max\ \alpha4\beta2}$ = 2 min, high ACh
E
F
normalized DA neuron activity v_D
time (min)
G, summary
H
experiments
model
wild type
β2 knock-out
α7 knock-out
relative change in DA activity
experiment
direct stimulation
disinhibition
direct stimulation
disinhibition
11.6 min
duration of elevated DA activity (min)
duration of Ni application T_{Ni} (min)

other hand, "disinhibition" refers to the scenario in which $\alpha 4\beta 2$ nAChRs mainly influence GABAergic activity (see Fig. 4.4B). Crucially, the conditions for the two mechanisms are different as we show below. In the model, the parameter r sets the balance of $\alpha 4\beta 2$ nAChR action between DA and GABA neurons, i.e. $r = 0.8$ is used for direct stimulation and $r = 0$ for disinhibition. We choose these two extreme values of r to illustrate the qualitative behavior of the model in the direct stimulation and the disinhibition case. The behavior of the model for intermediate values of r is examined in the next section. For simplicity, we utilize the same concentration of nicotine in this section as in the in vitro experiments, i.e. 1 µM.

Direct Stimulation Here, we identify the conditions for nicotine-evoked increase in DA activity if we assume that DA neurons predominantly express $\alpha 4\beta 2$ containing nAChRs. Based on the activation and desensitization properties of $\alpha 4\beta 2$-containing nAChRs, we derive two important requirements for such a scenario to boost DA activity. (i) The constant cholinergic drive to the VTA has to be low ($ACh \ll EC_{50}^{\alpha 4\beta 2}$ and $ACh \approx IC_{50}^{\alpha 4\beta 2}$) such that nicotine can further activate $\alpha 4\beta 2$-containing nAChRs. (ii) The duration of elevated DA activity cannot outlast the presence of nicotine, i.e. the increase of DA neuron activity for 10 min requires the presence of nicotine for the same amount of time (blue lines in Figs. 4.4C and E). In other words, DA activity stays elevated as long as nicotine is present and decays back to baseline level after removal of nicotine. The full lines in Figs. 4.4C and E show the temporal dynamics of the normalized GABAergic and dopaminergic neuron activity, respectively, for different durations of 1 µM nicotine being present (indicated by the bars on top of panel C) in case of direct stimulation (i.e. low constant ACh level, $ACh = 0.1$ µM). The activity of GABAergic neurons follows the same time course as DA neuron activity, i.e. increased activity during the presence of Nic, but with a smaller amplitude due to the smaller fraction of $\alpha 4\beta 2$ nAChR expressed on GABAergic cells ($r = 0.8$). Again, the biphasic nature of the DA and GABAergic neuron activity increases emerges from the slow build-up of Nic concentration in the model (see above). We furthermore find that when the tonic ACh drive is high, nicotine leads to a decrease in the DA activity for the direct excitation case ($r = 0.8$, $ACh = 1.77$ µM, dashed blue lines in Figs. 4.4C and E). The choice of $r = 0.8$ in the model represents an $\alpha 4\beta 2$ nAChR expression ratio of 80 to 20 on DA to GABA cells. In this case and as long as more $\alpha 4\beta 2$ nAChRs are expressed on DA cells in general ($r > 0.5$), the cholinergic drive has to be smaller than $ACh < 0.38$ µM in the model in order to observe an net increase of DA activity during and after nicotine application (we speak of an increase in DA activity as long as: $\int (v_D(t) - v_D(t = 0)) dt > 0$). Note that we do not chose the extreme case of $r = 1$, which would imply no $\alpha 4\beta 2$ nAChRs on GABA cells, since the existence of $\alpha 4\beta 2$ nAChRs on GABA cells has been shown experimentally (Klink et al. 2001; Mansvelder et al. 2002).

Disinhibition How can nicotine action through $\alpha 4\beta 2$ nAChRs expressed on GABAergic cells lead to an overall increase in DA activity? In case $r = 0$, we observe an increase in DA neuron activity in response to a 2 min exposure to 1 µM nicotine if the constant cholinergic drive to the VTA is bigger than $ACh \geq 0.17$ µM. The high cholinergic drive assures that nicotine mainly drives $\alpha 4\beta 2$ receptor desensitization after a short initial period of activation (Fig. 4.4D). Figure 4.4D shows that nicotine-induced desensitization removes cholinergic excitatory drive from the GABAergic cell population reducing their activity below baseline levels during and after the exposure to the drug (full blue, green and red lines). Return to baseline activity is determined by the maximal desensitization time constant of $\alpha 4\beta 2$ nAChRs (τ_{max}, see Table 4.2). The time course of GABAergic activity is depicted for three different values of for $\alpha 4\beta 2$ nAChRs in Fig. 4.4D. DA activity changes are solely governed by GABAergic cells in case of disinhibition since the activation of $\alpha 7$ nAChRs is negligible at 1 µM nicotine which is in the ballpark of realistic nicotine levels. The profile of DA activity is therefore a mirror-image of the GABAergic activity due to the inhibitory influence of the latter on DA cells (see Fig. 4.4F). In turn, the duration of boosted DA activity outlasts the presence of nicotine and depends on the recovery time constant from desensitization of $\alpha 4\beta 2$ nAChRs. We furthermore show in Fig. 4.4F that DA activity exhibits a drop below baseline if the constant cholinergic drive to the VTA is assumed to be low (dashed blue line, $ACh = 0.1$ µM), i.e. this drop stems from an increase of GABA cell activity (dashed blue line in Fig. 4.4D).

In summary, we show that the direct stimulation and the disinhibition scenarios can both potentially account for nicotine induced DA activity changes in wild type, $\alpha 4\beta 2$-, and $\alpha 7$-knockout mice (see summary in Fig. 4.4G). However, and this is a crucial point, these two mechanisms require different and distinct levels of afferent cholinergic input: Cholinergic drive has to be low for direct stimulation and high for disinhibition in order to lead to DA activity increases in response to nicotine. The model shows a second tell-tale difference in the relative observed time course of the DA activity: in case of direct stimulation the duration of DA response is determined directly by the duration of nicotine presence, whereas for disinhibition the recovery from desensitization of the $\alpha 4\beta 2$ nAChR defines the temporal scale of the DA activity (see Table 4.1 for an overview of the differences). The $\alpha 4\beta 2$ nAChR recovery time constant used in the model ($\tau_{max} = 10$ min) results in a ~ 12 min longer increase in DA activity with disinhibition as compared to direct stimulation for the same duration of nicotine application (see Fig. 4.4H).

4.9 Predictions of the VTA Model and Experimental Protocols to Pin Down the Major Nicotinic Pathway of Action

The experimental data available so far is not sufficient to determine conclusively to which extend direct stimulation or disinhibition of DA cells is at the origin of DA signal increases in response to nicotine. We now study the DA cell response in the model for different cholinergic input levels and for a range of nicotine concentrations applied in order to suggest concrete experiments to pinpoint the mechanisms of drug induced DA activity changes.

Table 4.1 Overview of the model results

Mechanism	Expression	Conditions to reproduce experimental data	Predictions
Disinhibition	$\alpha 4\beta 2$ nAChR-mediated action predominantly through GABA cells	– high afferent ACh input	– the higher the afferent ACh input level, the stronger the increase in DA activity in response to Nic
		– recovery from $\alpha 4\beta 2$ nAChR desensitization determines duration of elevated DA activity in response to Nic	– max. increase of DA activity saturates at low Nic
Direct stimulation	$\alpha 4\beta 2$ nAChR-mediated action predominantly through DA cells	– low afferent Ach input	– high ACh input levels turn the excitation of DA activity into inhibition in response to Nic
		– duration of Nic presence determines the duration of elevated DA activity in response to Nic	– higher Nic concentrations lead to stronger boost of DA

In our model the control parameter r, determining the distribution of $\alpha 4\beta 2$ nAChR on DA and GABAergic neuron populations, allows to change the balance of $\alpha 4\beta 2$ action, i.e. from direct stimulation to disinhibtion. For clarity, we chose extreme values of r for direct stimulation ($r = 0.8$) and disinhibition ($r = 0$). However, our analysis shows that the conditions for direct stimulation are met as long as $r > 0.5$ and for disinhibition with $r < 0.5$.

Figure 4.5A shows the temporal profile of DA activity in response to 1 µM nicotine for different values of r for low- (left panel, in vitro) and high afferent input strength (right panel, in vivo). Note that the DA activity is increased in the in vitro (in vivo) case as long as $r > 0.5$ ($r < 0.5$). The direct stimulation ($r = 0.8$) and disinhibition ($r = 0$) cases are highlighted by the red and the blue lines, respectively, for both input strength cases. Importantly, the extent to which the GABAergic neuron activity dominates DA cells in case of disinhibition ($r < 0.5$) depends on balancing the glutamatergic and GABAergic input to DA cells. In the model, we set the ratio of glutamatergic and GABAergic input current weights, w_{Glu}/w_G, to unity. If we chose $w_{Glu}/w_G = 100$, for example, the DA cell activity shows no inhibition in the in vitro scenario (results not shown). $w_{Glu}/w_G = 100$ means that DA cell activity is dominated by nicotine driven Glu input increases outweighing inhibitory input from GABA cells.

Our model predicts that the cholinergic input level (ACh) determines the extent of the delayed increase in DA activity in case of disinhibition ($r = 0$), as shown in Fig. 4.5B (left panel). The DA cells are initially inhibited due to fast activation of $\alpha 4\beta 2$ nAChR on GABAergic cell and subsequently disinhibited since those receptors are desensitized and GABAergic activity falls below baseline levels (see Fig. 4.4D). In turn, the amount of relief from inhibition of DA cells depends on the overall excitatory drive to GABAergic cells. This excitatory drive is directly related to the cholinergic input strength (Fig. 4.5B, left panel). Our analysis further shows differential effects on the two parts of the DA response to nicotine: the delayed increase (right panel in Fig. 4.5B, orange line) strongly depends on the constant cholinergic drive to the VTA, whereas the initial drop of DA activity (right panel in Fig. 4.5B, purple line) is less sensitive to the cholinergic input level.

How does the DA response for the direct stimulation and disinhibition depend on the applied nicotine concentration? We find that the relative maximal increase of DA activity augments with Nic in case of direct stimulation while this increase saturates at $\sim$0.5 µM nicotine for disinhibition (see Fig. 4.5C, right panel). The difference stems from the fact that the maximal increase for direct stimulation is due to fast activation of $\alpha 4\beta 2$ nAChRs, whereas in the disinhibition case, the increase arises from delayed desensitization of those receptors. Importantly, the two different mechanisms entrain different time scales of maximal DA activity (compare time points at which the maximal activity of DA neurons is attained in Fig. 4.5C, left panel). Higher nicotine levels result in stronger activation, i.e. the initial peak rises with Nic for direct stimulation (compare responses to 0.5 µM—full red line—and 3 µM nicotine—dashed red line—in the left panel of Fig. 4.5C; summary of results in the right panel). On the other hand, the fraction of receptors driven in the desensitized state cannot be increased with higher Nic concentrations for disinhibition since the maximal fraction ($IC_{50} \gg 0.5$ µM) and the minimal rate of desensitization, τ_0, ($K_\tau < 0.5$ µM) are already attained at $\sim$0.5 µM nicotine for $\alpha 4\beta 2$ nAChRs (see Table 4.2). In other words, 0.5 µM nicotine for 2 min is already effective in desensitizing the maximal amount of $\alpha 4\beta 2$ nAChRs. Note that the nicotine concentration is applied for 2 min in case of disinhibition and for 10 min in case of direct stimu-

Fig. 4.5 Predicted dynamics of the DA neuron population in the VTA in response to nicotine. **A**, ▶ Temporal dynamics of DA neuron activity in response to 1 µM nicotine for 2 min for different values of r and afferent input strengths. The *left panel* (*right panel*) shows the DA response in the presence of constant low (high) cholinergic and glutamatergic afferent input to the VTA, i.e. in vitro (in vivo) like conditions (in vitro: $ACh = 0.384$ µM, $\nu_{Glu} = 5.68 \cdot 10^{-4}$; $I_0 = 0.1$ in vivo: $ACh = 1.77$ µM, $\nu_{Glu} = 0.1$, $I_0 = 0.1$). The distribution of $\alpha 4\beta 2$ nAChRs is changed by varying the control parameter r as indicated in the panels. Direct stimulation (disinhibition) refers to the case with $\alpha 4\beta 2$ nAChRs dominantly on DA (GABAergic) cells, i.e. $r = 0.8$ and $r = 0$, respectively. **B**, Biphasic response of DA neuron activity in response to 1 µM nicotine for 2 min and for different cholinergic input levels. The temporal dynamics of ν_D is shown in the *left panel* for the disinhibition case ($r = 0$). The maximal initial inhibition of DA activity (marked by the *purple symbols*) and the maximal delayed excitation (marked by the *orange symbols*) is shown with respect to the cholinergic input strength in the *right panel* ($\nu_{Glu} = 0.1$, $I_0 = 0.1$). **C**, Different response profiles of the DA activity in the disinhibition and the direct stimulation case for varying nicotine concentrations. The *left panel* shows the time course of DA activity in response to 0.5 µM (*full lines*) and 3 µM nicotine (*dashed lines*) for 2 min, in case of disinhibition (*blue lines*; $r = 0$, $ACh = 1.77$ µM, $\nu_{Glu} = 0.1$, $I_0 = 0.1$), or 10 min, in case of direct stimulation (*inset, same axes, red lines*; $r = 0.8$, $ACh = 0.1$, $\nu_{Glu} = 0.1$, $I_0 = 0.0202$). The maximal DA response with respect to nicotine is depicted in the *right panel* for direct stimulation (*red line*) and disinhibition (*blue line*; data point adapted from Mameli-Engvall et al. 2006)

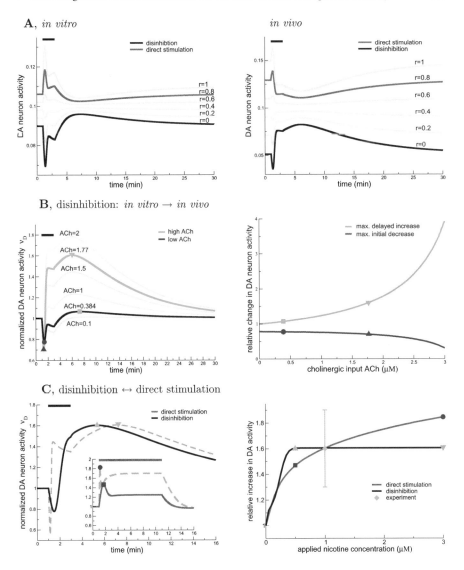

lation in order to achieve comparable durations of DA activity increases (compare Fig. 4.4H).

In summary, we suggest that changes of the cholinergic drive to the VTA during nicotine application, or changes in the nicotine concentration administered in vivo can potentially reveal the nature of the underlying mechanisms at the origin of DA activity changes. Decreasing the cholinergic drive to the VTA further boost DA activity increases for direct stimulation but diminishes that response for disinhibition (Figs. 4.5A and B). Furthermore, the maximal DA response saturates at low nicotine concentrations (~0.5 μM) for disinhibition but keeps rising with higher nicotine

Table 4.2 Parameters of nAChR activation and desensitization kinetics

Parameter	Definition	Value	Reference
$\alpha 7$-containing nAChR			
EC_{50}	half-maximum conc. of activation	80 µM	Peng et al. (1994), Gerzanich et al. (1995), Fenster et al. (1997), Papke (2006)
α	potency of nicotine to evoke response	~ 2	Peng et al. (1994), Gerzanich et al. (1995)
n_a	Hill coefficient of activation	1.73	Peng et al. (1994), Fenster et al. (1997), Papke (2006)
IC_{50}	half-maximum conc. of desensitization by Nic	1.3 µM	Peng et al. (1994), Fenster et al. (1997)
n_s	Hill coefficient of desensitization	2	Fenster et al. (1997)
τ_a	activation time constant	5 msec	Papke (2006)
K_τ	half-maximum conc. of desensitization time constant	1.73 µM	Fenster et al. (1997)
n_τ	Hill coefficient of des. time constant	2	Fenster et al. (1997)
τ_{max}	maximal des. time constant	2 min	Fenster et al. (1997)
τ_0	minimal des. time constant	50 msec	this study, Papke 2006
$\alpha 4\beta 2$-containing nAChR			
EC_{50}	half-maximum conc. of activation (ACh)	30 µM	Buisson and Bertrand (2001)
α	potency of Nic to evoke response	~ 3	Buisson and Bertrand (2001), Eaton et al. (2003)
N_A	Hill coefficient of activation	1.05	Fenster et al. (1997), Buisson and Bertrand (2001)
IC_{50}	half-maximum conc. of desensitization by Nic	0.061 µM	Fenster et al. (1997)
n_s	Hill coefficient of desensitization	0.5	Fenster et al. (1997)
τ_a	activation time constant	5 msec	Buisson and Bertrand (2001)
K_τ	half-maximum conc. of desensitization time constant	0.11 µM	Fenster et al. (1997)
n_τ	Hill coefficient of des. time constant	3	Fenster et al. (1997)
τ_{max}	maximal des. time constant	10 min	this study
τ_0	minimal des. time constant	500 msec	this study, Buisson and Bertrand (2001)

concentrations for direct stimulation (Fig. 4.5C). See Table 4.1 for a summary of the predictions for both mechanisms.

4.10 Discussion

Our major goals were to exhibit how the nicotine evoked responses at the receptor population level in the dopaminergic nucleus VTA can lead to the self-administration of nicotine and second to determine the dominant pathways of action for nicotine within the ventral tegmental area. In order to address the first issue we earlier proposed a novel neurodynamical modelling approach to drug-self-administration (Gutkin et al. 2006). In this multi-modular framework we managed to show that nicotine acting through nicotinic acetylcholine receptors, provoking activation/desensitization on the short time scale with upregulation on the longer time scales is able to drive the dopamine-dependent learning (read reward-modulated) in the action selection circuits and lead to the onset of nicotine self-administration. A slower opponent process that is engaged by nicotine then is shown to result in the drug-seeking being stamped-in and becoming a habitual behavior. This model made a number of predictions, e.g. changes in the general learning rates for addicted individuals, differential dose-availability thresholds for the initiation and habituation of nicotine-seeking. However it was of insufficient detail to tease out the precise receptor mechanisms governing the acute dopamine response to nicotine.

To address this issue we developed a novel approach allowing us to investigate the interplay of the pharmacodynamics of nicotine and the dopaminergic signal constructed in the VTA. The combination of a population activity model of the VTA with a detailed model of nAChR kinetics enables us to better understand the mechanisms of nicotine action on DA signaling. Our explorations of the model show that in vitro and in vivo data can be reconciled by taking into account the difference in afferent input strengths to the VTA in the two experimental settings. Hence, the differential activation and desensitization kinetics of $\alpha 7$- and $\alpha 4\beta 2$ nAChRs combined with different afferent input levels can explain the mechanism of nicotine action. Our approach confirms the previously expressed hypothesis that $\alpha 4\beta 2$ nAChRs predominantely mediate nicotine influence on DA signaling (Mansvelder et al. 2002; Champtiaux et al. 2003; Mameli-Engvall et al. 2006).

Available experimental data does not allow to pinpoint whether $\alpha 4\beta 2$ nAChRs on VTA DA—direct stimulation—or GABA cells—disinhibition—are the dominant site of nicotine action. We further demonstrate that disinhibition and direct stimulation of DA cells can potentially be at the origin of the experimentally observed nicotine induced boost of DA activity. In that sense, the here presented VTA model represents a necessary circuit to describe experimental data on nicotine applications yet the data is not sufficient to resolve unanimously the underlying mechanism. Using the model, we identify that the cholinergic input levels have to be low for direct stimulation, whereas high cholinergic input levels are crucial in the disinhibition case, to observe a DA activity increase. These results emerge directly from known activation and desensitization properties of $\alpha 4\beta 2$ nAChRs. Assuming that cortical glutamatergic and subcortical cholinergic as well as glutamatergic afferent activities are low in vitro and high in vivo suggests disinhibition of DA cells to be the candidate mechanism at the origin of increased DA activity. Several experimental results support the importance of $\alpha 4\beta 2$ nAChRs on GABAergics cells for nicotinic action: (i) Cholinergic axon terminals synapse selectively on non-DA neurons and on a subset of DA neurons in the VTA (Garzón et al. 1999). (ii) The α4-containing nAChR upregulation on VTA GABAergic neurons in response to chronic nicotine supports the functional importance of those receptors on GABAergic cells with respect to nicotine (Buisson and Bertrand 2001; Nashmi et al. 2007). (iii) A whole class of drugs— including opioids, cannabinoids, γ-hydroxy butyrate (GHB), benzodiazepines—has been identified which lead to inhibition of GABA neurons in the VTA and thereby disinhibition of DA neurons (e.g., Johnson and North 1992a, see Lüscher and Ungless 2006 for an overview). This together with the finding that the GABA antagonist bicuculline is self-administered by mice supports the hypothesis that reduced GABAergic input to DA cells induces addictive behavior (David et al. 1997; Ikemoto et al. 1997). (iv) The hyperexcitability of the VTA in response to nicotine (Sher et al. 2004) could be related to the higher abundance of GABAergic cells in the VTA as compared to the SN (GABA to DA ratio about 1/4 in the VTA, Johnson and North 1992b; and 1/19 in the substantia nigra, Lacey et al. 1989). We would like to point out that the disinhibition scenario emphasizes the role of local circuitry organization, i.e. the behavior of GABA cells is crucial for the augmentation of DA activity. This is opposed to the single cell mechanism associated with direct stimulation of DA cells.

Our model makes several predictions for the case that the reinforcing properties of nicotine are mediated through inhibition of GABA cells. The boost of DA activity induced by nicotine is preceded by a short-lasting inhibition of DA activity in the model stemming from fast activation of α4β2 nAChRs (see Figs. 4.4F and 4.5A). Independently obtained experimental results seem to support the finding that the response of DA neurons is biphasic, at least in some (but not all) cases of direct recordings from DA cells (private communications with Philippe Faure and Jie Wu; Erhardt et al. 2002). Another prediction of the model is the saturation of the nicotine induced DA boost at low ($\sim$500 nM) nicotine levels in case of disinhibition (see Fig. 4.5). Higher nicotine levels do not evoke further increases of DA activity since the maximal desensitization of α4β2 nAChR is already attained at low nicotine. Two interesting implications of this result are: (i) Nicotine elicits maximal α4β2 nAChR-mediated increase of DA activity at nicotine concentrations attained in the blood of smokers (Henningfield et al. 1993). (ii) Repetitive nicotine applications consistently evoke the same boost in DA activity even at high repetition rates, which would maintain the reinforcing properties of the drug even at high exposure frequencies.

We confirm previous findings suggesting that realistic doses of nicotine do not significantly desensitize α7-containing nACh receptors (Wooltorton et al. 2003). We extend this statement and propose that realistic concentrations of nicotine do not succeed to significantly activate α7-containing nAChRs. We consider it therefore unlikely that increased excitatory drive to DA cells in response to nicotine persistently augments their activity. It should however be noted that the mean-field approach presented here does not resolve the different firing modes of DA cells, i.e. bursting and regular firing. α7-containing nAChs could play a role in nicotine induced bursing since bursts have been shown to be induced by glutamatergic inputs to DA cells (Grenhoff et al. 1988; Chergui et al. 1993). One could imagine that the tonic inhibitory input from GABAergic cells sets the overall level of excitability of

DA cells. In case the DA cells are sufficiently disinhibited through desensitization of $\alpha4\beta2$ nAChRs, $\alpha7$ nAChR activation induces burst firing on top of elevated membrane depolarizations. Hence GABA cells would gate DA burst firing, as suggested in a biophysically detailed model by Komendantov et al. (2004). Furthermore, bursting induced by $\alpha7$ nAChR activation could be crucial for the induction of long-term potentiation of Glu synapses onto DA cells as proposed by Mansvelder and McGehee (2000). One future direction of our research is the investigation of the effect of nicotine in a spiking and bursting model of VTA DA cells in order to address the issue of the different DA cell activity patterns.

We note that we concentrated on the local VTA mechanisms and developed a minimal local-circuit model of the VTA that combined both neuronal population dynamics and receptor kinetics in a novel way. In particular, we focused only on feedforward afferent input (glutamatergic and cholinergic) and a simplified local circuitry of the VTA. We left aside the possible recurrent involvement of other neuronal structures participating in DA-signaling, such as GABAergic connections from the nucleus accumbens for example (Kalivas et al. 1993; Wu et al. 1996). Furthermore, we chose to not address the potential heterogeneity of the VTA itself (Garzón et al. 1999; Carr and Sesack 2000a, 2000b; Fagen et al. 2003). However, our proposed circuitry can be seen as a global description of the VTA or as a model of a local computational unit of neurons within the VTA. Whether the experimentally observed diversity of DA cell behavior could be explained by the coexistent presence of direct stimulation and disinhibition subcircuits in the VTA or whether recurrent inhibition has to be taken into account remains an area for future studies. We would however like to draw the attention to the fact that DA and GABAergic cells show a variety of temporal profiles in the model depending on their $\alpha4\beta2$ nAChR expression level and their cholinergic input level.

We show that the constant cholinergic tone crucially determines the phasic dopamine signal in response to nicotine. For disinhibition, we illustrate that the increase of the phasic DA activity grows with the tonic cholinergic input level to the VTA (see Fig. 4.5B). Furthermore data and theory suggest that phasic dopamine modulates learning (Reynolds and Wickens 2002; Dayan and Niv 2008). Our results together with this fact lead us to speculate that if salient characteristics of environmental cues are reflected in the overall cholinergic tone (Yu and Dayan 2005), the increase of phasic DA could explain the strong associations formed between these cues and the habit of smoking (Lichtenstein 1982; DiChiara 1999). Furthermore, taking diural rhythms of cholinergic signaling into account, the VTA may give rise to different DA output at different times of the day, e.g. the morning cigarette may deploy different mechanisms than an evening cigarette. However, how dynamic afferent input changes, possibly signaling behaviorally relevant features such as reward or expectation of reward, are translated into dopamine output is a future direction of the local VTA circuit modeling approach. The model would furthermore allow us to study how nicotine changes this input integration. Identifying the specific functional targets of nicotine driven reinforcement potentially has direct implication for developing nicotine addiction treatments, e.g. for designing replacement drugs. Overall, we suggest the dynamics of the local VTA circuit including GABA cells play a crucial role in constructing the dopamine signal in response to nicotine for the constant afferent input regime considered here.

Acknowledgements We thank Phillippe Faure, Huibert Mansvelder, Andrew Oster and Jie Wu for very helpful and fruitful discussions. This work is supported by CNRS, Collège de France, IST European consortium project BACS FP6-IST-027140 (MG and BG), École des Neurosciences de Paris Île-de-France (MG), and the Marie Curie Team of Excellence Grant BIND MECT-CT-20095-024831 (BG).

References

Andén N-E, Dahlström A, Fuxe K, Larsson K, Olson L, Ungersted U (1966) Ascending monoamine neurons to the telencephalon and diencephalon. Acta Physiol Scand 67(3–4):313–326

Beiser DG, Hua SE, Houk JC (1997) Network models of the basal ganglia. Curr Opin Neurobiol 7:185–190

Berger B, Thierry AM, Tassin JP, Moyne MA (1976) Dopaminergic innervation of the rat prefrontal cortex: a fluorescence histochemical study. Brain Res 106(1):133–145

Bobashev G, Costenbader E, Gutkin B (2007) Comprehensive mathematical modeling in drug addiction sciences. Drug Alcohol Depend 89(1):102–106

Bockstaele EJV, Pickel VM (1995) GABA-containing neurons in the ventral tegmental area project to the nucleus accumbens in rat brain. Brain Res 682(1–2):215–221

Bonci A, Malenka RC (1999) Properties and plasticity of excitatory synapses on dopaminergic and GABAergic cells in the ventral tegmental area. J Neurosci 19(10):3723–3730

Briggs CA, McKenna DG, Piattoni-Kaplan M (1995) Human alpha 7 nicotinic acetylcholine receptor responses to novel ligands. Neuropharmacology 34(6):583–590

Buisson B, Bertrand D (2001) Chronic exposure to nicotine upregulates the human (alpha)4(beta)2 nicotinic acetylcholine receptor function. J Neurosci 21(6):1819–1829

Calabresi P, Lacey MG, North RA (1989) Nicotinic excitation of rat ventral tegmental neurones in vitro studied by intracellular recording. Br J Pharmacol 98(1):135–140

Carr DB, Sesack SR (2000a) Gaba-containing neurons in the rat ventral tegmental area project to the prefrontal cortex. Synapse 38(2):114–123

Carr DB, Sesack SR (2000b) Projections from the rat prefrontal cortex to the ventral tegmental area: target specificity in the synaptic associations with mesoaccumbens and mesocortical neurons. J Neurosci 20(10):3864–3873

Champtiaux N, Gotti C, Cordero-Erausquin M, David DJ, Przybylski C, Lena C, Clementi F, Moretti M, Rossi FM, LeNovere N, McIntosh JM, Gardier AM, Changeux J-P (2003) Subunit composition of functional nicotinic receptors in dopaminergic neurons investigated with knock-out mice. J Neurosci 23:7820–7829

Changeux JP, Bertrand D, Corringer PJ, Dehaene S, Edelstein S, Léna C, Novère NL, Marubio L, Picciotto M, Zoli M (1998) Brain nicotinic receptors: structure and regulation, role in learning and reinforcement. Brain Res Brain Res Rev 26(2–3):198–216

Changeux JP, Devillers-Thiéry A, Chemouilli P (1984) Acetylcholine receptor: an allosteric protein. Science 225(4668):1335–1345

Chavez-Noriega LE, Crona JH, Washburn MS, Urrutia A, Elliott KJ, Johnson EC (1997) Pharmacological characterization of recombinant human neuronal nicotinic acetylcholine receptors h alpha 2 beta 2, h alpha 2 beta 4, h alpha 3 beta 2, h alpha 3 beta 4, h alpha 4 beta 2, h alpha 4 beta 4 and h alpha 7 expressed in xenopus oocytes. J Pharmacol Exp Ther 280(1):346–356

Chergui K, Charléty PJ, Akaoka H, Saunier CF, Brunet JL, Buda M, Svensson TH, Chouvet G (1993) Tonic activation of NMDA receptors causes spontaneous burst discharge of rat midbrain dopamine neurons in vivo. Eur J Neurosci 5(2):137–144

Chiara GD (2000) Role of dopamine in the behavioural actions of nicotine related to addiction. Eur J Pharmacol 393(1–3):295–314

Chiara GD, Imperato A (1988) Drugs abused by humans preferentially increase synaptic dopamine concentrations in the mesolimbic system of freely moving rats. Proc Natl Acad Sci USA 85(14):5274–5278

Cho RY, Nystrom LE, Brown ET, Jones AD, Braver TS, Holmes PJ, Cohen JD (2002) Mechanisms underlying dependencies of performance on stimulus history in a two-alternative forced-choice task. Cogn Affect Behav Neurosci 2(4):283–299

Christie MJ, Bridge S, James LB, Beart PM (1985) Excitotoxin lesions suggest an aspartatergic projection from rat medial prefrontal cortex to ventral tegmental area. Brain Res 333(1):169–172

Clarke P (1991) The mesolimbic dopamine system as a target for nicotine. In: Effects of nicotine on biological systems. Birkhäuser, Basel

Clements JR, Grant S (1990) Glutamate-like immunoreactivity in neurons of the laterodorsal tegmental and pedunculopontine nuclei in the rat. Neurosci Lett 120(1):70–73

Cornwall J, Cooper JD, Phillipson OT (1990) Afferent and efferent connections of the laterodorsal tegmental nucleus in the rat. Brain Res Bull 25(2):271–284

Corrigall WA, Coen KM (1991) Selective dopamine antagonists reduce nicotine self-administration. Psychopharmacology (Berlin) 104(2):171–176

Corrigall WA, Coen KM, Adamson KL (1994) Self-administered nicotine activates the mesolimbic dopamine system through the ventral tegmental area. Brain Res 653(1–2):278–284

Dani JA, Heinemann S (1996) Molecular and cellular aspects of nicotine abuse. Neuron 16(5):905–908

Dani JA, Ji D, Zhou FM (2001) Synaptic plasticity and nicotine addiction. Neuron 31(3):349–352

David V, Durkin TP, Cazala P (1997) Self-administration of the GABAA antagonist bicuculline into the ventral tegmental area in mice: dependence on D2 dopaminergic mechanisms. Psychopharmacology (Berlin) 130(2):85–90

Dayan P, Niv Y (2008) Reinforcement learning: the good, the bad and the ugly. Curr Opin Neurobiol 18(2):185–196

Dehaene S, Changeux J-P (2000) Reward-dependent learning in neuronal networks for planning and decision making. Prog Brain Res 126:217–229

DiChiara G (1999) Drug addiction as a dopamine-dependent associative learning disorder. Eur J Pharmacol 375:13–30

Eaton JB, Peng J-H, Schroeder KM, George AA, Fryer JD, Krishnan C, Buhlman L, Kuo Y-P, Steinlein O, Lukas RJ (2003) Characterization of human alpha 4 beta 2-nicotinic acetylcholine receptors stably and heterologously expressed in native nicotinic receptor-null SH-EP1 human epithelial cells. Mol Pharmacol 64(6):1283–1294

Erhardt S, Schwieler L, Engberg G (2002) Excitatory and inhibitory responses of dopamine neurons in the ventral tegmental area to nicotine. Synapse 43(4):227–237

Fagen ZM, Mansvelder HD, Keath JR, McGehee DS (2003) Short- and long-term modulation of synaptic inputs to brain reward areas by nicotine. Ann NY Acad Sci 1003:185–195

Feldberg W (1945) Present views on the mode of action of acetylcholine in the central nervous system. Physiol Rev 25:596–642

Fenster CP, Rains MF, Noerager B, Quick MW, Lester RA (1997) Influence of subunit composition on desensitization of neuronal acetylcholine receptors at low concentrations of nicotine. J Neurosci 17(15):5747–5759

Floresco SB, West AR, Ash B, Moore H, Grace AA (2003) Afferent modulation of dopamine neuron firing differentially regulates tonic and phasic dopamine transmission. Nat Neurosci 6(9):968–973

Forster GL, Blaha CD (2000) Laterodorsal tegmental stimulation elicits dopamine efflux in the rat nucleus accumbens by activation of acetylcholine and glutamate receptors in the ventral tegmental area. Eur J Neurosci 12(10):3596–3604

Garzón M, Vaughan RA, Uhl GR, Kuhar MJ, Pickel VM (1999) Cholinergic axon terminals in the ventral tegmental area target a subpopulation of neurons expressing low levels of the dopamine transporter. J Comp Neurol 410(2):197–210

Gerzanich V, Peng X, Wang F, Wells G, Anand R, Fletcher S, Lindstrom J (1995) Comparative pharmacology of epibatidine: a potent agonist for neuronal nicotinic acetylcholine receptors. Mol Pharmacol 48(4):774–782

Gotti C, Zoli M, Clementi F (2006) Brain nicotinic acetylcholine receptors: native subtypes and their relevance. Trends Pharmacol Sci 27(9):482–491

Gotti C, Moretti M, Gaimarri A, Zanardi A, Clementi F, Zoli M (2007) Heterogeneity and complexity of native brain nicotinic receptors. Biochem Pharmacol 74(8):1102–1111

Granon S, Faure P, Changeux J-P (2003) Executive and social behaviors under nicotine receptor regulation. Proc Natl Acad Sci USA 100:9596–9601

Grace AA, Onn SP (1989) Morphology and electrophysiological properties of immunocytochemically identified rat dopamine neurons recorded in vitro. J Neurosci 9(10):3463–3481

Grenhoff J, Tung CS, Svensson TH (1988) The excitatory amino acid antagonist kynurenate induces pacemaker-like firing of dopamine neurons in rat ventral tegmental area in vivo. Acta Physiol Scand 134(4):567–568

Gutkin BS, Dehaene S, Changeux J-P (2006) A neurocomputational hypothesis for nicotine addiction. Proc Natl Acad Sci USA 103(4):1106–1111

Hansel D, Sompolinsky H (1998) Modeling feature selectivity in local cortical circuits. In: Methods in neuronal modeling: from synapse to networks. MIT Press, Cambridge

Henningfield JE, Stapleton JM, Benowitz NL, Grayson RF, London ED (1993) Higher levels of nicotine in arterial than in venous blood after cigarette smoking. Drug Alcohol Depend 33(1):23–29

Hornykiewicz O (1966) Dopamine (3-hydroxytyramine) and brain function. Pharmacol Rev 18(2):925–964

Ikemoto S, Murphy JM, McBride WJ (1997) Self-infusion of GABA(A) antagonists directly into the ventral tegmental area and adjacent regions. Behav Neurosci 111(2):369–380

Johnson SW, North RA (1992a) Opioids excite dopamine neurons by hyperpolarization of local interneurons. J Neurosci 12(2):483–488

Johnson SW, North RA (1992b) Two types of neurone in the rat ventral tegmental area and their synaptic inputs. J Physiol 450:455–468

Kalivas PW, Churchill L, Klitenick MA (1993) GABA and enkephalin projection from the nucleus accumbens and ventral pallidum to the ventral tegmental area. Neuroscience 57(4):1047–1060

Katz B, Thesleff S (1957) A study of the desensitization produced by acetylcholine at the motor end-plate. J Physiol 138(1):63–80

Klink R, de Kerchove d'Exaerde A, Zoli M, Changeux JP (2001) Molecular and physiological diversity of nicotinic acetylcholine receptors in the midbrain dopaminergic nuclei. J Neurosci 21(5):1452–1463

Komendantov AO, Komendantova OG, Johnson SW, Canavier CC (2004) A modeling study suggests complementary roles for GABAA and NMDA receptors and the SK channel in regulating the firing pattern in midbrain dopamine neurons. J Neurophysiol 91(1):346–357

Lacey MG, Mercuri NB, North RA (1989) Two cell types in rat substantia nigra zona compacta distinguished by membrane properties and the actions of dopamine and opioids. J Neurosci 9(4):1233–1241

Lichtenstein E (1982) The smoking problem: a behavioral perspective. J Consult Clin Psychol 50(6):804–819

Lodge DJ, Grace AA (2006) The laterodorsal tegmentum is essential for burst firing of ventral tegmental area dopamine neurons. Proc Natl Acad Sci USA 103(13):5167–5172

Lüscher C, Ungless MA (2006) The mechanistic classification of addictive drugs. PLoS Med 3(11):e437

Mameli-Engvall M, Evrard A, Pons S, Maskos U, Svensson TH, Changeux J-P, Faure P (2006) Hierarchical control of dopamine neuron-firing patterns by nicotinic receptors. Neuron 50(6):911–921

Mansvelder HD, McGehee DS (2000) Long-term potentiation of excitatory inputs to brain reward areas by nicotine. Neuron 27(2):349–357

Mansvelder HD, Keath JR, McGehee DS (2002) Synaptic mechanisms underlie nicotine-induced excitability of brain reward areas. Neuron 33(6):905–919

Mereu G, Yoon KW, Boi V, Gessa GL, Naes L, Westfall TC (1987) Preferential stimulation of ventral tegmental area dopaminergic neurons by nicotine. Eur J Pharmacol 141(3):395–399

Nashmi R, Xiao C, Deshpande P, McKinney S, Grady SR, Whiteaker P, Huang Q, McClure-Begley T, Lindstrom JM, Labarca C, Collins AC, Marks MJ, Lester HA (2007) Chronic nicotine cell specifically upregulates functional alpha 4* nicotinic receptors: basis for both tolerance in midbrain and enhanced long-term potentiation in perforant path. J Neurosci 27(31):8202–8218

Nestler EJ, Aghajanian GK (1997) Molecular and cellular basis of addiction. Science 278(5335):58–63

Nisell M, Nomikos GG, Svensson TH (1994) Systemic nicotine-induced dopamine release in the rat nucleus accumbens is regulated by nicotinic receptors in the ventral tegmental area. Synapse 16(1):36–44

Oades RD, Halliday GM (1987) Ventral tegmental (A10) system: neurobiology. 1. Anatomy and connectivity. Brain Res 434(2):117–165

Oakman SA, Faris PL, Kerr PE, Cozzari C, Hartman BK (1995) Distribution of pontomesencephalic cholinergic neurons projecting to substantia nigra differs significantly from those projecting to ventral tegmental area. J Neurosci 15(9):5859–5869

Papke RL (2006) Estimation of both the potency and efficacy of alpha7 nAChR agonists from single-concentration responses. Life Sci 78(24):2812–2819

Papke RL, Papke JKP (2002) Comparative pharmacology of rat and human alpha7 nAChR conducted with net charge analysis. Br J Pharmacol 137(1):49–61

Peng X, Katz M, Gerzanich V, Anand R, Lindstrom J (1994) Human alpha7 acetylcholine receptor: cloning of the alpha7 subunit from the SH-SY5Y cell line and determination of pharmacological properties of native receptors and functional alpha7 homomers expressed in xenopus oocytes. Mol Pharmacol 45(3):546–554

Picciotto MR, Zoli M, Rimondini R, Lena C, Marubio LM, Pich EM, Fuxe K, Changeux J-P (1998) Subunit composition of functional nicotinic receptors in dopaminergic neurons investigated with knock-out mice. Nature 391:173–177

Pidoplichko VI, DeBiasi M, Williams JT, Dani JA (1997) Nicotine activates and desensitizes midbrain dopamine neurons. Nature 390(6658):401–404

Quick MW, Lester RAJ (2002) Desensitization of neuronal nicotinic receptors. J Neurobiol 53(4):457–478

Rahman S, Zhang J, Corrigall WA (2003) Effects of acute and chronic nicotine on somatodendritic dopamine release of the rat ventral tegmental area: in vivo microdialysis study. Neurosci Lett 348:1–4

Redish D (2004) Addiction as computation gone awry. Science 306:1944–1947

Reynolds JNJ, Wickens JR (2002) Dopamine-dependent plasticity of corticostriatal synapses. Neural Netw 15(4–6):507–521

Sesack SR, Pickel VM (1992) Prefrontal cortical efferents in the rat synapse on unlabeled neuronal targets of catecholamine terminals in the nucleus accumbens septi and on dopamine neurons in the ventral tegmental area. J Comp Neurol 320(2):145–160

Sher E, Chen Y, Sharples TJW, Broad LM, Benedetti G, Zwart R, McPhie GI, Pearson KH, Baldwinson T, Filippi GD (2004) Physiological roles of neuronal nicotinic receptor subtypes: new insights on the nicotinic modulation of neurotransmitter release, synaptic transmission and plasticity. Curr Top Med Chem 4(3):283–297

Steffensen SC, Svingos AL, Pickel VM, Henriksen SJ (1998) Electrophysiological characterization of GABAergic neurons in the ventral tegmental area. J Neurosci 18(19):8003–8015

Sugita S, Johnson SW, North RA (1992) Synaptic inputs to GABAA and GABAB receptors originate from discrete afferent neurons. Neurosci Lett 134(2):207–211

Swanson LW (1982) The projections of the ventral tegmental area and adjacent regions: a combined fluorescent retrograde tracer and immunofluorescence study in the rat. Brain Res Bull 9(1–6):321–353

Taber MT, Fibiger HC (1995) Electrical stimulation of the prefrontal cortex increases dopamine release in the nucleus accumbens of the rat: modulation by metabotropic glutamate receptors. J Neurosci 15(5 Pt 2):3896–3904

Thierry AM, Blanc G, Sobel A, Stinus L, Golwinski J (1973) Dopaminergic terminals in the rat cortex. Science 182(4111):499–501

Tong ZY, Overton PG, Clark D (1996) Stimulation of the prefrontal cortex in the rat induces patterns of activity in midbrain dopaminergic neurons which resemble natural burst events. Synapse 22(3):195–208

Tong ZY, Overton PG, Martinez-Cué C, Clark D (1998) Do non-dopaminergic neurons in the ventral tegmental area play a role in the responses elicited in A10 dopaminergic neurons by electrical stimulation of +the prefrontal cortex? Exp Brain Res 118(4):466–476

Ungerstedt U (1971) Stereotaxic mapping of the monoamine pathways in the rat brain. Acta Physiol Scand Suppl 367:1–48

Usher M, McClelland JL (2001) The time course of perceptual choice: the leaky, competing accumulator model. Psychol Rev 108:550–592

Wilson HR, Cowan JD (1972) Excitatory and inhibitory interactions in localized populations of model neurons. Biophys J 12(1):1–24

Wooltorton JRA, Pidoplichko VI, Broide RS, Dani JA (2003) Differential desensitization and distribution of nicotinic acetylcholine receptor subtypes in midbrain dopamine areas. J Neurosci 23(8):3176–3185

Wu M, Hrycyshyn AW, Brudzynski SM (1996) Subpallidal outputs to the nucleus accumbens and the ventral tegmental area: anatomical and electrophysiological studies. Brain Res 740(1–2):151–161

Yu AJ, Dayan P (2005) Uncertainty, neuromodulation, and attention. Neuron 46(4):681–692

Chapter 5
Dual-System Learning Models and Drugs of Abuse

Dylan A. Simon and Nathaniel D. Daw

Abstract Dual-system theories in psychology and neuroscience propose that a deliberative or goal-directed decision system is accompanied by a more automatic or habitual path to action. In computational terms, the latter is prominently associated with model-free reinforcement learning algorithms such as temporal-difference learning, and the former with model-based approaches. Due in part to the close association between drugs of abuse and dopamine, and also between dopamine, temporal-difference learning, and habitual behavior, addictive drugs are often thought to specifically target the habitual system.

However, although many drug-taking behaviors are well explained under such a theory, evidence suggests that drug-seeking behaviors must leverage a goal-directed controller as well. Indeed, one exhaustive theoretical account proposed that drugs may have numerous, distinct impacts on both systems as well as on other processes.

Here, we seek a more parsimonious account of these phenomena by asking whether the apparent profligacy of drugs' effects might be explained by a single mechanism of action. In particular, we propose that the pattern of effects observed under drug abuse may reveal interactions between the two controllers, which have typically been modeled as separate and parallel. We sketch several different candidate characterizations and architectures by which model-free effects may impinge on a model-based system, including sharing of cached values through truncated tree search and bias of transition selection for prioritized value sweeping.

5.1 Introduction

Dual-system theories of decision making—involving, for instance, a deliberative "goal-directed" controller and a more automatized or "habitual" one—are ubiq-

D.A. Simon
Department of Psychology, New York University, New York, NY, USA

N.D. Daw (✉)
Center for Neural Science and Department of Psychology, New York University, New York, NY, USA
e-mail: nathaniel.daw@nyu.edu

uitous across the behavioral sciences (Blodgett and McCutchan 1947; Dickinson 1985; Verplanken et al. 1998; Kahneman and Frederick 2002; Loewenstein and O'Donoghue 2004; Daw et al. 2005; Wood and Neal 2007). Many theories of drug abuse draw on this sort of framework, proposing that the compulsive nature of abuse reflects a transition of behavioral control from the voluntary system to the habitual one (Tiffany 1990; Ainslie 2001; Everitt et al. 2001; Vanderschuren and Everitt 2004; Everitt and Robbins 2005; Bechara 2005). Such a characterization may explain many drug-taking behaviors that become stereotyped and automatic, and dovetails naturally with models of the function of the neuromodulator dopamine (a ubiquitous target of drugs of abuse) suggesting a specific role for this neuromodulator in reinforcing habits (Di Chiara 1999; Redish 2004). However, the view of abusive behaviors as excessively automatized stimulus-response habits cannot easily explain many sorts of drug-*seeking* behaviors, which can involve novel and often increasingly inventive goal-directed acquisition strategies (Tiffany 1990). Such theories also do not speak to more cognitive phenomena such as craving.

Drug abuse is a dysfunction of decision making, acquired through learning. In this domain, theories are often formalized in terms of reinforcement learning (RL) algorithms from artificial intelligence (Sutton and Barto 1998). By providing a quantitative characterization of decision problems, RL theories have enjoyed success in behavioral neuroscience as methods for direct analysis and interpretation of trial-by-trial decision data, both behavioral and neural (Schultz et al. 1997; Daw and Doya 2006). Importantly, these theories also offer a putative computational counterpart to the goal-directed vs. habitual distinction, which may be useful for characterizing either system's role in drug abuse. In these terms, the more automatic, habitual behaviors are typically associated with so-called *model-free* RL, notably temporal-difference (TD) methods such as the actor/critic, in which successful actions are reinforced so that they may be repeated in the future. However, it has more recently been proposed that goal-directed behaviors can be captured with a categorically distinct type of RL known as *model-based*, in which actions may be planned based on a learned associative model of the environment (Doya 1999; Daw et al. 2005; Tanaka et al. 2006; Hampton et al. 2006; Pan et al. 2007; Redish and Johnson 2007; Rangel et al. 2008; Gläscher et al. 2010).

Such theories hypothesize that goal-directed and habitual behaviors arise from largely separate and parallel RL systems in the brain: model-based and model-free. Model-free RL forms the basis for a well-known account of dopamine neurons in the midbrain, as well as BOLD activity in dopamine targets in the basal ganglia (Houk et al. 1994; Schultz et al. 1997; Berns et al. 2001; O'Doherty et al. 2003; McClure et al. 2003). Since it is well established that drugs of abuse affect the function of these systems, and that other problem behaviors such as compulsive gambling show evidence of related effects, it has been a natural and fruitful line of research to apply TD-like theories to drug abuse (Di Chiara 1999; Redish 2004). However, these more computational theories pose the same puzzle as their psychological counterparts: how to account for the role of more flexible, drug-seeking behaviors apparently associated with goal-directed (in this case, model-based) control. Here, we consider how these behaviors might be understood in terms of the less well characterized model-based system. We follow Redish et al. (2008) in this endeavor, but focus more on what drug abuse phenomena suggest about potential variants or elaborations of the standard model-based account. In particular, we relate these issues to a range of other data suggesting that the two hypothesized RL systems are not as separate as they have been envisioned, but may instead interact in some respects. We consider how different sorts of interaction might be captured in modified forms of these theories in order to extend the computational account of drug abuse.

5.2 Background: Reinforcement Learning and Behavior

As a framework for formalizing theories of drug abuse, this section lays out the basics of RL, the study of learning optimal decisions through trial and error. For a more detailed description of this branch of computer science, see Sutton and Barto (1998) or, for its applications to psychology, Balleine et al. (2008).

5.2.1 The Markov Decision Process

Most decision problems in RL are based on Markov decision processes (MDPs), which formalize real-world problems as a sequence of steps, each of which involves a choice between actions affecting the resulting reward and the situation going forward. Formally, an MDP is a set of *states*, $\mathcal{S}$, and *actions*, $\mathcal{A}$, which occur in some sequence, s_t and a_t over timesteps t, such that s_{t+1} depends stochastically on s_t and a_t, but on no other information. This dependence is described by a *transition function* specifying the probability distribution over possible next states given the current state and chosen action:

$$T(s, a, s') = \mathrm{P}\left[s_{t+1} = s' | s_t = s, a_t = a\right]$$

Rewards are similarly described by a stochastic *reward function* mapping each state to the quantity of reward received in that state: $R : \mathcal{S} \to \mathbb{R}$, such that $r_t = R(s_t)$. The transition and reward functions thus fully describe the process.

5.2.2 Values and Policies

The goal of an agent in an MDP is to select actions so as to maximize its reward, and more specifically, to *learn* to do so by trial and error, using only information about the underlying process observed during behavior (i.e., samples from the transition and reward functions). Specifically, at a state, s, an agent aims to pick the action,

a, that will maximize the cumulative, temporally discounted rewards that will be received in the future, in expectation over future states and actions:

$$Q(s,a) = \mathrm{E}\left[\sum_{i=1} \gamma^{i-1} r_{t+i} | s_t = s, a_t = a\right]$$

where $\gamma < 1$ is an exponential time discounting factor. This quantity is known as the state-action value function, and many approaches to RL involve estimating it, either directly or indirectly, so as to choose the action maximizing it at each state.

A key aspect of MDPs (indeed, what makes them difficult), is their sequential nature. An agent's future value prospects depend not only on the current state and action, but on future choices as well. Formally, consider a *policy* by which an agent selects actions, that is, a (possibly stochastic) function describing the action to take in each state: $\pi : \mathcal{S} \to \mathcal{A}$. From this, we can define the expected value of taking action a in state s, and then following policy π thereafter:

$$Q^\pi(s,a) = \sum_{s'} T(s,a,s') \left[\mathrm{E}[R(s')] + \gamma \sum_{s''} T(s', \pi(s'), s'') [\mathrm{E}[R(s'')] + \cdots] \right] \tag{5.1}$$

This value depends on the sequence of future expected rewards that will be obtained, averaged over all possible future trajectories of states, $s, s', s'', \ldots$, according to the policy and transition function. One way of framing the goal, then, is to determine the optimal policy, known as π^*, that will maximize $Q^{\pi^*}(s_t, \pi^*(s_t))$ at each step.

A key insight relevant to solving this problem is that the state-action value may be written recursively:

$$Q^\pi(s,a) = \sum_{s'} T(s,a,s') [R(s') + \gamma Q^\pi(s', \pi(s'))]$$

Since the optimal policy must maximize Q at each step, the optimal value satisfies:

$$Q^*(s,a) = \sum_{s'} T(s,a,s') \left[R(s') + \gamma \max_{a'} Q^*(s',a')\right] \tag{5.2}$$

This is known as the Bellman equation, which provides a recursive relationship between all the action values in the MDP.

The optimal policy can be extracted directly from the optimal value function, if it is known. That is, an agent can achieve maximal expected reward by simply choosing the maximally valued action at each step: $\pi^*(s) = \mathrm{argmax}_a Q^*(s,a)$. Accordingly, we next consider two different approaches to learning to choose actions, which each work via learning to estimate $Q^*(s,a)$.

5.2.3 Algorithms for RL

5.2.3.1 Model-Free RL

The recursive nature of the state-action value motivates one approach to RL, often exemplified by temporal-difference learning (Sutton 1988). Here, an agent attempts directly to estimate the optimal value function Q^*. (A closely related variant, the actor/critic algorithm, estimates the policy π^* itself using similar methods.)

The recursion in Eq. (5.2) shows how such an estimate may be updated, by changing it so as to reduce the observed deviation between the left and right hand sides of the equation, known as the prediction error. Specifically, consider any step at which an action, a, is taken in state s, and a new state, s', and reward, r, are observed. From Eq. (5.2), it can be seen that the quantity $r + \gamma \max_{a'} Q(s', a')$ is a *sample* of the value of the preceding state and action, $Q(s, a)$, where the state s' samples the transition distribution $T(s, a, s')$, and the agent's own estimate of the new state's value, $Q(s', a')$, stands in for the true Q^*. We can then update the estimated value toward the observed value, with learning rate α:

$$Q(s,a) \leftarrow Q(s,a) + \alpha \underbrace{\big(\overbrace{R(s') + \gamma \max_{a'} Q(s',a')}^{Q \text{ sample}} - Q(s,a) \big)}_{\text{prediction error, } \delta}$$

Algorithms of this sort are known as model-free approaches because they do not directly represent or make use of the underlying MDP transition or reward functions, but instead learn the relevant summary quantity directly: the state-action value function.

5.2.3.2 Model-Based RL

A second approach to RL is model-based learning. Here, representations of the transition and reward functions are themselves learned, which function as a model of the MDP, thus giving rise to the name. This is quite straightforward; for instance, the transition function may be estimated simply by counting state-action-state transitions. Given any estimate of these functions, the state-action value function may be computed directly, for example, through the iterative expansion of Eq. (5.1) to explicitly compute the expected rewards over different possible trajectories.

Such a recursive computation can be laborious, in contrast to and thus motivating model-free methods which involve minimal computation at choice time (e.g., simply comparing learned state-action values). The flip side of this trade-off is that computing these values on the basis of a full world model, rather than simply relying on a previously learned summary, offers more flexible possibilities for combining information learned at different times, and enables the agent to respond more dynamically under changing situations.

5.2.4 RL and Behavioral Neuroscience

These two frameworks for solving an MDP, one (model-free) computationally fast and reactive, and the other (model-based) involving more deliberative or proactive consideration of possibilities, are closely related to the psychological concepts of habits and goal-directed actions, respectively.

In psychology, these two sorts of instrumental behavior are envisioned as relying on different underlying representations (Balleine and Dickinson 1998). Goal-directed actions are supposed to be based on a representation of the action-outcome contingency (e.g., that pressing a lever produces a certain amount of cheese; or, in a spatial task, a 'cognitive map' of the maze), allowing deliberative choice by examining the consequences of different possible actions. Habits are instead assumed to be based on direct stimulus-response associations, which may be learned by a simple reinforcement rule (i.e., if a response in the presence of some stimulus is followed by reward, strengthen it, as proposed by Thorndike 1898) and embody a very simple, switchboard-like choice strategy.

However, since the stimulus-response association lacks any representation of the specific outcome (e.g., cheese) that originally reinforced it, a choice mechanism of this sort predicts odd inflexibilities and insensitivities to certain shifts in circumstances. For instance, it predicts that a rat who is trained to lever-press for food while hungry, but then fed to satiety, will continue to work for food given the opportunity, at least until given enough experience to unlearn or relearn the association. In contrast, since choosing a goal-directed action involves examining the action-outcome association, this approach can adjust behavioral preferences instantly to comply with new situations such as changes in outcome values. Another important capability of a goal-directed approach is the ability to plan novel actions to obtain new goals or react to new information. For instance, in a maze, an animal might use a cognitive map to plan a route not previously followed, such as a shortcut between two locations (Tolman 1948). Such flexibility is not possible using only stimulus-response associations (since such a route will not have previously been reinforced).

All this motivates standard experimental procedures, such as outcome devaluation, for distinguishing these two sorts of behaviors. The results of such tests (specifically, whether actions are or are not sensitive to devaluation under different circumstances) indicate that the brain uses both approaches (Dickinson and Balleine 2002).

The two sorts of RL algorithms directly mimic these psychological theories in key respects (Daw et al. 2005; Balleine et al. 2008). Like habits, model-free approaches support easy choices by relying on a summary representation of an immediately relevant decision variable: the value function or policy. For the same reason, these representations lack information about outcome identity and are insensitive to changes; they can be updated only following additional experience with the consequences of a state and action, and often through its repetition. Conversely, model-based algorithms formalize the idea of an associative or cognitive search in which possible outcomes are explicitly considered in relation to their likelihood of achieving some goal (i.e., reward). These forms of reasoning depend on representations of outcomes and state transitions (analogous to a cognitive map or action-outcome association), and, like their psychological counterparts, can adjust rapidly to changes in the worth or availability of outcomes and can combine previously experienced sequences of actions in novel ways to reach goals.

RL approaches have also been associated with specific neural systems. Model-free algorithms in particular have been a valuable tool for explaining the function of the dopamine system, as the firing rates of midbrain dopamine neurons closely match the error signals predicted by these algorithms (Houk et al. 1994; Schultz et al. 1997). There is also evidence for representation of state-action values in other areas of the brain, including prefrontal cortex, striatum, and parietal regions (Delgado et al. 2000; Arkadir et al. 2004; Tanaka et al. 2004; Samejima et al. 2005; Plassmann et al. 2007; Tom et al. 2007; Kable and Glimcher 2007; Hare et al. 2008; Kim et al. 2009; Wunderlich et al. 2009; Chib et al. 2009).

Less is known about the neural substrate for model-based or goal-directed actions, though there are now a number of reports of potentially model-related activity throughout the brain (Hampton et al. 2006, 2008; Pan et al. 2007; Bromberg-Martin et al. 2010). In general, these actions are not envisioned to involve dopamine, since model-based approaches rely on quite different learning mechanisms with error signals that do not match the dopaminergic response (Gläscher et al. 2010) and because lesions of the dopaminergic system appear to spare goal-directed action while affecting habits (Faure et al. 2005). More generally, the use of the reward devaluation procedure together with numerous brain lesions has allowed the demonstration of an anatomical double-dissociation, wherein different areas of striatum (and associated parts of cortex and thalamus) support each learning strategy even under circumstances when the other would be observed in intact animals (Killcross and Coutureau 2003; Yin et al. 2004, 2005; Balleine et al. 2007). These findings have suggested that the brain implements both model-based and model-free approaches as parallel and, to some extent, independent systems.

5.2.5 RL and Drugs of Abuse

In this light it seems natural to interpret the strongly habitual behaviors associated with drug taking as an effect of drug abuse specifically on model-free valuations (Redish 2004; Redish et al. 2008; Schultz 2011). In particular, compulsive behaviors have been attributed to overly strong habitual responses (or state-action values), whereby learned responses persist despite contrary evidence of their value available to a contemplative, model-based system (Everitt and Robbins 2005). A candidate mechanism for such uncontrolled reinforcement is effects of drugs on the dopaminergic signal carrying the reward prediction error supposed to train model-free values or policies (Redish 2004; Panlilio et al. 2007; Redish et al. 2008). This interpretation is consistent with the fact that most if not all drugs of abuse share effects on dopamine as a common mechanism of their reinforcing action.

However, it has also been pointed out that such an account is necessarily incomplete, and in particular that drug abusers demonstrate highly elaborate and often novel drug-seeking behaviors (Tiffany 1990; Olmstead et al. 2001; Kalivas and Volkow 2005; Root et al. 2009). Just as with short cuts in mazes, such flexible planning cannot be explained by the model-free repetition of previously reinforced actions. Therefore, the remainder of this chapter considers algorithmic possibilities for ways a model-based system could be affected by drug abuse. These considerations have consequences for theories of appetitively motivated behavior more generally, since they strongly suggest some sort of integration or cooperation between the systems in commonly valuing drug outcomes.

5.3 Drugs and Model-Based RL

The problem facing us is that, under a standard theory (e.g., Daw et al. 2005), drugs of abuse affect valuations only in the model-free system, via effects on a dopaminergic prediction error. Valuations in a model-based system have been presumed to be entirely separate and independent, and in particular, to be unaffected by manipulations of dopamine. However, if the effects of drugs are isolated to a model-free system (and drugs are not, by comparison, disproportionately valued in a model-based system) then actions motivated by drugs should exclusively constitute simple repetitions of previously reinforced actions. Such a system has no mechanism for planning novel drug-seeking actions.

In this section, we consider a number of potential solutions to this issue, focusing on effects either via inflation of values per se, or biasing them via changes in the search process by which they are computed.

5.3.1 Drugs and Model-Based Reward

A typical application of model-free theories to drug abuse depends on drugs affecting the learned value or policy function, for example, by inflating the state-action values leading to drug rewards. Is there some simple analogy in a model-based system for such inflation? While model-based systems typically construct a value function on demand, rather than maintaining a representation of one, they do maintain a representation of rewards in some other form, often as an approximation to the state reward function. This reward function could theoretically be learned through prediction errors just as state-action values are, and similarly be inflated as an effect of drug abuse (Redish et al. 2008; Schultz 2011). In this case, an increased reward associated with the attainment state would flexibly elicit a wide range of goal-directed behaviors, as any actions likely to eventually reach that state would themselves have a higher computed action value, even along novel paths. However, this explanation raises a problematic question: what is the process by which drugs of abuse could inflate the reward function?

By analogy with the TD account, the natural answer would seem to be that the inflation happens in much the same way as model-free value inflation is supposed to occur: via effects on dopaminergic responses effectively exaggerating the prediction error used to learn these representations. However, as previously mentioned, the representations learned in a model-based system (notably, the reward function, R) require different sorts of prediction errors (Gläscher et al. 2010). On available evidence, the responses of dopaminergic neurons appear consistent with a prediction error appropriate for training future (discounted) value (Q), not immediate reward (R). In particular, the signature phenomenon whereby dopamine responses transfer with training to cues predicting upcoming reward is inconsistent with a prediction error for the one-step reward R: there are no immediate rewards and no errors in their predictions tied to this event (Schultz et al. 1997). Moreover, although reward values for the model-based system are likely represented in a dissociable location in the brain from model-free values, it is unlikely that this learning is driven by some atypical dopaminergic signal, since reports suggest at least anecdotally that dopamine neurons are consistent in this respect, regardless of where they project (Schultz 1998).

If dopamine controls these secondary incentives or motivational values and not representations of one-step rewards, then the latter are unlikely to be a mechanism by which drugs of abuse impact model-based valuations.

5.3.2 Drugs and Model-Based Value

In order to solve this problem, we return to the Bellman equation (5.2) which connects model-free and model-based approaches by defining the state-action value that they both compute in different ways. A key claim of the model-free approaches is that the brain maintains internal ("cached" or stored) estimates of the state-action values, which are updated in place by prediction error and are putatively inflated by drugs of abuse via their effects on this prediction error signaling. The model-based approach is assumed instead to compute the state-action values anew at decision time by evaluating the Bellman equation, deriving them from more elemental information (the reward and transition functions).

If indeed both systems operate in the brain and aim to compute equivalently defined state-action values, then the Bellman equation suggests an obvious possibility for their interaction: a model-based system could make use of the cached state-action values maintained by the model-free system. In particular, because of the recursive form of the Bellman equation, at any point in its iterative, tree-structured expansion, it is possible to substitute a cached (e.g., model-free) estimate of the right-hand value, Q, to terminate the expansion. One motivation for this "partial evaluation" is that the full reevaluation of the Bellman equation at each decision step is computationally laborious; moreover, repeating this computation each step may have diminishing returns if, for instance, the learned estimates of transition and reward functions change little between each evaluation (Moore and Atkeson 1993).

If a model-based search immediately terminated with cached action values (i.e., on the first step) it would simply revert to a model-free system, while each additional step of evaluation using the model's transition and reward functions would provide a view of value which is model-based out to a horizon extended one step further into the future, at the cost of additional computation. Thus, if a model-based system engaged in such partial evaluation by terminating its search at states with model-free state-action values inflated by the theorized dopamine mechanisms (such as states associated with drug attainment), the model-based system would be similarly compromised, with this exaggeration carried back to other computed action values that may reach such a state. The combination of the two sorts of evaluation would allow the model-based system to plan novel action trajectories aimed at attaining states with high (potentially drug-inflated) value in the model-free system's estimates. In this sense, the model-free estimates can serve as secondary incentives for guiding the model-based system's preferences, an idea reminiscent of "incentive salience" accounts of drug motivation (Robinson and Berridge 2008).

The foregoing considerations suggest a new perspective on the joint contribution of model-based and model-free evaluations to behavior. Whereas previous work (Daw et al. 2005) envisioned that the brain must select between separate model-based and model-free values, the partial evaluation approach suggests that the key question is instead where to integrate the values: at each step, whether to further evaluate a decision branch or to truncate the trajectory using cached values. With this extension, the traditional story of a shift from goal-directed to automatic processing can make a broader range of behavioral predictions as a shift towards more limited searches under model-based evaluation (Nordquist et al. 2007).

Also, interacting architectures of this broad sort may help to explain numerous indications from the neuroscientific literature that model-free and model-based evaluation may be more interacting than separate. For instance, goal-directed learning appears to involve a subregion of striatum, dorsomedial, which is adjacent to the part apparently responsible for habits, and which also receives heavy dopaminergic innervation (Yin et al. 2005). Moreover, indications of model-based computations (such as devaluation sensitivity) have been observed throughout areas of the brain traditionally thought to be part of the model-free system including ventral striatum (Daw et al. 2011; Simon and Daw 2011; van der Meer et al. 2010), downstream ventral pallidum (Tindell et al. 2009), and even dopaminergic neurons (Bromberg-Martin et al. 2010).

In the drug context, this view also raises a new set of questions, surrounding how drugs might affect search termination. For instance, if drug-inflated estimates of state-action values serve as secondary incentives for model-based search, why would the model-based system terminate with them, rather than planning *past* the contaminated states?

5.3.3 Drugs and Model-Based Search

If a model-based search process were biased at search time to adopt exaggerated cached values rather than pursuing further evaluation, the resulting behavior would show strong preferences for actions (even novel ones) that tend to lead to such outcomes. The question is why such a bias would arise. That is, the concept of partial evaluation explains how inflated values in the model-free system could affect the model-based system, but may not adequately account for the particular fixations drugs of abuse engender, whereby goal-directed behaviors may operate to fulfill the craving to the exclusion of other goals.

To begin to address this question, we consider how search progress and search termination might be affected by drugs of abuse. A more general and flexible framework for reasoning about these issues is Sutton's Dyna architecture (Sutton 1990), which provides a framework by which model-based and model-free RL can coexist and dynamically trade-off their contributions to learning. This architecture has also been employed in theories of model-based learning in the brain (Johnson and Redish 2005). The Dyna-Q algorithm envisions that an agent will maintain a single set of cached state-action values, but that these can be updated by both model-based and model-free updates in any mixture. As with standard model-free learning, state-action values may be updated directly by prediction errors according to actual experience. A learned world model can also be used to produce simulated experience (i.e., state, action and reward trajectories sampled from the modeled transition and reward functions), which can train the cached state-action values in the same way as real experience. Full model-based value updates (i.e., averaging rather than sampling over possible successor states for an action using the Bellman equation) can also be applied in place.

As opposed to the traditional view of a tree-structured search, Dyna-Q has the freedom to apply these model-based updates in arbitrary orders. All these updates may be interleaved during behavior, at decision time, or off-line. Given sufficient updates, the values learned will approach the same model-based values a fully expanded search would. Because of the possibility of learning from simulated sample trajectories, the theory also exposes the connection between model-based valuation and simulation. Intuitively this idea comports well with ideas that search may be implemented by cognitive simulation (Buckner and Carroll 2007; Buckner 2010) as well as evidence for various sorts of on- and off-line replay or preplay over spatial trajectories in hippocampal place cells (Johnson and Redish 2005; Foster and Wilson 2006; Hasselmo 2008; Koene and Hasselmo 2008; Davidson et al. 2009; Lansink et al. 2009; Derdikman and Moser 2010; Carr et al. 2011; Dragoi and Tonegawa 2011).

The question of drug abuse now can be further refined to which trajectories are simulated, as well as where these trajectories are terminated. One principled approach to this question is the prioritized sweeping algorithm (Moore and Atkeson 1993). In its original form, it is fully model-based (i.e., no direct TD updates from experience are used) but the same principle is equally applicable within Dyna. The general idea is that if new experience or computation changes the value (or transition and reward functions) at a state, then these changes will have the most extreme effects on the state-action values for actions leading up to those states, and so those predecessors should have the highest priority for simulated updates. For example, if a novel reward is experienced following an action, with the standard TD algorithm, this reward will not have an effect on other actions that may lead to the reward state until those actions are taken, while a model-based system will be able to update other action values accordingly, but only with extensive computation. Under a Dyna algorithm, however, this reward value could be propagated to other cached, model-free action values through simulated sampling of actions. By sampling states in reverse order along trajectories leading to the reward state, for instance, 'backing up' the values to more distant states, this can happen quite efficiently without requiring any additional real experience (Foster and Wilson 2006).

A neural system that implements such an algorithm suggests a mechanism for exploitation by drugs of abuse, whereby values inflated by distorted prediction errors could preferentially be selected for backing up. In particular, the principle that model-based updates are prioritized toward areas of the state space with new learning will be directly compromised by inflated prediction errors, since these will drive new learning and thereby attract more priority for model-based updates. Thus, the standard dopamine-mediated drug abuse story, whereby effective prediction errors are enhanced by drug experiences even when no new reward information is available, now cleanly predicts such prioritized model-based value updates as well. The action values associated with drug-taking would continue to increase in such a scenario, and thus always be given high priority for backups. As a result, these inflated values would propagate throughout the model, even to actions not previously resulting in drug attainment that have some probability of leading to other inflated states, to the exclusion of other potential goals or even negative experiences that may occur subsequent to fulfillment. This may constitute a computational description of phenomena associated with drug abuse, such as salience-driven sensitization or motivational magnets (Di Ciano 2008; Robinson and Berridge 2008), and can also explain suggestions that even goal-directed drug-seeking actions are insensitive to devaluation (Root et al. 2009). Here, the high priority given to such continually changing values is analogous to high salience for drug-associated stimuli.

Finally, a related phenomenon observed in drug abuse that might be similarly explained in this framework is cue-specific craving, in which stimuli associated with drug-taking result in increased drug-seeking motivation (Meil and See 1996; Garavan et al. 2000; Bonson et al. 2002; See 2005; Volkow et al. 2008). A potentially related effect in psychology is known as outcome-specific Pavlovian-instrumental transfer (PIT), in which presentation of cues associated with a particular reward increase the preference for instrumental actions associated with the same reward (Lovibond 1983; Rescorla 1994). A pure model-free learning system has no way to explain these effects, as action values abstract specific outcomes, and so while cues could generally enhance motivation, they cannot do so in an outcome-specific way. Further, it is unclear why cues in themselves should change an agent's action preferences or valuations, since the cues do not in fact carry information relevant to action valuation. A model-based system, however, stores specific outcomes as part of the reward function. These, in the Dyna framework, may be used to drive simulation priorities for value updates. Through a priority mechanism, and since this approach allows on-the-fly updating of model-free values based on model-driven updates, it could theoretically drive updates preferentially toward a cued goal, ignoring other rewards to effect an updated value map more biased toward that outcome. Similarly, a drug-associated cue could simply trigger further updates back from objective states, pushing the values for related actions higher.

5.4 Conclusion

Drug abuse is a disorder of decision making, and as such its phenomena are relevant to and can be informed by the established computational theories of the domain. Building on two-system theories of learned decision making (Dickinson 1985; Balleine and Dickinson 1998; Poldrack et al. 2001; Daw et al. 2005; Wood and Neal 2007) and on the broad taxonomy of their potential vulnerabilities to drugs of abuse by Redish et al. (2008), we have considered the implications of drug-seeking behavior for algorithms and architectures hypothesized to comprise such a system. Drugs of abuse are commonly thought to target a habit learning system, specifically via their effects on dopamine and resultant amplification of model-free prediction errors. That they appear to serve as incentives for goal-directed behavior as well strongly suggests that the two decision systems interchange information rather than operating independently. We suggest this interchange might be captured within a modified architecture, such as Dyna or tree search with partial evaluation, allowing model-free and model-based influences to converge within a single representation. Importantly, such a mechanism, coupled with a scheduling principle for model-based searches like prioritized-sweeping, allows the single, ubiquitous, model-free mechanism of drug action to account for the range of behavioral phenomena.

The implications of these hypothesized mechanisms for decision making theories more generally remain to be developed. In particular, previous work has addressed a range of data on how animals' behaviors are differentially sensitive to devaluation in different circumstances by assuming two separate RL algorithms whose preferences were arbitrated according to relative uncertainty (Daw et al. 2005). It remains to be seen whether the same phenomena can be understood in the more integrated architectures suggested here, either in terms of prioritized sweeping heuristics or, alternatively, by developing the uncertainty explanation in this setting. That said, indications are accumulating rapidly, beyond the context of drugs of abuse, that the systems are more interactive than was assumed in previous theories (Root et al. 2009; Bromberg-Martin et al. 2010; van der Meer et al. 2010; Daw et al. 2011; Simon and Daw 2011). This accumulation of evidence strongly motivates the investigation of hybrid algorithms and interacting architectures of the type discussed here to expand our understanding of the range of strategies by which humans make decisions.

Acknowledgements The authors are supported by a Scholar Award from the McKnight Foundation, a NARSAD Young Investigator Award, Human Frontiers Science Program Grant RGP0036/2009-C, and NIMH grant 1R01MH087882-01, part of the CRCNS program.

References

Ainslie G (2001) Breakdown of will. Cambridge University Press, Cambridge
Arkadir D, Morris G, Vaadia E, Bergman H (2004) Independent coding of movement direction and reward prediction by single pallidal neurons. J Neurosci 24(45):10047–10056
Balleine BW, Daw ND, O'Doherty JP (2008) Multiple forms of value learning and the function of dopamine. In: Glimcher PW, Camerer CF, Fehr E, Poldrack RA (eds) Neuroeconomics: decision making and the brain. Academic Press, London, pp 367–387
Balleine BW, Delgado MR, Hikosaka O (2007) The role of the dorsal striatum in reward and decision-making. J Neurosci 27(31):8161–8165
Balleine BW, Dickinson A (1998) Goal-directed instrumental action: contingency and incentive learning and their cortical substrates. Neuropharmacology 37(4–5):407–419
Bechara A (2005) Decision making, impulse control and loss of willpower to resist drugs: a neurocognitive perspective. Nat Neurosci 8(11):1458–1463
Berns GS, McClure SM, Pagnoni G, Montague PR (2001) Predictability modulates human brain response to reward. J Neurosci 21(8):2793–2798
Blodgett HC, McCutchan K (1947) Place versus response learning in the simple T-maze. J Exp Psychol 37(5):412–422
Bonson KR, Grant SJ, Contoreggi CS, Links JM, Metcalfe J, Weyl HL et al (2002) Neural systems and cue-induced cocaine craving. Neuropsychopharmacology 26(3):376–386
Bromberg-Martin ES, Matsumoto M, Hong S, Hikosaka O (2010) A pallidus-habenula-dopamine pathway signals inferred stimulus values. J Neurophysiol 104(2):1068–1076
Buckner RL (2010) The role of the hippocampus in prediction and imagination. Annu Rev Psychol 61:27–48, C1-8
Buckner RL, Carroll DC (2007) Self-projection and the brain. Trends Cogn Sci 11(2):49–57
Carr MF, Jadhav SP, Frank LM (2011) Hippocampal replay in the awake state: a potential substrate for memory consolidation and retrieval. Nat Neurosci 14(2):147–153
Chib VS, Rangel A, Shimojo S, O'Doherty JP (2009) Evidence for a common representation of decision values for dissimilar goods in human ventromedial prefrontal cortex. J Neurosci 29(39):12315–12320
Davidson TJ, Kloosterman F, Wilson MA (2009) Hippocampal replay of extended experience. Neuron 63(4):497–507
Daw ND, Doya K (2006) The computational neurobiology of learning and reward. Curr Opin Neurobiol 16(2):199–204
Daw ND, Niv Y, Dayan P (2005) Uncertainty-based competition between prefrontal and dorsolateral striatal systems for behavioral control. Nat Neurosci 8(12):1704–1711
Daw ND, Gershman SJ, Seymour B, Dayan P, Dolan R (2011) Model-based influences on humans' choices and striatal prediction errors. Neuron 69(6):1204–1215
Delgado MR, Nystrom LE, Fissell C, Noll DC, Fiez JA (2000) Tracking the hemodynamic responses to reward and punishment in the striatum. J Neurophysiol 84(6):3072–3077
Derdikman D, Moser M-B (2010) A dual role for hippocampal replay. Neuron 65(5):582–584
Di Chiara G (1999) Drug addiction as dopamine-dependent associative learning disorder. Eur J Pharmacol 375(1–3):13–30
Di Ciano P (2008) Facilitated acquisition but not persistence of responding for a cocaine-paired conditioned reinforcer following sensitization with cocaine. Neuropsychopharmacology 33(6):1426–1431
Dickinson A (1985) Actions and habits: The development of behavioural autonomy. Philos Trans R Soc Lond B, Biol Sci 308:67–78

Dickinson A, Balleine B (2002) The role of learning in the operation of motivational systems. In: Stevens' handbook of experimental psychology. Wiley, New York

Doya K (1999) What are the computations of the cerebellum, the basal ganglia and the cerebral cortex? Neural Netw 12(7–8):961–974

Dragoi G, Tonegawa S (2011) Preplay of future place cell sequences by hippocampal cellular assemblies. Nature 469(7330):397–401

Everitt BJ, Robbins TW (2005) Neural systems of reinforcement for drug addiction: from actions to habits to compulsion. Nat Neurosci 8(11):1481–1489

Everitt BJ, Dickinson A, Robbins TW (2001) The neuropsychological basis of addictive behaviour. Brains Res Rev 36(2–3):129–138

Faure A, Haberland U, Condé F, Massioui NE (2005) Lesion to the nigrostriatal dopamine system disrupts stimulus-response habit formation. J Neurosci 25(11):2771–2780

Foster DJ, Wilson MA (2006) Reverse replay of behavioural sequences in hippocampal place cells during the awake state. Nature 440(7084):680–683

Garavan H, Pankiewicz J, Bloom A, Cho JK, Sperry L, Ross TJ et al (2000) Cue-induced cocaine craving: neuroanatomical specificity for drug users and drug stimuli. Am J Psychiatry 157(11):1789–1798

Gläscher J, Daw ND, Dayan P, O'Doherty JP (2010) States versus rewards: Dissociable neural prediction error signals underlying model-based and model-free reinforcement learning. Neuron 66(4):585–595

Hampton AN, Bossaerts P, O'Doherty JP (2006) The role of the ventromedial prefrontal cortex in abstract state-based inference during decision making in humans. J Neurosci 26(32):8360–8367

Hampton AN, Bossaerts P, O'Doherty JP (2008) Neural correlates of mentalizing-related computations during strategic interactions in humans. Proc Natl Acad Sci 105(18):6741–6746

Hare TA, O'Doherty JP, Camerer CF, Schultz W, Rangel A (2008) Dissociating the role of the orbitofrontal cortex and the striatum in the computation of goal values and prediction errors. J Neurosci 28(22):5623–5630

Hasselmo ME (2008) Temporally structured replay of neural activity in a model of entorhinal cortex, hippocampus and postsubiculum. Eur J Neurosci 28(7):1301–1315

Houk JC, Adams JL, Barto AG (1994) A model of how the basal ganglia generate and use neural signals that predict reinforcement. In: Houk JC, Davis JL, Beiser DG (eds) Models of information processing in the basal ganglia. MIT Press, Cambridge, pp 249–270

Johnson A, Redish AD (2005) Hippocampal replay contributes to within session learning in a temporal difference reinforcement learning model. Neural Netw 18(9):1163–1171

Kable JW, Glimcher PW (2007) The neural correlates of subjective value during intertemporal choice. Nat Neurosci 10(12):1625–1633

Kahneman D, Frederick S (2002) Representativeness revisited: Attribute substitution in intuitive judgment. In: Gilovich T, Griffin DW, Kahneman D (eds) Heuristics and biases: the psychology of intuitive judgement. Cambridge University Press, New York, pp 49–81

Kalivas PW, Volkow ND (2005) The neural basis of addiction: a pathology of motivation and choice. Am J Psychiatry 162(8):1403–1413

Killcross S, Coutureau E (2003) Coordination of actions and habits in the medial prefrontal cortex of rats. Cereb Cortex 13(4):400–408

Kim H, Sul JH, Huh N, Lee D, Jung MW (2009) Role of striatum in updating values of chosen actions. J Neurosci 29(47):14701–14712

Koene RA, Hasselmo ME (2008) Reversed and forward buffering of behavioral spike sequences enables retrospective and prospective retrieval in hippocampal regions CA3 and CA1. Neural Netw 21(2–3):276–288

Lansink CS, Goltstein PM, Lankelma JV, McNaughton BL, Pennartz CMA (2009) Hippocampus leads ventral striatum in replay of place-reward information. PLoS Biol 7(8):e1000173

Loewenstein G, O'Donoghue T (2004) Animal spirits: Affective and deliberative processes in economic behavior (Working Papers Nos. 04–14). Cornell University, Center for Analytic Economics

Lovibond PF (1983) Facilitation of instrumental behavior by a Pavlovian appetitive conditioned stimulus. J Exp Psychol, Anim Behav Processes 9(3):225–247

McClure SM, Berns GS, Montague PR (2003) Temporal prediction errors in a passive learning task activate human striatum. Neuron 38(2):339–346

van der Meer MAA, Johnson A, Schmitzer-Torbert NC, Redish AD (2010) Triple dissociation of information processing in dorsal striatum, ventral striatum, and hippocampus on a learned spatial decision task. Neuron 67(1):25–32

Meil W, See R (1996) Conditioned cued recovery of responding following prolonged withdrawal from self-administered cocaine in rats: an animal model of relapse. Behav Pharmacol 7(8):754–763

Moore AW, Atkeson CG (1993) Prioritized sweeping: Reinforcement learning with less data and less time. Mach Learn 13:103–130. (10.1007/BF00993104)

Nordquist RE, Voorn P, de Mooij-van Malsen JG, Joosten RNJMA, Pennartz CMA, Vanderschuren LJMJ (2007) Augmented reinforcer value and accelerated habit formation after repeated amphetamine treatment. Eur Neuropsychopharmacol 17(8):532–540

O'Doherty JP, Dayan P, Friston K, Critchley H, Dolan RJ (2003) Temporal difference models and reward-related learning in the human brain. Neuron 38(2):329–337

Olmstead MC, Lafond MV, Everitt BJ, Dickinson A (2001) Cocaine seeking by rats is a goal-directed action. Behav Neurosci 115(2):394–402

Pan X, Sawa K, Sakagami M (2007) Model-based reward prediction in the primate prefrontal cortex. Neurosci Res 58(Suppl 1):229

Panlilio LV, Thorndike EB, Schindler CW (2007) Blocking of conditioning to a cocaine-paired stimulus: testing the hypothesis that cocaine perpetually produces a signal of larger-than-expected reward. Pharmacol Biochem Behav 86(4):774–777

Plassmann H, O'Doherty J, Rangel A (2007) Orbitofrontal cortex encodes willingness to pay in everyday economic transactions. J Neurosci 27(37):9984–9988

Poldrack RA, Clark J, Paré-Blagoev EJ, Shohamy D, Creso Moyano J, Myers C et al (2001) Interactive memory systems in the human brain. Nature 414(6863):546–550

Rangel A, Camerer C, Montague P (2008) A framework for studying the neurobiology of value-based decision making. Nat Rev, Neurosci 9(7):545–556

Redish AD (2004) Addiction as a computational process gone awry. Science 306(5703):1944–1947

Redish AD, Johnson A (2007) A computational model of craving and obsession. Ann NY Acad Sci 1104(1):324–339

Redish AD, Jensen S, Johnson A (2008) Addiction as vulnerabilities in the decision process. Behav Brain Sci 31(04):461–487

Rescorla RA (1994) Control of instrumental performance by Pavlovian and instrumental stimuli. J Exp Psychol, Anim Behav Processes 20(1):44–50

Robinson TE, Berridge KC (2008) The incentive sensitization theory of addiction: some current issues. Philos Trans R Soc Lond B, Biol Sci 363(1507):3137–3146

Root DH, Fabbricatore AT, Barker DJ, Ma S, Pawlak AP, West MO (2009) Evidence for habitual and goal-directed behavior following devaluation of cocaine: a multifaceted interpretation of relapse. PLoS ONE 4(9):e7170

Samejima K, Ueda Y, Doya K, Kimura M (2005) Representation of action-specific reward values in the striatum. Science 310(5752):1337–1340

Schultz W (1998) Predictive reward signal of dopamine neurons. J Neurophysiol 80(1):1–27

Schultz W (2011) Potential vulnerabilities of neuronal reward, risk, and decision mechanisms to addictive drugs. Neuron 69(4):603–617

Schultz W, Dayan P, Montague PR (1997) A neural substrate of prediction and reward. Science 275(5306):1593–1599

See RE (2005) Neural substrates of cocaine-cue associations that trigger relapse. Eur J Pharmacol 526(1–3):140–146

Simon DA, Daw ND (2011) Neural correlates of forward planning in a spatial decision task in humans. J Neurosci 31(14):5526–5539

Sutton RS (1988) Learning to predict by the methods of temporal differences. Mach Learn 3(1):9–44
Sutton RS (1990) Integrated architectures for learning, planning, and reacting based on approximating dynamic programming. In: Proceedings of the seventh International Conference on Machine Learning. Morgan Kaufmann, San Mateo, pp 216–224
Sutton RS, Barto AG (1998) Reinforcement learning. MIT Press, Cambridge
Tanaka SC, Doya K, Okada G, Ueda K, Okamoto Y, Yamawaki S (2004) Prediction of immediate and future rewards differentially recruits cortico-basal ganglia loops. Nat Neurosci 7(8):887–893
Tanaka SC, Samejima K, Okada G, Ueda K, Okamoto Y, Yamawaki S et al (2006) Brain mechanism of reward prediction under predictable and unpredictable environmental dynamics. Neural Netw 19(8):1233–1241
Thorndike EL (1898) Animal intelligence: An experimental study of the associative processes in animals. Psychol Rev Monogr Suppl 2(4):1–8
Tiffany ST (1990) A cognitive model of drug urges and drug-use behavior: Role of automatic and nonautomatic processes. Psychol Rev 97(2):147–168
Tindell AJ, Smith KS, Berridge KC, Aldridge JW (2009) Dynamic computation of incentive salience: "wanting" what was never "liked". J Neurosci 29(39):12220–12228
Tolman EC (1948) Cognitive maps in rats and men. Psychol Rev 55:189–208
Tom SM, Fox CR, Trepel C, Poldrack RA (2007) The neural basis of loss aversion in decision-making under risk. Science 315(5811):515–518
Vanderschuren LJMJ, Everitt BJ (2004) Drug seeking becomes compulsive after prolonged cocaine self-administration. Science 305(5686):1017–1019
Verplanken B, Aarts H, van Knippenberg AD, Moonen A (1998) Habit versus planned behaviour: a field experiment. Br J Soc Psychol 37(1):111–128
Volkow ND, Wang G-J, Telang F, Fowler JS, Logan J, Childress A-R et al (2008) Dopamine increases in striatum do not elicit craving in cocaine abusers unless they are coupled with cocaine cues. NeuroImage 39(3):1266–1273
Wood W, Neal DT (2007) A new look at habits and the habit goal interface. Psychol Rev 114(4):843–863
Wunderlich K, Rangel A, O'Doherty JP (2009) Neural computations underlying action-based decision making in the human brain. Proc Natl Acad Sci 106(40):17199–17204
Yin HH, Knowlton BJ, Balleine BW (2004) Lesions of dorsolateral striatum preserve outcome expectancy but disrupt habit formation in instrumental learning. Eur J Neurosci 19(1):181–189
Yin HH, Ostlund SB, Knowlton BJ, Balleine BW (2005) The role of the dorsomedial striatum in instrumental conditioning. Eur J Neurosci 22(2):513–523

Chapter 6
Modeling Decision-Making Systems in Addiction

Zeb Kurth-Nelson and A. David Redish

Abstract This chapter describes addiction as a failure of decision-making systems. Existing computational theories of addiction have been based on temporal difference (TD) learning as a quantitative model for decision-making. In these theories, drugs of abuse create a non-compensable TD reward prediction error signal that causes pathological overvaluation of drug-seeking choices. However, the TD model is too simple to account for all aspects of decision-making. For example, TD requires a state-space over which to learn. The process of acquiring a state-space, which involves both situation classification and learning causal relationships between states, presents another set of vulnerabilities to addiction. For example, problem gambling may be partly caused by a misclassification of the situations that lead to wins and losses. Extending TD to include state-space learning also permits quantitative descriptions of how changing representations impacts patterns of intertemporal choice behavior, potentially reducing impulsive choices just by changing cause-effect beliefs. This approach suggests that addicts can learn healthy representations to recover from addiction. All the computational models of addiction published so far are based on learning models that do not attempt to look ahead into the future to calculate optimal decisions. A deeper understanding of how decision-making breaks down in addiction will certainly require addressing the interaction of drugs with model-based look-ahead decision mechanisms, a topic that remains unexplored.

Decision-making is a general process that applies to all the choices made in life, from which ice cream flavor you want to whether you should use your children's college savings to buy drugs. Neural systems evolved to make decisions about what actions to take to keep an organism alive, healthy and reproducing. However, the same decision-making processes can fail under particular environmental or pharmacological conditions, leading the decision-maker to make pathological choices.

Z. Kurth-Nelson · A.D. Redish (✉)
Department of Neuroscience, University of Minnesota, 6-145 Jackson Hall, 321 Church St. SE, Minneapolis, MN 55455, USA
e-mail: redish@umn.edu

Z. Kurth-Nelson
e-mail: kurt0073@umn.edu

Both substance addiction and behavioral addictions such as gambling can be viewed in this framework, as failures of decision-making.

The simplest example of a failure in decision-making is in response to situations that are engineered to be disproportionately rewarding. In the wild, sweetness is a rare and useful signal of nutritive value, but refined sugar exploits this signal, and given the opportunity, people will often select particularly sweet foods over more nutritive choices. A more dangerous failure mode can be found in drugs of abuse. These drugs appear to directly modulate elements of the decision-making machinery in the brain, such that the system becomes biased to choose drug-seeking actions.

There are three central points in this chapter. First, a mathematical language of decision-making is developed based on *temporal difference (TD)* algorithms applied to *reinforcement learning (RL)* (Sutton and Barto 1998). Within this mathematical language, we review existing quantitative theories of addiction, most of which are based on identified failure modes within that framework (Redish 2004; Gutkin et al. 2006; Dezfouli et al. 2009). However, we will also discuss evidence that the framework is incomplete and that there are decision-making components that are not easily incorporated into the TD-RL framework (Dayan and Balleine 2002; Daw et al. 2005; Balleine et al. 2008; Dayan and Seymour 2008; Redish et al. 2008). Second, an organism's understanding of the world is central to its decision-making. Two organisms that perceive the contingencies of an experiment differently will behave differently. We extend quantitative decision-making theories to account for ways that organisms identify and utilize structure in the world to make decisions (Redish et al. 2007; Courville 2006; Gershman et al. 2010), which may be altered in addiction. Third, decision-making models naturally accommodate a description of how future rewards can be compared to immediate ones (Sutton and Barto 1998; Redish and Kurth-Nelson 2010). Both drug and behavioral addicts often exhibit impulsive choice, where a small immediate reward is preferred over a large delayed reward (Madden and Bickel 2010). There is evidence that impulsivity is both cause and consequence of addiction (Madden and Bickel 2010; Rachlin 2000). In particular, a key factor in recovery from addiction seems to be the ability to take a longer view on one's decisions and the ability to construct representations that support healthy decision-making (Ainslie 2001; Heyman 2009; Kurth-Nelson and Redish 2010).

6.1 Multiple Decision-Making Systems, Multiple Vulnerabilities to Addiction

Organisms use a combination of decision-making strategies. When faced with a choice, a human or animal may employ one or more of these strategies to produce a decision. The strategies used may also change with experience. For example, a classic experiment in rodent navigation involves a plus-shaped maze with four arms. On each trial, a food reward is placed in the east arm of the maze and the animal is placed in the south arm. The animal quickly learns to turn right to the east arm to reach the food. On a probe trial, the animal can be placed in the north arm instead of the south arm. If these probe trials are conducted early in the course of learning, the animal turns left to the east arm, indicating that the animal is following a *location-based strategy* that dynamically calculates appropriate actions based on new information. On the other hand, if probe trials are conducted after the animal has been overtrained on the original task, the animal turns right into the west arm of the maze, indicating that it is following a *response strategy* where actions are precalculated and stored (Tolman 1948; Restle 1957; Packard and McGaugh 1996).

These different decision-making systems have different neuroanatomical substrates. In the rodent navigation example, the location-based strategy requires hippocampal integrity (Barnes 1979; Packard and McGaugh 1996), while the response strategy is dependent on the integrity of lateral aspects of striatum (Packard and McGaugh 1996; Yin et al. 2004). The location-based system is more computationally intensive but is more flexible to changing environments, while the response-based system is quick to calculate but inflexible to changing environments (O'Keefe and Nadel 1978; Redish 1999).

How the results of these different decision-making systems are integrated into a final decision remains an important open question. Obviously, if the two predicted actions are incompatible (as in the example above where one system decides to turn right while the other decides to turn left) and the animal takes an action, then the results must be integrated by the time the signals reach the muscles to perform the action. For example, an oversight system could enable or disable the place and response strategies, or could decide between the suggested actions provided by the two systems. However, economic theory implies the results are integrated much sooner (Glimcher et al. 2008). In neuroeconomic theory, every possible outcome is assumed to have a *utility*. The utilities of any possible outcome can be represented in a *common currency*, allowing direct comparison of the expected utilities to select a preferred action. In between the two extremes of common currency and muscle-level integration, there is a wide range of possibilities for how different decision-making systems could interact to produce a single decision. For example, a location-based strategy and a response strategy could each select an action (e.g., "turn left" or "turn right"), and these actions could compete to be transformed into a motor pattern.

In the following sections, we will develop a theoretical description of the brain's decision-making systems and show how drugs of abuse can access specific failure modes that lead to addictive choice. Addictive drugs have a variety of pharmacological effects on the brain, ranging from blockade of dopamine transporters to agonism of μ-opioid receptors to antagonism of adenosine receptors. Fundamentally, the common effect of addictive drugs is to cause pathological over-selection of the drug-taking decision, but this may be achieved in a variety of ways by accessing vulnerabilities in the different decision-making systems. This theory suggests that addicts may use and talk about drugs differently depending on which vulnerability the drugs access, and that appropriate treatment will likely differ depending on how the decision-making system has failed (Redish et al. 2008). For example, craving and relapse are separable entities in addictive processes—overvaluation in a stimulus-response based system could lead to relapse of the action of drug-taking even in the absence of explicit craving, while overvaluation in the value system could lead to explicit identifiable desires for drug, but may not necessarily lead to relapse (Redish and Johnson 2007; Redish et al. 2008; Redish 2009).

6.1.1 Temporal Difference Reinforcement Learning and the Dopamine Signal

To explain why reward learning seems to occur only when an organism is confronted with an unexpected reward, Rescorla and Wagner (1972) introduced the idea of a *reward learning prediction error*. In their model, an agent (i.e., an organism or a computational model performing decision-making) learns how much reward is predicted by each cue, and generates a prediction error if the actual reward received does not match the net prediction of the cues they experienced. The prediction error is then used to update the reward prediction. To a first approximation, the fast phasic firing of midbrain dopamine neurons matches the Rescorla-Wagner prediction error signal (Ljungberg et al. 1992; Montague et al. 1996; Schultz 2002): when an animal is presented with an unexpected reward, dopamine neurons fire in a phasic burst of activity. If the reward is preceded by a predictive cue, the phasic firing of dopamine neurons gradually diminishes over several trials. The loss of dopamine firing at reward matches the loss of Rescorla-Wager prediction error, as the reward is no longer unpredicted.

However, there are several phenomena that the Rescorla-Wagner model does not account for. First, in animal behavior, conditioned stimuli can also act as reinforcers (Domjan 1998), and this shift is also reflected in the dopamine signals (Ljungberg et al. 1992). The Rescorla-Wagner model cannot accommodate this shift in reinforcement (Niv and Montague 2008). Second, a greater latency between stimulus and reward slows learning, reduces the amount of responding at the stimulus, and reduces dopamine firing at the stimulus (Mackintosh 1974; Domjan 1998; Bayer and Glimcher 2005; Fiorillo et al. 2008). The Rescorla-Wagner model does not represent time and cannot account for any effects of timing. Third, the Rescorla-Wagner model is a model of Pavlovian prediction and does not address instrumental action-selection. A generalized version of the Rescorla-Wagner model that accounts for stimulus chaining, temporal effects and action-selection is temporal difference reinforcement learning (TDRL).

Reinforcement learning is the general problem of how to learn what actions to take in order to maximize reward. Temporal difference learning is a common theoretical approach to solving the problem of reinforcement learning (Sutton and Barto 1998). Although the agent may be faced with a complex sequence of actions and observations before receiving a reward, temporal difference learning allows the agent to assign a value to each action along the way.

In order to apply a mathematical treatment, TDRL formalizes the learning problem as a set of states and transitions that define the situation of the animal and how that situation can change (for example, see the very simple state-space in Fig. 6.1A). This collection of states and transitions is called a *state-space*, and defines the cause-effect relationships of the world that pertain to the agent. The agent maintains an estimate, for each state, of the reward it expects to receive in the future of that state. This estimate of future reward is called *value*, or V. We will use S_t to refer to the state of the agent at time t; $V(S_t)$ is the value of this state.

When the agent receives reward, it compares this reward with the amount of reward it expected to receive at that moment. Any difference is an error signal, called δ, which represents how incorrect the prior expectation was.

$$\delta = (R_t + V(S_t)) \cdot disc(d) - V(S_{t-1}) \qquad (6.1)$$

where R_t is the reward at time t, d is the time spent in state S_{t-1}, and *disc* is a monotonically decreasing temporal discounting function with a range from 0 to 1. (Note that in the *semi-Markov* formulation of temporal difference learning (Daw 2003; Si et al. 2004; Daw et al. 2006), which we use here, the world can dwell in each state for an extended period of time.) A commonly used discounting function is

$$disc(d) = \gamma^d \qquad (6.2)$$

where $\gamma \in [0, 1]$ is the exponential discounting rate. δ (Eq. (6.1)) is zero if the agent correctly estimated the value of state S_{t-1}; that is, it correctly identified the discounted future reward expected to follow that state. The actual reward received immediately following S_{t-1} is R_t, and the future reward expected after S_t is $V(S_t)$. Together, $R_t + V(S_t)$ is the future reward expected following S_{t-1}. This is discounted by the delay between S_{t-1} and S_t. The difference between this and the prior expectation $V(S_{t-1})$ is the value prediction error δ.

The estimated value of state S_{t-1} is updated proportional to δ, so that the expectation is brought closer to reality.

$$V(S_{t-1}) \leftarrow V(S_{t-1}) + \delta \cdot \alpha \qquad (6.3)$$

where $\alpha \in (0, 1)$ is a learning rate. With appropriate exploration parameters and unchanging state space and reward contingencies, this updating process is guaranteed to converge on the correct expectation of discounted future reward for each state (Sutton and Barto 1998). Once reward expectations are learned, the agent can choose the actions that lead to the states with highest expected reward.

6.1.2 Value Prediction Error as a Failure Mode

The psychostimulants, including cocaine and amphetamine, directly increase dopamine action at the efferent targets of dopaminergic neurons (Ritz et al. 1987; Phillips et al. 2003; Aragona et al. 2008). The transient, or *phasic*, component of dopamine neuron firing appears to carry a reward prediction error signal like δ

(Montague et al. 1996; Schultz et al. 1997; Tsai et al. 2009). Thus, the psychostimulant drugs may act by pharmacologically increasing the δ signal (di Chiara 1999; Bernheim and Rangel 2004; Redish 2004).

Redish (2004) implemented this hypothesis in a computational model. Drug delivery was simulated by adding a non-compensable component to δ,

$$\delta = \max(D_t,\ D_t + (R_t + V(S_t)) \cdot disc(d) - V(S_{t-1})) \tag{6.4}$$

This is the same as Eq. (6.1) with the addition of a D_t term representing the drug delivered at time t. The value of δ cannot be less than D_t, due to the max function. The effect of D_t is that even after $V(S_{t-1})$ has reached the correct estimation of future reward, $V(S_{t-1})$ will keep growing without bound. In other words, D_t can never be compensated for by increasing $V(S_{t-1})$, so δ is never driven to zero. If there is a choice between a state that leads to drugs and a state that does not, the state leading to drugs will eventually (after a sufficient number of trials) have a higher value and thus be preferred.

This model exhibits several features of real drug addiction. The degree of preference for drugs over natural rewards increases with drug experience. Further, drug use is less sensitive to costs (i.e., drugs are less elastic) than natural rewards, and the elasticity of drug use decreases with experience (Christensen et al. 2008). Like other neuroeconomic models of addiction (e.g., Becker and Murphy (1988)), the Redish (2004) model predicts that even highly addicted individuals will still be sensitive to drug costs, albeit less sensitive than non-addicts, and less sensitive than to natural reward costs. (Even though they are willing to pay remarkably high costs to feed their addiction, addicts remain sensitive to price changes in drugs (Becker et al. 1994; Grossman and Chaloupka 1998; Liu et al. 1999).) The Redish (2004) model achieves inelasticity due to overvaluation of drugs of abuse.

The hypotheses that phasic dopamine serves as a value prediction error signal in a Rescorla-Wagner or TDRL-type learning system and that cocaine increases that phasic dopamine signal imply that Kamin blocking should not occur when cocaine is used as a reinforcer. In Kamin blocking (Kamin 1969), a stimulus X is first paired with reward until the X→reward association is learned. (The existence of a learned association is measured by testing whether the organism will respond to the stimulus.) Then stimuli X and Y are together paired with reward. In this case, no association between Y and reward is learned. The Rescorla-Wagner model explains this result by saying that because X already fully predicts reward, there is no prediction error and thus no learning when X and Y are paired with reward. Consistent with the dopamine-as-δ hypothesis, phasic dopamine signals do not appear in response to the blocked stimuli (Waelti et al. 2001). However, if the blocking experiment is performed with cocaine instead of a natural reinforcer, the hypothesis that cocaine produces a non-compensable δ signal predicts that the δ signal should still occur when training XY→cocaine, so the organism should learn to respond for Y. Contrary to this prediction, Panlilio et al. (2007) recently provided evidence that blocking does occur with cocaine in rats, implying that either the phasic dopamine signal is not equivalent to the δ signal, or cocaine does not boost phasic dopamine. Recently, Jaffe et al. (2010) presented data that a subset of high-responding animals did not show Kamin blocking when faced with nicotine rewards, suggesting that the lack of Kamin blocking may produce overselection of drug rewards in a subset of subjects. An extension to the Redish model to produce overselection of drug rewards while still accounting for blocking with cocaine is given by Dezfouli et al. (2009) (see also Chap. 8 in this book). In this model, new rewards are compared against a long-term average reward level. Drugs increase this average reward level, so the effect of drugs is compensable and the δ signal goes to zero with long-term drug exposure. If this model is used to simulate the blocking experiment with cocaine as the reinforcer, then during the X→cocaine training, the average reward level is elevated, so that when XY→cocaine occurs, there is no prediction error signal and Y does not acquire predictive value.

Other evidence also suggests that the Redish (2004) model is not a complete picture. First, the hypotheses of the model imply that continued delivery of cocaine will eventually overwhelm any reinforcer whose prediction error signal is compensable (such as a food reward). Recent data (Lenoir et al. 2007) suggest that this is not the case, implying that the Redish (2004) model is not a complete picture. Second, the Redish (2004) model is based on the assumption that addiction arises from the action of drugs on the dopamine system. Many addictive drugs do not act directly on dopamine (e.g., heroin, which acts on μ-opioid receptors (Nestler 1996)), and some drugs that boost dopamine are not addictive (e.g., bupropion (Stahl et al. 2004)). Most psychostimulant drugs also have other pharmacological effects; for example, cocaine also has an action on the norepinephrine and serotonin systems (Kuhar et al. 1988). Norepinephrine has been implicated in signaling uncertainty (Yu and Dayan 2005) and attention (Berridge et al. 1993), while serotonin has other effects on decision-making structures in the brain (Tanaka et al. 2007). All of these actions could also potentially contribute to the effects of cocaine on decision-making.

Action selection can be performed in a variety of ways. When multiple actions are available, the agent may choose the action leading to the highest valued state. Alternatively, the benefit of each action may be learned separately from state values. Separating *policy learning* (i.e., learning the benefit of each action) from value learning has the theoretical advantage of being easier to compute when there are many available actions (for example, if the action space is continuous, Sutton and Barto 1998). In this case, the policy learning system is called the *actor* and the value learning system is called the *critic*. The actor and critic systems have been proposed to correspond to different brain structures (Barto 1994; O'Doherty et al. 2004; Daw and Doya 2006). The dopamine-as-δ hypothesis can provide another explanation for drug addiction if learning in the critic system is saturable. During actor learning, feedback from the critic is required to calculate how much unexpected reinforcement occurred, and thus how much the actor should learn. If drugs produce a large increase in δ that cannot be compensated for by the saturated critic, then the actor will over-learn the benefit of the action leading to this drug-delivery (see Chap. 8 in this book).

The models we have discussed so far use the assumption that decision-making is based on learning, for each state, an expectation of future value that can be expressed in a common currency. There are many experiments that show that not all decisions are explicable in this way (Balleine and Dickinson 1998; Dayan 2002; Daw et al. 2005; Dayan and Seymour 2008; Redish et al. 2008; van der Meer and Redish 2010). The limitations of the temporal difference models can be addressed by incorporating additional learning and decision-making algorithms (Pavlovian systems, deliberative systems) and by addressing the representations of the world over which these systems work.

6.1.3 Pavlovian Systems

Unconditioned stimuli can provoke an approach or avoidance response that does not depend on the instrumental contingencies of the experiment (Mackintosh 1974; Dayan and Seymour 2008). These Pavlovian systems can produce non-optimal decisions in some animals under certain conditions (Breland and Breland 1961; Balleine 2001, 2004; Dayan et al. 2006; Uslaner et al. 2006; Flagel et al. 2008; Ostlund and Balleine 2008). For example, in a classic experiment, birds were placed on a linear track, near a cup of food that was mechanically designed to move in the same direction as the bird, at twice the bird's speed. The optimal strategy for the bird was to move away from the food until the food reached the bird, but in the experiment, birds never learned to move away; instead always chasing the food to a greater distance (Hershberger 1986). Theories of Pavlovian influence on decision-making suggest that the food-related cues provoked an approach response (Breland and Breland 1961; Dayan et al. 2006). Similarly, if animals are trained that a cue predicts a particular reward in a Pavlovian conditioning task, later presenting that cue during an instrumental task in which one of the choices leads to that reward will increase preference for that choice (Pavlovian-instrumental transfer (Estes 1943; Kruse et al. 1983; Lovibond 1983; Talmi et al. 2008)). Although models of Pavlovian systems exist (Balleine 2001, 2004; Dayan et al. 2006) as do suggestions that Pavlovian failures underlie aspects of addiction (Robinson and Berridge 1993, 2001, 2004; Berridge 2007), computational models of addiction taking into account interactions between Pavlovian effects and temporal difference learning are still lacking.

6.1.4 Deliberation, Forward Search and Executive Function

During a decision, the brain may explicitly consider alternatives in order to predict outcomes (Tolman 1939; van der Meer and Redish 2010). This process allows evaluation of those outcomes in the light of current goals, expectations, and values (Niv et al. 2006). Therefore part of the decision-making process plausibly involves predicting the future situation that will arise from taking a choice and accessing the reinforcement associations that are present in that future situation. This stands in contrast to decision-making strategies that use only the value associations present in the current situation.

When rats running in a maze come to an important choice-point where they could go right or left and possibly receive reward, they will sometimes pause and turn their head from side to side as if to sample the options. This is known as vicarious trial and error (VTE) (Muenzinger 1938; Tolman 1938, 1939, 1948). VTE behavior is correlated to hippocampal activity and is reduced by hippocampal lesions (Hu and Amsel 1995; Hu et al. 2006). During most behavior, cells in the hippocampus encode the animal's location in space (O'Keefe and Dostrovsky 1971; O'Keefe and Nadel 1978; Redish 1999). But during VTE, this representation sometimes projects forward in one direction and then the other (Johnson and Redish 2007). Johnson and Redish (2007) proposed that this "look-ahead" that occurs during deliberation may be part of the decision making process. By imagining the future, the animal may be attempting to determine whether each choice is rewarded (Tolman 1939, 1948). Downstream of the hippocampus, reward-related cells in the ventral striatum also show additional activity during this deliberative process (van der Meer and Redish 2009), which may be evidence for prediction and calculation of expectancies (Daw et al. 2005; Redish and Johnson 2007; van der Meer and Redish 2010).

Considering forward search as part of the decision making process permits a computational explanation for the phenomena of craving and obsession in drug addicts (Redish and Johnson 2007). Craving is the recognition of a high-value outcome, and obsession entails constraint of searches to a single high-value outcome. Current theories suggest that endogenous opioids signal the hedonic value of received rewards (Robinson and Berridge 1993). If these endogenous opioids also signal imagined rewards, then opioids may be a key to craving (Redish and Johnson 2007). This fits data that opioid antagonists reduce craving (Arbisi et al. 1999; Levine and Billington 2004). Under this theory, an opioidergic signal at the time of reward or drug delivery may cause neural plasticity in such a way that the dynamics of the forward search system become biased to search toward the outcome linked to the opioid signal. Activation of opioid receptors is known to modulate synaptic plasticity in structures such as the hippocampus (Liao et al. 2005), suggesting a possible physiological basis for altering forward search in the hippocampus.

6.2 Temporal Difference Learning in a Non-stationary Environment

Temporal difference learning models describe how to learn an expectation of future reward over a known state-space. In the real world, the state-space itself is not known a priori. It must be learned and may even change over time. This is illustrated by the problem of extinction and reinstatement. After a cue-reinforcer association is learned, it can be extinguished by presenting the cue alone (Domjan 1998). Over time, animals will learn to stop responding for the cue. If extinction is done in a different environment from the original learning, placing the animal back in the original environment causes responding to start again immediately (Bouton and Swartzentruber 1989). Similarly, even if acquisition and extinction occur in the same environment, a single presentation of the reinforcer following extinction can cause responding to start again (Pavlov 1927; McFarland and Kalivas 2001; Bouton 2002). This implies that the original association was not unlearned during extinction. A similar phenomenon occurs in abstaining human drug addicts, where drug-related cues can trigger relapse to full resumption of drug-seeking behavior much faster than the original development of addiction (Jaffe et al. 1989; Childress et al. 1992). In extinction paradigms, the world is non-stationary: a cue that used to lead to a reward or drug-presentation now no longer does. Thus, a decision-making system trying to accurately predict the world requires a mechanism to construct state-spaces flexibly from the observed dynamics of the world. This mechanism does not exist in standard TDRL models.

To explain the phenomenon of renewal of responding after extinction, a recent model extended temporal difference learning by adding state-classification (Redish et al. 2007). In this model, the total information provided from the world to the agent at each moment was represented as an n-dimensional sensory cue. The model classified cue vectors into the same state if they were similar, or into different states if they were sufficiently dissimilar. During acquisition of a cue-reinforcer association, the model grouped these similar observations (many trials with the same cue) into a state representing "cue predicts reward". The model learned to associate the value of the reward with instrumental responding in this "cue predicts reward" state. This learning occurred at the learning rate of the model. During extinction, as the model accumulated evidence that a cue did not predict reward in a new context, these observations were classified into a new state representing "cue does not predict reward", from which actions had no value. When returned to the original context, the model switched back to classifying cue observations into the "cue predicts reward" state. Because instrumental responding in the "cue predicts reward" state had already been associated with reward during acquisition, no additional learning was needed, and responding immediately resumed at the pre-extinction rate.

This situation-classification component may be vulnerable to its own class of failures in decision-making. Based on vulnerabilities in situation-classification, Redish et al. (2007) were also able to simulate behavioral addiction to gambling. These errors followed both from over-separation of states, in which two states that were not actually different were identified as different due to unexpected consistencies in noise, and from over-generalization of states, in which two states that were different were not identified as different due to the similarities between them. The first process is similar to that of "the illusion of control" in which subjects misperceive that they have control of random situations, producing superstition (Langer and Roth 1975; Custer 1984; Wagenaar 1988; Elster 1999). The illusion of control can be created by having too many available cues, particularly when combined with the identification of near-misses (Cote et al. 2003; Parke and Griffiths 2004). The phenomenon of "chasing", in which subjects continue to place deeper and deeper losing bets, may arise because gamblers over-generalize a situation in which they received a large win, to form a belief that gambling generally leads to reward (Custer 1984; Wagenaar 1988;

Elster 1999). We suggest this is a problem of state-classification: the gamblers classify the generic gambling situation as leading to reward.

In the Redish et al. (2007) model, states were classified from sensory and reinforcement experience, but the transition structure of the world was not learned. Smith et al. (2006) took the converse approach. Here the algorithm started with a known set of states, each with equal temporal extent, and learned the transition probability matrix based on observed transitions. A "surprise" factor measured the extent to which a reinforcer was unpredicted by previous cues, also allowing the model to reproduce the Kamin blocking effect (Kamin 1969) and the reduction of latent inhibition by amphetamine (Weiner et al. 1988).

Both the Redish et al. (2007) and Smith et al. (2006) models are special cases of the more general *latent cause theory*, in which the agent attempts to identify hidden causes underlying sets of observations (Courville 2006; Gershman et al. 2010). In these models, agents apply an approximation of Bayesian statistical inference to all observations to infer hidden causes that could underlie correlated observations. Because latent cause models take into account any change in stimulus–stimulus or stimulus–outcome contingencies, these models are able to accommodate any nonstationary environment.

The ability of the brain to dynamically construct interpretations of the causal structure of the world is likely seated in frontal cortex and hippocampus. Hippocampus is involved in accommodating cue-reward contingency changes (Hirsh 1974; Isaacson 1974; Hirsh et al. 1978; Nadel and Willner 1980; Corbit and Balleine 2000; Fuhs and Touretzky 2007). Returning to a previously reinforced context no longer triggers renewal of extinguished responding if hippocampus is lesioned (Bouton et al. 2006). Medial prefrontal cortex appears to be required for learning the relevance of new external cues that signal altered reinforcement contingencies (Lebron et al. 2004; Milad et al. 2004; Quirk et al. 2006; Sotres-Bayon et al. 2006). Classification and causality representations in hippocampus and frontal cortex may form a cognitive input to the basal ganglia structures that perform reinforcement learning. Drugs of abuse that negatively impact the function of hippocampal or cortical structures could inhibit the formation of healthy state-spaces, contributing to addiction. Alcohol, for example, has been hypothesized to preferentially impair both hippocampal and prefrontal function (Hunt 1998; Oscar-Berman and Marinkovic 2003; White 2003).

In general, if the brain constructs state-spaces that do not accurately reflect the world but instead overemphasize the value of the addictive choice, this constitutes an addiction vulnerability. Behavioral addiction to gambling may arise from a failure of state classification as described above. Addiction to drugs could result from state-spaces that represent only the immediate choice and not the long-range consequences. This would suggest that training new state-space constructions, and mechanisms designed to prevent falling back into old state-spaces, may improve relapse outcomes in addicts.

6.3 Discounting and Impulsivity

In this section we will discuss the phenomenon of intertemporal choice (how the delay to a reward influences decisions), and show how changes in the agent's state-space can change the intertemporal decisions made by an organism.

If offered a choice between $10 right now and $11 tomorrow, many people will feel it is not worth waiting one day for that extra dollar, and choose the $10 now. When offered a choice between a small immediate reward and a large delayed reward, *impulsivity* is the extent to which the agent prefers the small immediate reward, being unwilling to wait for the future reward. This is sometimes viewed as a special case of temporal discounting, which is the general problem of how the value of rewards diminishes as they recede into the future.[1] As discussed above, a discounting function $disc(d)$ maps a delay d to a number in [0, 1] specifying how much a reward's value is attenuated due to being postponed by time d. The impulsive decision to take a smaller-sooner reward rather than a larger-later one can be studied in the context of temporal difference learning.

Addicts tend to be more impulsive than non-addicts. It is easy to see why impulsivity could lead to addiction: the benefit of drug-taking tends to be more immediate than the benefits of abstaining. It is also possible that drugs increase impulsivity. Smokers discount faster than those who have never smoked, but ex-smokers discount at a rate similar to those who have never smoked (Bickel et al. 1999). In the Dezfouli et al. (2009) model, simulations show that choice for non-drug rewards becomes more impulsive following repeated exposure to drugs. Although the causal relationship between drug-taking and impulsivity is difficult to study in humans, animal data show that chronic drug-taking increases impulsivity (Paine et al. 2003; Simon et al. 2007).

If offered a choice between $10 right now and $11 tomorrow, many people will choose $10; however, if offered a choice between $10 in a year and $11 in a year and a day, the same people often prefer the $11 (Ainslie 2001). This is an example of *preference reversal*. Economically, the two decisions are equivalent and, under simple assumptions of stability, it should not matter if the outcomes are each postponed by a year. But in practice, many experiments have found that the preferred option changes as the time of the present changes relative to the outcomes (Madden and Bickel 2010).

In principle, any monotonically decreasing function with a range from 0 to 1 could make a reasonable discounting function. Exponential discounting (as in Eq. (6.2)) is often used in theoretical models because it is easy to calculate and matches economic assumptions of behavior. However, preference reversal does not occur in exponential discounting, but does occur with any non-exponential

---

[1] There are multiple decision factors often referred to as "impulsivity", including the inability to inhibit a pre-potent response, the inability to inhibit an over-learned response, and an over-emphasis on immediate versus delayed rewards (which we are referring to here). These multiple factors seem to be independent (Reynolds et al. 2006) and to depend on different brain structures (Isoda and Hikosaka 2008) and we will not discuss the other factors here.

discounting function (Frederick et al. 2002). Discounting data in humans and animals generally does show preference reversal (Chung and Herrnstein 1967; Baum and Rachlin 1969; Mazur 1987; Kirby and Herrnstein 1995), indicating that organisms are not performing exponential discounting. Human and animal discounting data are often best fit by a hyperbolic discount function (Ainslie 2001):

$$disc(d) = \frac{1}{1 + kd} \quad (6.5)$$

where $k \in [0, \infty)$ is the discount rate. It is therefore important to consider how hyperbolic discounting can fit into reinforcement learning models.

Hyperbolic discounting is empirically a good fit to human and animal discounting data, but it also has a theoretical basis in uncertain hazard rates. Agents are assumed to discount future rewards because there is some risk that the reward will never be received, and this risk grows with temporal distance (but see Henly et al. 2008). Events that would prevent reward receipt, such as death of the organism, are called *interruptions*. If interruptions are believed to occur randomly at some rate (i.e., the hazard rate), then the economically optimal policy is exponential discounting at that rate. However, if the hazard rate is not known a priori, it could be taken to be a uniform distribution over the possible rates (ranging from 1 where interruptions never occur to 0 where interruptions occur infinitely fast). Under this assumption, the economically optimal policy is hyperbolic discounting (Sozou 1998). Using the data from a large survey, it was found that factoring out an individual's expectation and tolerance of risk leaves individuals with a discounting factor well-fit by an exponential discounting function (Andersen et al. 2008). This function was correlated with the current interest rate, suggesting that humans may be changing their discounting rates to fit the expected hazard functions. Studies in which subjects could maximize reward by discounting exponentially at particular rates have found that humans can match their discounting to those exponential functions (Schweighofer et al. 2006). However, neurological studies have found that risk and discounted rewards may be utilizing different brain structures (Preuschoff et al. 2006).

Semi-Markov temporal difference models, such as those described above, can represent varying time intervals within a single state, permitting any discount function to be calculated across a single state-transition. However, the value of a state is still calculated recursively using the discounted value of the next state (rather than looking ahead all the way to the reward). Thus, across multiple state-transitions, the discounting of semi-Markov models depends on the way that the total temporal interval between now and reward is divided between states. With exponential discounting, the same percent reduction in value occurs for a given delay, regardless of the absolute distance in the future. Because of this, exponential discounting processes convolve appropriately; that is, the discounted value of a reward R is independent of whether the transition is modeled as one state with delay d or two states with delay $d/2$. In contrast, hyperbolic discounting functions do not convolve to produce hyperbolic discounting across a sequence of multiple states, and the discounted value of a reward R depends on the number of state transitions encompassing the delay.

As a potential explanation for how hyperbolic discounting could be calculated in a way that is not dependent on the division of time into states, Kurth-Nelson and Redish (2009) noted that a hyperbolic discount function is mathematically equivalent to the sum of exponential discounting functions with a range of exponential discount factors.

$$\int_0^1 \gamma^x d\gamma = \frac{1}{1+x} \tag{6.6}$$

Kurth-Nelson and Redish extended TDRL using a population of "micro-agents", each of which independently performed temporal difference learning using exponential discounting. Each micro-agent used a different discount rate. Actions were selected in the model by a simple voting process among the micro-agents. The overall model exhibited hyperbolic discounting that did not depend on the division of time into states (Fig 6.1).

There is evidence that a range of discounting factors are calculated in the striatum, with a gradient from faster discount rates represented in ventral striatum to slower rates in dorsal striatum (Tanaka et al. 2004). Doya (2000) proposed that serotonin levels regulate which of these discounting rates are active. Tanaka et al. (2007) and Schweighofer et al. (2007) showed that changing serotonin levels (by loading/unloading the serotonin precursor tryptophan) produced changes in which components of striatum were active in a given task. Drugs of abuse could pharmacologically modulate different aspects of striatum (Porrino et al. 2004). Kurth-Nelson and Redish (2009) predicted that drugs of abuse may change the distribution of discount factors and thus speed discounting. The multiple-discount hypothesis predicts that if the distribution of discount rates is altered by drugs, the shape of the discounting curve will be altered as well.

6.3.1 Seeing Across the Intertrial Interval

Discounting is often operationally measured by offering the animal a choice between a smaller reward available sooner or a larger reward available later (Mazur 1987). In the mathematical language used in this chapter, this experiment can be modeled as a reinforcement learning state-space (Fig. 6.2). The discount rate determines whether the smaller-sooner or larger-later reward will be preferred by a temporal difference model.

Rather than running a single trial, the animal is usually required to perform multiple trials in sequence. In these experiments the total trial length is generally held constant (i.e. the intertrial interval following the smaller-sooner choice is longer than the intertrial interval following the larger-later choice) so that smaller-sooner does not become the superior choice simply by hastening the start of the next trial. This creates a theoretical paradox. On any individual trial, the animal may prefer the smaller-sooner option because of its discount rate. But consistently choosing smaller-sooner over larger-later only changes the phase of reward delivery and decreases the overall reward magnitude.

6 Modeling Decision-Making Systems in Addiction

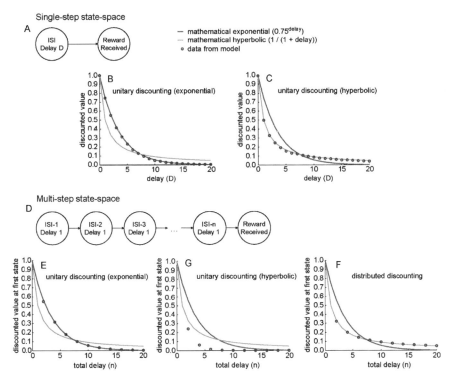

Fig. 6.1 Distributed discounting permits hyperbolic discounting across multiple state transitions. **A,** All delay between stimulus and reward is represented in a single state, permitting any discount function to be calculated over this delay, including exponential (**B**) or hyperbolic (**C**). (**D**) The delay between stimulus and reward is divided into multiple states. Exponential discounting (**E**) can still be calculated recursively across the entire delay (because $\gamma^a \gamma^b = \gamma^{a+b}$), but if hyperbolic discounting is calculated at each state transition, the net discounting at the stimulus is not hyperbolic (**G**). However, if exponential discounting is performed in parallel at many different rates, the average discounting across the entire time interval is hyperbolic (**F**). [From Kurth-Nelson and Redish (2009).]

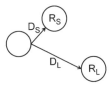

Fig. 6.2 A state-space representing intertemporal choice. From the initial state, a choice is available between a smaller reward (of magnitude R_S) available after a shorter delay (of duration D_S), or a larger reward (R_L) after a longer delay (D_L)

This suggests that there are two different potential state-space representations to describe this experiment. In one description, each trial is seen independently (Fig. 6.3, top); this is the standard approach in TDRL. In the other description,

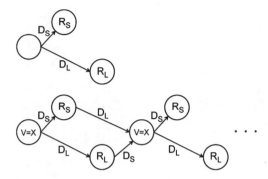

Fig. 6.3 Allowing the agent to see across the inter-trial interval changes the state-space representation of the task. *Top*, A state-space in which each trial is independent from the next. *Bottom*, A state-space in which the end of one trial has a transition to the beginning of the next trial, allowing the value estimates to include expectation of reward from future trials. The delays following the rewards are set to keep the total trial length constant. Note that the states are duplicated for illustrative purposes; an equivalent diagram would have only three states, with arrows wrapping back from R_S and R_L states to the initial choice state

the end of the last trial has a transition to the beginning of the next trial (Fig. 6.3, bottom). By adding this transition (which we will call a *wrap-around* transition), the algorithm can integrate expectation of future reward across all future trials. The total expectation is still convergent because future trials are discounted increasingly with temporal distance.

Adding a wrap-around transition to the state-space has the effect of slowing the apparent rate of discounting. Without wrap-around, the value of the smaller-sooner option is $R_S \cdot disc(D_S)$, and the value of the larger-later option is $R_L \cdot disc(D_L)$. With wrap-around, the smaller-sooner option becomes $R_S \cdot disc(D_S) + X$, and the larger-later option becomes $R_L \cdot disc(D_L) + X$, where X is the value of the initial state in which the choices are available. In other words, wrap-around adds the same constant to the reward expectation for each choice. Thus, if the smaller-sooner option was preferred without wrap-around, with wrap-around it is still preferred but to a lesser degree. Because additional delay devalues the future reward less (proportional to its total value), the apparent rate of discounting is reduced. Note that adding a wrap-around transition does not change the underlying discount function $disc(d)$, but the agent's behavior changes as if it were discounting more slowly. Also, because X is a constant added to both choices, X can change the degree to which the smaller-sooner option is preferred to the larger-later, but it cannot reverse the preference order. Thus, if the agent prefers the smaller-sooner option without a wrap-around state transition, adding wrap-around cannot cause the agent to switch to prefer the larger-later option.

If addicts could be influenced to change their state-space to see across the inter-trial interval, they should exhibit slower discounting. Heyman (2009) observes that recovered addicts have often made the time-course at which they view their lives more global. An interesting question is whether this reflects a change in state-space in the individuals.

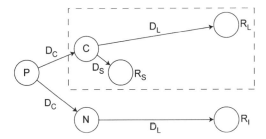

Fig. 6.4 A state-space in which the agent can make a precommitment to avoid having access to a smaller-sooner reward option. The portion of the state-space inside the *dashed box* is the smaller-sooner versus larger-later choice state-space shown in Fig. 6.2. Now a prechoice is available to enter the smaller-sooner versus larger-later choice, or to enter a situation from which only larger-later is available. Following the prechoice is a delay D_C

6.3.2 Precommitment and Bundling

The phenomenon of preference reversal suggests that an agent who can predict their own impulsivity may prefer to remove the future impulsive choice if given an opportunity (Strotz 1956; Ainslie 2001; Gul and Pesendorfer 2001; Heyman 2009; Kurth-Nelson and Redish 2010). For example, an addict may decline to visit somewhere drugs are available. When the drug-taking choice is viewed from a temporal distance, he prefers not to take drugs. But he knows that if faced with drug-taking as an immediate option, he will take it, so he does not wish to have the choice. Precommitment to larger-later choices by eliminating future smaller-sooner choices is a common behavioral strategy seen in successful recovery from addiction (Rachlin 2000; Ainslie 2001; Dickerson and O'Connor 2006; Heyman 2009).

Kurth-Nelson and Redish (2010) showed that precommitment behavior can be modeled with reinforcement learning. The reinforcement learning state-space for precommitment is represented in Fig. 6.4. The agent is given a choice to either enter a smaller-sooner versus larger-later choice, or to enter a situation where only the larger-later option is available. Because the agent discounts hyperbolically, the agent can prefer the smaller-sooner option when making the choice at C, but also prefer the larger-later option when making the earlier choice at P. Mathematically, when the agent is in state C, it is faced with a choice between two options with values $R_S \cdot disc(D_S)$ and $R_L \cdot disc(D_L)$. But when the agent is in state P, the choice is between two options with values $R_L \cdot disc(D_C + D_L)$ and $R_S \cdot disc(D_C + D_S)$. In hyperbolic discounting, the rate of discounting slows as rewards recede into the future, so $\frac{disc(D_S)}{disc(D_L)} > \frac{disc(D_C+D_S)}{disc(D_C+D_L)}$, meaning that the extra delay D_C makes the smaller-sooner choice relatively less valuable. This experiment has been performed in pigeons, and some pigeons consistently elected to take away a future impulsive choice from themselves, despite preferring that choice when it was available (Rachlin and Green 1972; Ainslie 1974). However, to our knowledge this experiment has not yet been run in humans or other species.

In order for a reinforcement learning agent to exhibit precommitment in the state-space in Fig. 6.4, it must behave in state P as if it were discounting R_S across the entire time interval $D_C + D_S$, and discounting R_L across the entire interval $D_C + D_L$. As noted earlier (cf. Fig. 6.1), hyperbolic discounting across multiple states cannot be done with a standard hyperbolic discounting model (Kurth-Nelson and Redish 2010). It requires a model such as the distributed discounting model (Kurth-Nelson and Redish 2009) described above. In this model, each μAgent has a different exponential discounting rate and has a different value estimate for each state. This model performs hyperbolic discounting across multi-step state-spaces (cf. Fig. 6.1) by not collapsing future reward expectation to a single value for each state. Thus, if the distributed discounting model is trained over the state-space of Fig. 6.4, it prefers the smaller-sooner option from state C, but from state P prefers to go to state N (Kurth-Nelson and Redish 2010).

Another way for an impulsive agent to regulate its future choices is with bundling (Ainslie 2001). In bundling, an agent reduces a sequence of future decisions to a single decision. For example, an alcoholic may recognize that having one drink is not a choice that can be made in isolation, because it will lead to repeated impulsive choice. Therefore the choice is between being an alcoholic or never drinking.

Consider the state-spaces in Fig. 6.5. If each choice is treated as independent, the value of the smaller-sooner choice is $R_S \cdot disc(D_S)$ and the value of the larger-later choice is $R_L \cdot disc(D_L)$. However, if making one choice is believed to also determine the outcome of the subsequent trial, then the value of smaller-sooner is $R_S \cdot disc(D_S) + R_S \cdot disc(D_S + D_L + D_S)$ and the value of larger-later is $R_L \cdot disc(D_L) + R_L \cdot disc(D_L + D_S + D_L)$. In an agent performing hyperbolic discounting, the attenuation of value produced by the extra $D_S + D_L$ delay is less if this delay comes later relative to the present. Thus bundling can change the agent's preferences so that the larger-later choice is preferred from the initial state. Like precommitment, bundling can be modeled with reinforcement learning, but only if the model correctly performs hyperbolic discounting across multiple state transitions (Kurth-Nelson and Redish 2010).

It is interesting to note that the agent can represent a given choice in a number of ways: existing in isolation (Fig. 6.3, top), leading to subsequent choices (Fig. 6.3, bottom), viewed in advance (Fig. 6.4), or viewed as a categorical choice (Fig. 6.5, bottom). These four different state-spaces are each reasonable representations of the same underlying choice, but produce very different behavior in reinforcement learning models. This highlights the importance of constructing a state-space for reinforcement learning. If state-space construction is a cognitive operation, it is possible that it can be influenced by semantic inputs. For example, perhaps by verbally suggesting to someone that the decision to have one drink cannot be made in isolation, they are led to create a state-space that reflects this idea.

Throughout these examples in which state-space construction has influenced the apparent discount rate, the *underlying* discount rate (the function $disc(d)$) is unaffected. The difference is in the agent's choice behavior, from which discounting is inferred. Since state-space construction in temporal difference models affects apparent discount rates, it may be that discounting in the brain is modulated by the capacity of the organism to construct state-spaces. This suggests that a potential treatment

Fig. 6.5 Bundling two choices. *Top*, Each choice is made independently. *Bottom*, One choice commits the agent to make the same choice on the next trial

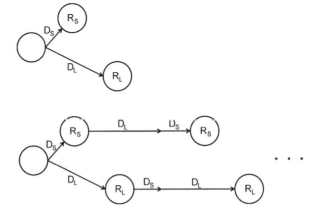

for addiction may lie in the creation of better state-spaces. Gershman et al. (2010) proposed that a limited ability to infer causal relations in the world explains the fact that young animals exhibit less context-dependence in reinforcement learning. This matches the data that people with higher cognitive skills exhibit slower discounting (Burks et al. 2009). It is also consistent with the emphasis of addiction treatment programs (such as 12-step programs) on cognitive strategies that alter the perceived contingencies of the world.

However, it is not clear that the learning systems for habitual or automatic behaviors always produce impulsive choice, or that the executive systems always produce non-impulsive choice. For example, smokers engage in complex planning to find the cheapest cigarettes, in line with the economic view that addicts should be sensitive to cost (Becker and Murphy 1988; Redish 2004). Addicts can perform very complex planning in order to get their drugs (Goldman et al. 1987; Goldstein 2000; Jones et al. 2001; Robinson and Berridge 2003). Thus it does not appear that the problem of addiction is simply a case of the habitual system pharmacologically programmed to carry out drug-seeking behaviors (as arises from the Redish (2004), Gutkin et al. (2006), or Dezfouli et al. (2009) models discussed above; see also Chap. 8 in this book). Rather, addictive drugs seem to have the potential to access vulnerabilities in multiple decision-making systems, including cognitive or executive systems. These different vulnerabilities are likely accessed by different drugs and have differentiable phenotypes (Redish et al. 2008).

6.4 Decision-Making Theories and Addiction

We have seen examples of how decision-making models exhibit vulnerabilities to addictive choice. Another important question is how people actually made decisions in the real-world. There is a key aspect of addiction that does not fit easily into current theories of addiction: the high rate of remission. Current theories of addiction generally account for the development and escalation of addiction by supposing that drugs have a pharmacological action that cumulatively biases the decision-making system of the brain toward drug-choice. These models do not account for cases of spontaneous (untreated) remission, such as a long-term daily drug user who suddenly realizes that she would rather support her children than use drugs, and stops her drug use (Heyman 2009).

Approaches like the 12-step programs (originally Alcoholics Anonymous) have a high success rate in achieving lasting abstinence (Moos and Moos 2004, 2006a, 2006b). These programs use a variety of strategies to encourage people to give up their addictive behavior. These strategies may be amenable to description in the framework of decision-making modeling. For example, one effective strategy is to offer addicts movie rental vouchers in exchange for one week of abstinence (McCaul and Petry 2003; Higgins et al. 2004). If an addict is consistently making decisions that prefer having a gram of cocaine over having $60, why would the addict prefer a movie rental worth $3 over a week of drug taking? This is, as yet, an unanswered question which may require models that include changes in state-space representation, more complex forward-modeling, and more complex evaluation mechanisms than those currently included in computational models of addiction.

References

Ainslie G (1974) Impulse control in pigeons. J Exp Anal Behav 21:485
Ainslie G (2001) Breakdown of will. Cambridge University Press, Cambridge
Andersen S, Harrison GW, Lau MI, Rutström EE (2008) Eliciting risk and time preferences. Econometrica 76:583
Aragona BJ, Cleaveland NA, Stuber GD, Day JJ, Carelli RM, Wightman RM (2008) Preferential enhancement of dopamine transmission within the nucleus accumbens shell by cocaine is attributable to a direct increase in phasic dopamine release events. J Neurosci 28:8821
Arbisi PA, Billington CJ, Levine AS (1999) The effect of naltrexone on taste detection and recognition threshold. Appetite 32:241
Balleine BW (2001) Incentive processes in instrumental conditioning. In: Handbook of contemporary Learning Theories, p 307
Balleine BW (2004) Incentive behavior. In: The behavior of the laboratory rat: a handbook with tests, p 436
Balleine BW, Dickinson A (1998) Goal-directed instrumental action: contingency and incentive learning and their cortical substrates. Neuropharmacology 37:407
Balleine BW, Daw ND, O'Doherty JP (2008) Multiple forms of value learning and the function of dopamine. In: Neuroeconomics: decision making and the brain, p 367
Barnes CA (1979) Memory deficits associated with senscence: A neurophysiological and behavioral study in the rat. J Comp Physiol Psychol 93:74
Barto AG (1994) Adaptive critics and the basal ganglia. In: Models of information processing in the basal ganglia, p 215
Baum W, Rachlin H (1969) Choice as time allocation. J Exp Anal Behav 12:861
Bayer HM, Glimcher P (2005) Midbrain dopamine neurons encode a quantitative reward prediction error signal. Neuron 47:129
Becker GS, Murphy KM (1988) A theory of rational addiction. J Polit Econ 96:675
Becker GS, Grossman M, Murphy KM (1994) An empirical analysis of cigarette addiction. Am Econ Rev 84:396
Bernheim BD, Rangel A (2004) Addiction and cue-triggered decision processes. Am Econ Rev 94:1558

Berridge KC (2007) The debate over dopamine's role in reward: the case for incentive salience. Psychopharmacology 191:391

Berridge CW, Arnsten AF, Foote SL (1993) Noradrenergic modulation of cognitive function: clinical implications of anatomical, electrophysiological and behavioural studies in animal models. Psychol Med 23:557

Bickel WK, Odum AL, Madden GJ (1999) Impulsivity and cigarette smoking: delay discounting in current, never, and ex-smokers. Psychopharmacology (Berlin) 146:447

Bouton ME (2002) Context, ambiguity, and unlearning: sources of relapse after behavioral extinction. Biol Psychiatry 52:976

Bouton ME, Swartzentruber D (1989) Slow reacquisition following extinction: context, encoding, and retrieval mechanisms. J Exp Psychol, Anim Behav Processes 15:43

Bouton ME, Westbrook RF, Corcoran KA, Maren S (2006) Contextual and temporal modulation of extinction: behavioral and biological mechanisms. Biol Psychiatry 60:352

Breland K, Breland M (1961) The misbehavior of organisms. Am Psychol 16:682

Burks SV, Carpenter JP, Goette L, Rustichini A (2009) Cognitive skills affect economic preferences, strategic behavior, and job attachment. Proc Natl Acad Sci 106:7745

Childress AR, Ehrman R, Rohsenow DJ, Robbins SJ, O'Brien CP (1992) Classically conditioned factors in drug dependence. In: Substance abuse: a comprehensive textbook, p 56

Christensen CJ, Silberberg A, Hursh SR, Roma PG, Riley AL (2008) Demand for cocaine and food over time. Pharmacol Biochem Behav 91:209

Chung SH, Herrnstein RJ (1967) Choice and delay of reinforcement. J Exp Anal Behav 10:67

Corbit LH, Balleine BW (2000) The role of the hippocampus in instrumental conditioning. J Neurosci 20:4233

Cote D, Caron A, Aubert J, Desrochers V, Ladouceur R (2003) Near wins prolong gambling on a video lottery terminal. J Gambl Stud 19:433

Courville AC (2006) A latent cause theory of classical conditioning. Doctoral dissertation, Carnegie Mellon University

Custer RL (1984) Profile of the pathological gambler. J Clin Psychiatry 45:35

Daw ND (2003) Reinforcement learning models of the dopamine system and their behavioral implications. Doctoral dissertation, Carnegie Mellon University

Daw ND, Doya K (2006) The computational neurobiology of learning and reward. Curr Opin Neurobiol 16:199

Daw ND, Kakade S, Dayan P (2002) Opponent interactions between serotonin and dopamine. Neural Netw 15:603

Daw ND, Niv Y, Dayan P (2005) Uncertainty-based competition between prefrontal and dorsolateral striatal systems for behavioral control. Nat Neurosci 8:1704

Daw ND, Courville AC, Touretzky DS (2006) Representation and timing in theories of the dopamine system. Neural Comput 18:1637

Dayan P (2002) Motivated reinforcement learning. Advances in neural information processing systems: proceedings of the 2002 conference

Dayan P, Balleine BW (2002) Reward, motivation, and reinforcement learning. Neuron 36:285

Dayan P, Seymour B (2008) Values and actions in aversion. In: Neuroeconomics: decision making and the brain, p 175

Dayan P, Niv Y, Seymour B, Daw ND (2006) The misbehavior of value and the discipline of the will. Neural Netw 19:1153

Dezfouli A, Piray P, Keramati MM, Ekhtiari H, Lucas C, Mokri A (2009) A neurocomputational model for cocaine addiction. Neural Comput 21:2869

di Chiara G (1999) Drug addiction as dopamine-dependent associative learning disorder. Eur J Pharmacol 375:13

Dickerson M, O'Connor J (2006) Gambling as an addictive behavior. Cambridge University Press, Cambridge

Domjan M (1998) The principles of learning and behavior. Brooks/Cole

Doya K (2000) Metalearning, neuromodulation, and emotion. In: Affective minds, p 101

Elster J (1999) Gambling and addiction. In: Getting hooked: rationality and addiction, p 208

Estes WK (1943) Discriminative conditioning. I. A discriminative property of conditioned anticipation. J Exp Psychol 32:150

Fiorillo CD, Newsome WT, Schultz W (2008) The temporal precision of reward prediction in dopamine neurons. Nat Neurosci 11:966

Flagel SB, Watson SJ, Akil H, Robinson TE (2008) Individual differences in the attribution of incentive salience to a reward-related cue: Influence on cocaine sensitization. Behav Brain Res 186:48

Frederick S, Loewenstein G, O'Donoghue T (2002) Time Discounting and time preference: A critical review. J Econ Lit 40:351

Fuhs MC, Touretzky DS (2007) Context learning in the rodent hippocampus. Neural Comput 19:3172

Gershman SJ, Blei DM, Niv Y (2010) Context, learning, and extinction. Psychol Rev 117:197

Glimcher PW, Camerer C, Fehr E, Poldrack RA (2008) Neuroeconomics: decision making and the brain. Elsevier/Academic Press, London

Goldman MS, Brown SA, Christiansen BA (1987) Expectancy theory: thinking about drinking. In: Psychological theories of drinking and alcoholism, p 181

Goldstein A (2000) Addiction: from biology to drug policy. Oxford University Press, Oxford

Grossman M, Chaloupka FJ (1998) The demand for cocaine by young adults: a rational addiction approach. J Health Econ 17:427

Gul F, Pesendorfer W (2001) Temptation and self-control. Econometrica 69:1403

Gutkin BS, Dehaene S, Changeux JP (2006) A neurocomputational hypothesis for nicotine addiction. Proc Natl Acad Sci USA 103:1106

Henly SE, Ostdiek A, Blackwell E, Knutie S, Dunlap AS, Stephens DW (2008) The discounting-by-interruptions hypothesis: model and experiment. Behav Ecol 19:154

Hershberger WA (1986) An approach through the looking-glass. Anim Learn Behav 14:443

Heyman GM (2009) Addiction: a disorder of choice. Harvard University Press, Cambridge

Higgins ST, Heil SH, Lussier JP (2004) Clinical implications of reinforcement as a determinant of substance use disorders. Annu Rev Psychol 55:431

Hirsh R (1974) The hippocampus and contextual retrieval of information from memory: A theory. Behav Biol 12:421

Hirsh R, Leber B, Gillman K (1978) Fornix fibers and motivational states as controllers of behavior: A study stimulated by the contextual retrieval theory. Behav Biol 22:463

Hu D, Amsel A (1995) A Simple Test of the Vicarious Trial-and-Error Hypothesis of Hippocampal Function. Proc Natl Acad Sci USA 92:5506

Hu D, Xu X, Gonzalez-Lima F (2006) Vicarious trial-and-error behavior and hippocampal cytochrome oxidase activity during Y-maze discrimination learning in the rat. Int J Neurosci 116:265

Hunt WA (1998) Pharmacology of alcohol. In: Tarter RE, Ammerman RT, Ott PJ (eds) Handbook of substance abuse: Neurobehavioral pharmacology. Plenum, New York, pp 7–22

Isaacson RL (1974) The limbic system. Plenum, New York

Isoda M, Hikosaka O (2008) Role for subthalamic nucleus neurons in switching from automatic to controlled eye movement. J Neurosci 28:7209

Jaffe JH, Cascella NG, Kumor KM, Sherer MA (1989) Cocaine-induced cocaine craving. Psychopharmacology (Berlin) 97:59

Jaffe A, Gitisetan S, Tarash I, Pham AZ, Jentsch JD (2010) Are nicotine-related cues susceptible to the blocking effect? Society for Neuroscience Abstracts, Program Number 268.4

Johnson A, Redish AD (2007) Neural ensembles in CA3 transiently encode paths forward of the animal at a decision point. J Neurosci 27:12176

Jones BT, Corbin W, Fromme K (2001) A review of expectancy theory and alcohol consumption. Addiction 96:57

Kamin LJ (1969) Predictability, surprise, attention, and conditioning. In: Learning in animals, p 279

Kirby KN, Herrnstein RJ (1995) Preference reversals due to myopic discounting of delayed reward. Psychol Sci 6:83

Kruse JM, Overmier JB, Konz WA, Rokke E (1983) Pavlovian conditioned stimulus effects upon instrumental choice behavior are reinforcer specific. Learn Motiv 14:165

Kuhar MJ, Ritz MC, Sharkey J (1988) Cocaine receptors on dopamine transporters mediate cocaine-reinforced behavior. In: Mechanisms of cocaine abuse and toxicity, p 14

Kurth-Nelson Z, Redish AD (2009) Temporal-difference reinforcement learning with distributed representations. PLoS ONE 4:e7362

Kurth-Nelson Z, Redish AD (2010) A reinforcement learning model of precommitment in decision making. Frontiers Behav Neurosci 4:184

Langer EJ, Roth J (1975) Heads I win, tails it's chance: The illusion of control as a function of the sequence of outcomes in a purely chance task. J Pers Soc Psychol 32:951

Lebron K, Milad MR, Quirk GJ (2004) Delayed recall of fear extinction in rats with lesions of ventral medial prefrontal cortex. Learn Mem 11:544

Lenoir M, Serre F, Cantin L, Ahmed SH (2007) Intense sweetness surpasses cocaine reward. PLoS ONE 2:e698

Levine AS, Billington CJ (2004) Opioids as agents of reward-related feeding: a consideration of the evidence. Physiol Behav 82:57

Liao D, Lin H, Law PY, Loh HH (2005) Mu-opioid receptors modulate the stability of dendritic spines. Proc Natl Acad Sci USA 102:1725

Liu J-, Liu J-, Hammit JK, Chou S- (1999) The price elasticity of opium in Taiwan, 1914–1942. J Health Econ 18:795

Ljungberg T, Apicella P, Schultz W (1992) Responses of monkey dopamine neurons during learning of behavioral reactions. J Neurophysiol 67:145

Lovibond PF (1983) Facilitation of instrumental behavior by a Pavlovian appetitive conditioned stimulus. J Exp Psychol Anim Behav Process 9:225

Mackintosh NJ (1974) The psychology of animal learning. Academic Press, San Diego

Madden GJ, Bickel WK (2010) Impulsivity: the behavioral and neurological science of discounting. American Psychological Association, Washington, DC

Mazur J (1987) An adjusting procedure for studying delayed reinforcement. In: Quantitative analyses of behavior, p 55

McCaul ME, Petry NM (2003) The role of psychosocial treatments in pharmacotherapy for alcoholism. Am J Addict 12:S41

McFarland K, Kalivas PW (2001) The circuitry mediating cocaine-induced reinstatement of drug-seeking behavior. J Neurosci 21:8655

Milad MR, Vidal-Gonzalez I, Quirk GJ (2004) Electrical stimulation of medial prefrontal cortex reduces conditioned fear in a temporally specific manner. Behav Neurosci 118:389

Montague PR, Dayan P, Sejnowski TJ (1996) A framework for mesencephalic dopamine systems based on predictive Hebbian learning. J Neurosci 16:1936

Moos RH, Moos BS (2004) Long-term influence of duration and frequency of participation in alcoholics anonymous on individuals with alcohol use disorders. J Consult Clin Psychol 72:81

Moos RH, Moos BS (2006a) Participation in treatment and Alcoholics Anonymous: a 16-year follow-up of initially untreated individuals. J Clin Psychol 62:735

Moos RH, Moos BS (2006b) Rates and predictors of relapse after natural and treated remission from alcohol use disorders. Addiction 101:212

Muenzinger KF (1938) Vicarious trial and error at a point of choice. I. A general survey of its relation to learning efficiency. J Genet Psychol 53:75

Nadel L, Willner J (1980) Context and conditioning: A place for space. Physiol Psychol 8:218

Nestler EJ (1996) Under siege: The brain on opiates. Neuron 16:897

Niv Y, Montague PR (2008) Theoretical and empirical studies of learning. In: Neuroeconomics: decision making and the brain, p 331

Niv Y, Daw ND, Dayan P (2006) Choice values. Nat Neurosci 9:987

O'Doherty J, Dayan P, Schultz J, Deichmann R, Friston K, Dolan RJ (2004) Dissociable roles of ventral and dorsal striatum in instrumental conditioning. Science 304:452

O'Keefe J, Dostrovsky J (1971) The hippocampus as a spatial map. Preliminary evidence from unit activity in the freely moving rat. Brain Res 34:171

O'Keefe J, Nadel L (1978) The hippocampus as a cognitive map. Clarendon, Oxford
Oscar-Berman M, Marinkovic K (2003) Alcoholism and the brain: an overview. Alcohol Res Health 27(2):125–134
Ostlund SB, Balleine BW (2008) The disunity of Pavlovian and instrumental values. Behav Brain Sci 31:456
Packard MG, McGaugh JL (1996) Inactivation of hippocampus or caudate nucleus with lidocaine differentially affects expression of place and response learning. Neurobiol Learn Mem 65:65
Paine TA, Dringenberg HC, Olmstead MC (2003) Effects of chronic cocaine on impulsivity: relation to cortical serotonin mechanisms. Behav Brain Res 147:135
Panlilio LV, Thorndike EB, Schindler CW (2007) Blocking of conditioning to a cocaine-paired stimulus: Testing the hypothesis that cocaine perpetually produces a signal of larger-than-expected reward. Pharmacol Biochem Behav 86:774
Parke J, Griffiths M (2004) Gambling addiction and the evolution of the near miss. Addict Res Theory 12:407
Pavlov I (1927) Conditioned reflexes. Oxford Univ Press, Oxford
Phillips PEM, Stuber GD, Heien MLAV, Wightman RM, Carelli RM (2003) Subsecond dopamine release promotes cocaine seeking. Nature 422:614
Porrino LJ, Lyons D, Smith HR, Daunais JB, Nader MA (2004) Cocaine self-administration produces a progressive involvement of limbic, association, and sensorimotor striatal domains. J Neurosci 24:3554
Preuschoff K, Bossaerts P, Quartz SR (2006) Neural differentiation of expected reward and risk in human subcortical structures. Neuron 51:381
Quirk GJ, Garcia R, González-Lima F (2006) Prefrontal mechanisms in extinction of conditioned fear. Biol Psychiatry 60:337
Rachlin H (2000) The science of self-control. Harvard University Press, Cambridge
Rachlin H, Green L (1972) Commitment, choice, and self-control. J Exp Anal Behav 17:15
Redish AD (1999) Beyond the cognitive map: from place cells to episodic memory. MIT Press, Cambridge
Redish AD (2004) Addiction as a computational process gone awry. Science 306:1944
Redish AD (2009) Implications of the multiple-vulnerabilities theory of addiction for craving and relapse. Addiction 104:1940
Redish AD, Johnson A (2007) A computational model of craving and obsession. Ann NY Acad Sci 1104:324
Redish AD, Kurth-Nelson Z (2010) Neural models of temporal discounting. In: Impulsivity: the behavioral and neurological science of discounting, p 123
Redish AD, Jensen S, Johnson A, Kurth-Nelson Z (2007) Reconciling reinforcement learning models with behavioral extinction and renewal: implications for addiction, relapse, and problem gambling. Psychol Rev 114:784
Redish AD, Jensen S, Johnson A (2008) A unified framework for addiction: vulnerabilities in the decision process. Behav Brain Sci 31:415
Rescorla RA, Wagner AR (1972) A theory of Pavlovian conditioning: Variations in the effectiveness of reinforcement and nonreinforcement. In: Classical conditioning II, p 64
Restle F (1957) Discrimination of cues in mazes: A resolution of the 'place-vs-response' question. Psychol Rev 64:217
Reynolds B, Ortengren A, Richards JB, de Wit H (2006) Dimensions of impulsive behavior: personality and behavioral measures. Pers Individ Differ 40:305
Ritz MC, Lamb RJ, Goldberg SR, Kuhar MJ (1987) Cocaine receptors on dopamine transporters are related to self-administration of cocaine. Science 237:1219
Robinson TE, Berridge KC (1993) The neural basis of drug craving: An incentive-sensitization theory of addiction. Brains Res Rev 18:247
Robinson TE, Berridge KC (2001) Mechanisms of action of addictive stimuli: Incentive-sensitization and addiction. Addiction 96:103
Robinson TE, Berridge KC (2003) Addiction. Annu Rev Psychol 54:25
Robinson TE, Berridge KC (2004) Incentive-sensitization and drug 'wanting'. Psychopharmacology 171:352

Schultz W (2002) Getting formal with dopamine and reward. Neuron 36:241

Schultz W, Dayan P, Montague R (1997) A neural substrate of prediction and reward. Science 275:1593

Schweighofer N, Shishida K, Han CE, Yamawaki S, Doya K (2006) Humans can adopt optimal discounting strategy under real-time constraints. PLoS Comput Biol 2:e152

Schweighofer N, Tanaka SC, Doya K (2007) Serotonin and the evaluation of future rewards. Theory, experiments, and possible neural mechanisms. Ann NY Acad Sci 1104:289

Si J, Barto AG, Powell WB, Wunsch D (2004) Handbook of learning and approximate dynamic programming. Wiley/IEEE Press, New York

Simon NW, Mendez IA, Setlow B (2007) Cocaine exposure causes long-term increases in impulsive choice. Behav Neurosci 121:543

Smith A, Li M, Becker S, Kapur S (2006) Dopamine, prediction error and associative learning: a model-based account. Network: Comput Neural Syst 17:61

Sotres-Bayon F, Cain CK, LeDoux JE (2006) Brain mechanisms of fear extinction: historical perspectives on the contribution of prefrontal cortex. Biol Psychiatry 60:329

Sozou PD (1998) On hyperbolic discounting and uncertain hazard rates. R Soc Lond B 265:2015

Stahl SM, Pradko JF, Haight BR, Modell JG, Rockett CB, Learned-Coughlin S (2004) A review of the neuropharmacology of bupropion, a dual norepinephrine and dopamine reuptake inhibitor. Prim Care Companion J Clin Psychiat 6:159

Strotz RH (1956) Myopia and inconsistency in dynamic utility maximization. Rev Econ Stud 23:165

Sutton RS, Barto AG (1998) Reinforcement learning: an introduction. MIT Press, Cambridge

Talmi D, Seymour B, Dayan P, Dolan RJ (2008) Human Pavlovian instrumental transfer. J Neurosci 28:360

Tanaka SC, Doya K, Okada G, Ueda K, Okamoto Y, Yamawaki S (2004) Prediction of immediate and future rewards differentially recruits cortico-basal ganglia loops. Nat Neurosci 7:887

Tanaka SC, Schweighofer N, Asahi S, Shishida K, Okamoto Y, Yamawaki S, Doya K (2007) Serotonin differentially regulates short- and long-term prediction of rewards in the ventral and dorsal striatum. PLoS ONE 2:e1333

Tolman EC (1938) The determiners of behavior at a choice point. Psychol Rev 45:1

Tolman EC (1939) Prediction of vicarious trial and error by means of the schematic sowbug. Psychol Rev 46:318

Tolman EC (1948) Cognitive maps in rats and men. Psychol Rev 55:189

Tsai HC, Zhang F, Adamantidis A, Stuber GD, Bonci A, de Lecea L, Deisseroth K (2009) Phasic firing in dopaminergic neurons is sufficient for behavioral conditioning. Science 324:1080

Uslaner JM, Acerbo MJ, Jones SA, Robinson TE (2006) The attribution of incentive salience to a stimulus that signals an intravenous injection of cocaine. Behav Brain Res 169:320

van der Meer MA, Redish AD (2009) Covert expectation-of-reward in rat ventral striatum at decision points. Frontiers Integr Neurosci 3:1

van der Meer MA, Redish AD (2010) Expectancies in decision making, reinforcement learning, and ventral striatum. Front Neurosci 4:29

Waelti P, Dickinson A, Schultz W (2001) Dopamine responses comply with basic assumptions of formal learning theory. Nature 412:43

Wagenaar WA (1988) Paradoxes of gambling behavior. Erlbaum, London

Weiner I, Lubow RE, Feldon J (1988) Disruption of latent inhibition by acute administration of low doses of amphetamine. Pharmacol Biochem Behav 30:871

White AM (2003) What happened? Alcohol, memory blackouts, and the brain. Alcohol Res Health 27(2):186–196

Yin HH, Knowlton B, Balleine BW (2004) Lesions of dorsolateral striatum preserve outcome expectancy but disrupt habit formation in instrumental learning. Eur J Neurosci 19:181

Yu AJ, Dayan P (2005) Uncertainty, neuromodulation, and attention. Neuron 46:681

Chapter 7
Computational Models of Incentive-Sensitization in Addiction: Dynamic Limbic Transformation of Learning into Motivation

Jun Zhang, Kent C. Berridge, and J. Wayne Aldridge

Abstract Incentive salience is a motivational magnet property attributed to reward-predicting conditioned stimuli (cues). This property makes the cue and its associated unconditioned reward 'wanted' at that moment, and pulls an individual's behavior towards those stimuli. The incentive-sensitization theory of addiction posits that permanent changes in brain mesolimbic systems in drug addicts can amplify the incentive salience of Pavlovian drug cues to produce excessive 'wanting' to take drugs. Similarly, drug intoxication and natural appetite states can temporarily and dynamically amplify cue-triggered 'wanting', promoting binge consumption. Finally, sensitization and drug intoxication can add synergistically to produce especially strong moments of urge for reward. Here we describe a computational model of incentive salience that captures all these properties, and contrast it to traditional cache-based models of reinforcement and reward learning. Our motivation-based model incorporates dynamically modulated physiological brain states that change the ability of cues to elicit 'wanting' on the fly. These brain states include the presence of a drug of abuse and longer-term mesolimbic sensitization, both of which boost mesocorticolimbic cue-triggered signals. We have tested our model by recording neuronal activity from mesolimbic output signals for reward and Pavlovian cues in the ventral pallidum (VP), and a novel technique for analyzing neuronal firing "profile", presents evidence in support of our dynamic motivational account of incentive salience.

Definition Box:
Incentive salience: Also called 'wanting', incentive salience represents motivation for reward (UCS), and is typically triggered in anticipation by a reward-related cue (Pavlovian CS) when the cue is encountered by an individual whose brain mesocorticolimbic circuits are in a highly reactive state (determined by a modulation parameter kappa in our model). Attribution of incentive salience to the cue or reward representations make them more attractive, sought after, and likely to be consumed. Brain mesolimbic systems, especially those involving dopamine, are espe-

---

J. Zhang · K.C. Berridge (✉) · J.W. Aldridge
Department of Psychology, University of Michigan, Ann Arbor, MI, USA
e-mail: berridge@umich.edu

cially important to "wanting." Ordinarily "wanting" occurs together with other reward components of "liking" and learning, but can be dissociated both from other components and subjective desire under some conditions. Incentive salience may occur in the absence of conscious, declarative goal in the ordinary sense of the word wanting. This cognitive form of wanting involves additional cortical brain mechanisms beyond the mesolimbic systems that mediate "wanting" as incentive salience. The difference between conscious desire (want) and incentive salience ('want') can sometimes confer an irrational feature on the excessive urges of sensitized addicts who are not in withdrawal yet still 'want' to take a drug that they know will not give much pleasure.

7.1 Introduction

Incentive salience is a psychological process and neural mechanism to explain the acquisition and expression of motivational values of conditioned stimuli (Berridge and Robinson 1998; Berridge 2007). Incentive salience arises typically as a consequence of reward learning (Bindra 1978; Toates 1986; Berridge 2004), and involves a fundamental dissociation in brain mechanisms of learning, "liking" (hedonic impact or pleasure associated with the receipt of a primary reward) and "wanting" (incentive salience itself; features that makes a stimulus a desirable and attractive goal), see Berridge and Robinson (2003); Robinson and Berridge (2003). Incentive salience is attributed to a sensory stimulus after prior learning of cue-reward associations (between a Pavlovian cue for reward [CS], or the reward itself [UCS]), and transforms it from a sensory representation into a salient and attractive goal representation capable of grabbing the animal's attention and motivating the animal's approach and consumption behaviors.

Beyond learning, physiological brain states relevant to drugs and addiction for the relevant reward, such as activation of mesocorticolimbic dopamine circuits or their regulatory inputs, also modulate attributions of incentive salience on a moment-to-moment basis, in part via alteration in mesolimbic dopamine activation. The incentive salience hypothesis specifically suggests Pavlovian-guided attribution of incentive salience to be dynamically modulated by physiological states that impact NAcc-related circuitry, including dopamine neurotransmission. Regarding addiction, the incentive-sensitization hypothesis suggests that drugs of abuse induce compulsion to take drugs by hijacking neural circuits of incentive salience that evolved to motivate behavior for natural rewards (Robinson and Berridge 1993, 2003, 2008). The hypothesis is not exclusive: it does not deny an important role for drug pleasure or drug withdrawal in motivating drug taking behavior (Koob and Le Moal 2006; Gutkin et al. 2006; Redish et al. 2008). But it suggests that the compulsive and persistent nature of addiction, may be best explained by the concept of a sensitized 'wanting' systems in susceptible individuals, mediated by long-term neuroadaptations that may involve alterations in gene expression, neurotransmitter release and receptor levels, and dendritic sprouting patterns in mesocorticolimbic structures. Incentive-sensitization can create addiction even to drugs that are not particularly pleasant, and still produce cue-triggered relapse back into drug taking even long after recovery from withdrawal.

The relation to mesocorticolimbic modulation makes incentive salience particularly influenced by natural appetite states, by psychostimulant drugs that promote dopamine and by enduring neural sensitization of mesolimbic NAc-VP systems. Previous computational models have suggested that incentive salience can be construed purely by reinforcement learning mechanisms such as the temporal difference or prediction error model (McClure et al. 2003; Redish 2004). Such models account for incentive salience in terms of dopamine-based learning mechanisms (Schultz et al. 1997; Schultz 2002), without invoking a role for physiological modulation of motivation after learning. For example, McClure et al. identified incentive salience with the "(state) value function" or expected total discounted reward. Redish (2004) identified drug sensitization with a mechanism that amplifies the temporal difference prediction error signal itself. This is different from our view, which posts incentive salience to be dynamically modulated from moment to moment, based on inputs from current physiological/brain states as well as from learned associations to a reward cue (Zhang et al. 2009). In this chapter, we first review those standard learning models, and then contrast them to data that indicate incentive salience involves more than merely learning. Namely, cue-triggered 'wanting' also involves an additional physiological factor that dynamically transforms transforms static learned values into a flexible level of motivation appropriate to the moment (Tindell et al. 2009). Rather than simply reflecting a previously-learned value, our model proposes a specific gain-control mechanism for modulating on the fly the expected values of reward-predicting stimuli to dynamically compute incentive salience.

7.1.1 Dopamine and Reinforcement Learning: The "Standard" Model and Critique

Contemporary reinforcement learning theory posits an actor-critic architecture for reward prediction and action control. The critic computes the error in reward-prediction—the discrepancy between the reward expected from a stimulus (technically, a state) and the reward actually received. The temporal-difference (TD) method provides an explicit formula for calculating such expected reward through incorporating the subsequent prediction made by the same reward-predicting system as a part of predicted reward of the current state, thereby allowing a refined estimate of the value of a state in the sequential context. The actor, on the other hand, evaluates the merits of policies and selects for each state an action associated with highest long-term reward values. Critic and actor are often discussed as potential functions of ventral striatum and its mesolimbic dopamine inputs from ventral tegmental area (VTA) and substantia nigra pars compacta (SNc), respectively (O'Doherty et al. 2004).

There is growing consensus (though not without controversy, see Redgrave et al. 1999; Redgrave and Gurney 2006), that the predictive error signal, which lies at the core of temporal difference learning, is carried by the firing of dopaminergic neurons projecting to nucleus accumbens and neostriatum (Schultz 1998; Schultz et al. 1997). Phasic firing in midbrain dopaminergic neurons (Schultz et al. 1997; Schultz 1998) has been suggested to express the TD error in reward prediction. Such signal has also been posited to update working memory and goal stack representations in prefrontal cortex (Montague et al. 2004; O'Reilly 2006), consistent with neuropsychological proposals that the prefrontal cortex (PFC) controls goals and goal-directed action (Miller and Cohen 2001).

The actor-critic architecture and TD-based learning rule derives its computational power from its consistent and effective scheme for optimizing sequential decision-making in a stationary Markov environment (e.g., Puterman 1994), without the need of an elaborate model of the world. However, an important missing piece for this framework is how learned predictions are used to generate motivation on a moment-to-moment basis in a way that incorporates the current physiological state such as hunger, satiety, psychostimulant sensitization, or the immediate impact of drugs. This was pointed out by Dayan (2009) and Dayan and Balleine (2002). Attempts to grapple with this issue have been made via tree-search devaluations of goals (Daw et al. 2005a, 2005b) or via satiation decrements that reduce motor arousal or limit generalization (Niv et al. 2006). A problem that has remained unaddressed is how the motivation value of specific reward stimuli may be dynamically *increased* in targeted fashion by physiological states involving mesolimbic activation, including neural sensitization states relevant to addiction (Robinson and Berridge 2003; Berridge 2007).

Going beyond the act-outcome ("A-O") devaluation that can be successfully modeled by the cognitively-based tree-search mechanism (Daw et al. 2005a, 2005b) the concept of incentive salience posits an additional Pavlovian-based cue-triggered ("S-S") motivation process (i.e., incentive salience attribution), which can be either increased or decreased phasically by physiological hunger-satiety states that are relevant to the brain's calculation of hedonic value for particular sensory goals (for example, sweet versus salty tastes during caloric hunger versus salt appetite) (Berridge and Valenstein 1991; Robinson and Berridge 1993; Berridge and Robinson 1998; Tindell et al. 2006). Incentive salience takes Pavlovian associations as its primary learned input, but also takes a separate input in the form of current mesolimbic states that amplify or dampen 'wanting' for specific cues and their rewards. Further, the incentive-sensitization theory of addiction posits that mesolimbic activation, drugs of abuse and persisting sensitization can tap into those motivation-amplifying brain circuits to incrementally raise the incentive salience carried by particular reward stimuli, so that cues may trigger compulsive motivation for their rewards by the same S-S incentive modulation mechanism (Robinson and Berridge 1993, 2003) The incentive salience attribution mechanism is thus essentially Pavlovian-guided, but calculates value anew on a moment-by-moment basis that is influenced by mesolimbic dopamine states (see Berridge 2001 chapter). This view assigns a very different role to dopamine function from traditional learning models, but one that we suggest below can be made compatible with the existing computational theories of reinforcement learning.

7.2 Previous Computational Approaches to Learning and Incentive Salience

Several models of dopamine function with respect to computation of incentive salience and incentive sensitization have been proposed, including our own. They are all anchored on the now-standard reinforcement learning framework along with the temporal difference (TD) learning method.

Reinforcement learning theory provides a mathematical framework to model action control and reward prediction by an agent in a stable Markov environment. In temporal difference models (Sutton and Barto 1981), the expected total future discounted reward V associated with an environmental state s (i.e., the conditioned stimulus [CS] associated with reward) is

$$V(s_t) = \left\langle \sum_{i=0} \gamma^i r_{t+i} \right\rangle = \langle r_t \rangle + \gamma \langle r_{t+1} \rangle + \gamma^2 \langle r_{t+2} \rangle + \cdots, \quad (7.1)$$

where $\gamma \in [0, 1)$ is the discount factor, $r_t, r_{t+1}, r_{t+2}, \ldots$, representing the sequence of primary rewards (UCS) starting from the current state (subscripted t, predictive CS), and the expectation $\langle \cdot \rangle$ is taken over generally stochastic state transition and reward delivery. The estimated value of reward prediction $\hat{V}$ (denoted with a hat) is a cached value that becomes gradually established through temporal difference learning over past instances in which r and s are paired. On each trial, specifically, a prediction error δ concerning deviation from consistent successive predictions is calculated, based on instantaneous reward r_t (which might be stochastic)

$$\delta(s_t) = r_t + \gamma \hat{V}(s_{t+1}) - \hat{V}(s_t), \quad (7.2)$$

and is used to update $\hat{V}$ via $\delta \hat{V}(s_t) \propto \delta(s_t)$. After learning has completed, $\delta(s_t) = 0$, so

$$\hat{V}(s_t) = \langle r_t \rangle + \gamma \hat{V}(s_{t+1}). \quad (7.3)$$

In the early application to 'wanting' mentioned above, McClure et al. (2003) proposed that the notion of incentive salience be mapped directly to the computational concept of total expected future discounted reward, namely V (see Eq. (7.1)). In TD learning theory, V is a cached, incrementally-learnt value function. However, a difficulty arises from identifying incentive salience with the value function V, as V is usually defined in TD models. That difficulty is that V can change if a reward is revalued only after further relearning about the new prediction error introduced by re-encounters with the revalued reward. Thus, to change incentive salience of a CS requires further pairing with its revalued UCS, according to such a pure learning model based on re-training via new prediction errors. That contradicts our idea described above that CS incentive salience is also modulated on the fly by relevant physiological states that alter mesolimbic function, which produces a synergistic interaction between prior learning and current mesolimbic reactivity in determining the current level of CS-triggered motivation (Robinson and Berridge 1993; Berridge and Robinson 1998).

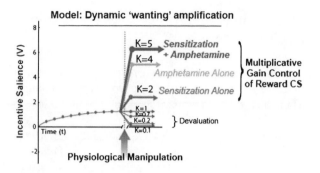

Fig. 7.1 Simulations of dynamic shifts in incentive salience. Initial learning is assumed to proceed by a TD type of rule initially. At time step $t = 11$, a new mesolimbic activation introduced via amphetamine administration, sensitization, or both. The change in incentive salience occurs as indicated by the *arrows*, multiplicatively ($V \cdot \kappa$). Modified from Zhang et al. (2009)

A variant on prediction-error approach applied to addiction has been to posit that sensitization elevates learning itself (e.g., Redish 2004). For example, by magnifying the drug-elicited prediction error signal carried by dopamine neurons, this has been suggested to result in a surge in the value of δ itself (Redish 2004). Such a change would then induce an increase in the learned value function V via the standard, δ-driven TD learning mechanism on the next trial that paired the initial CS with the UCS. An exaggerated TD error signal, repeated again and again in drug addicts who continue to take drugs, has been posited to increase V without upper bound, hence explaining addiction as over-learning of V prediction (Redish 2004). Such ideas are elegant applications of TD learning theory to addiction, but nevertheless rely exclusively on the assumption that dopamine causes a predictive error that functions as a teaching signal in the TD framework. There are reasons to question that assumption (e.g., Berridge 2007).

7.3 Our Dynamic Model of Incentive Salience: Integrating Learning with Current State

To incorporate motivational modulation of previously learned values (Fig. 7.1), we propose that the *incentive salience* or motivational value $\tilde{V}(s_t)$ of a reward-predicting CS be described as

$$\tilde{V}(s_t) = \tilde{r}(r_t, \kappa) + \gamma V(s_{t+1}), \qquad (7.4)$$

where the value of the UCS (i.e., the primary reward value r_t) is modulated by a factor κ reflecting current physiological state, such as hunger, thirst, salt appetite, amphetamine administration, drug sensitization, etc. Two specific forms of r were postulated in Zhang et al. (2009): an additive mechanism and a multiplicative mechanism. Only the multiplicative mechanism is required regarding addiction in terms of drug activation of dopamine systems and long-term sensitization of those systems, and so in this chapter, we concentrate on the multiplicative form

$$\tilde{r}(r_t, \kappa) = \kappa \langle r_t \rangle. \tag{7.5}$$

In our model, the multiplicative constant κ can be specific to a particular appetitive system. Incentive salience can be either increased or decreased, with $\kappa < 1$ representing decrease (such as satiation or devaluation) and $\kappa > 1$ representing increase (such as hunger or sensitization). Equation (7.4) along with (7.5) suggests that moment-to-moment evaluation of the incentive value of the goal associated with the current state s_t, is contributed to by two parts: a gain-controlled evaluation of the immediately available reward (first term), and a γ-discounted evaluation of future rewards based on stored prediction value (second term). Physiological state κ factors may couple with geometric discounting under γ, in that satiation ($\kappa < 1$) may increase the temporal horizon γ, whereas sensitization or an increased physiological appetite (κ becomes greater than 1) may decrease γ, and disproportionately raise the motivational value of temporal proximity to reward UCS (see Giordano et al. 2002). The incentive value of a state s_t is the motivationally-modulated value of the immediate reward r_t plus the discounted value of the expected reward in the next state s_{t+1}; both are loaded into the goal representation as s_t is presented.

The multiplicative modulation (i.e., the calculation of $\tilde{V}$) is the key distinction from learning-based models of incentive salience—it corresponds to model-based action control via goal representation versus model-free action control using stored predictions. Subtracting both sides of Eq. (7.5) from those of Eq. (7.3), and substituting in the multiplicative relation (7.5), we obtain

$$\tilde{V}(s_t) - V(s_t) = (\kappa - 1)\langle r_t \rangle, \tag{7.6}$$

that is, the incentive salience $\tilde{V}(s_t)$ reduces to $V(s_t)$ in the absence of devaluation/sensitization manipulation ($\kappa = 1$). We believe, however, that the κ parameter does not act as an indiscriminate tide that floats all boats to raise the incentive salience of all CSs equally. Rather, specific physiological appetite states, such as drug addiction, caloric hunger or salt appetite, each amplify the hedonic value of their own reward (drugs, sweets, salty foods), and hence specifically amplifies the incentive salience of particular CSs related to that UCS (Bindra 1978; Toates 1986; Berridge 2001, 2004; Dickinson and Balleine 2002), presumably each modulated by its own κ parameters.

It follows from Eqs. (7.1), (7.4) and (7.5) that

$$\tilde{V}(s_t) = \kappa \langle r_t \rangle + \gamma \left(\left\langle \sum_{i=0} \gamma^i r_{t+1+i} \right\rangle \right). \tag{7.7}$$

In essence, Eq. (7.7) is an expression of what is known as the "quasi-hyperbolic" discounting model (Laibson 1997; Frederick et al. 2002). Hence, our model provides an incentive salience explanation of why mesolimbic NAc-VP systems may sometimes be activated by an immediately available reward more than by temporally distant reward (McClure et al. 2004), and suggests that the degree of discounting in such situations will be modulated by current mesolimbic states.

Note that our postulated gain-control mechanism effectively modulates the tradeoff of immediate primary reward and future expected reward. Recent neuro-imaging studies implicating ventral striatum and parietal cortex in mediating the relative weighting of immediate versus future rewards in humans (McClure et al. 2004; Glimcher and Kable 2005) and in monkeys (Louie and Glimcher 2005), with mesolimbic neural activation specifically potentiating the motivational value of short-term rewards at the expense of long-term rewards, are consistent with our proposed gain-control function for dopamine.

In short, our model proposes that incentive salience requires (1) an online gain control (gating) mechanism κ that dynamically modulates CS reward value according to changes in reward-relevant physiological or neurobiological states, including mesolimbic dopamine activation or sensitization, along with (2) potential adjustment of the temporal horizon γ for evaluating stored prediction values; both motivational consequences are adjustable on-the-fly without the need for (re)learning.

7.4 Testing Model Predictions Using a Serial Conditioning Task

The above-mentioned models of incentive salience and mesolimbic dopamine function were teased apart previously by our colleagues and us in an electrophysiological recording study that employed post-learning mesolimbic modulation in a Pavlovian conditioning task (Tindell et al. 2004). Let us consider the serial conditioning paradigm involving two conditioned stimuli in sequence, CS1 and CS2, that predict a terminal reward UCS: the full series is CS1 $\rightarrow$ CS2 $\rightarrow$ UCS (Fig. 7.2). After the rat is trained on this sequential paradigm, it learns the values associated with V_1 (of CS1) and V_2 (of CS2), as well as with the value r of the terminal reward. In later tests, the rat's mesolimbic systems may be activated by amphetamine administration; or between the training and the test, the rat may be sensitized by a hefty regimen of exposure to psychostimulant drugs. In all test cases, the first CS still predicts all following stimuli, and because of temporal discounting, their magnitude will be in descending order: $V_1 < V_2 < r$. So a pure TD value-coding model would predict that neuronal coding of incentive salience should follow the same ordering, with activation to UCS being the largest. Under sensitization manipulation, the primary UCS reward values will be magnified, $r \rightarrow \kappa r$.

A TD error model by comparison would predict that, after learning is complete in the CS1/CS2/UCS paradigm, $\delta_2 = \delta_3 = 0$, whereas $\delta_1 > 0$. Allowing the possibility for incomplete learning, one still has $\delta_1 > \delta_2 > \delta_3$, where the ordering reflects the propagation of learning gradient from reward-distal to reward-proximal direction. Assuming the effect of sensitization or acute amphetamine challenge to be either additive or multiplicative on the existing δ signal, it follows from the above line of reasoning that the response of TD error-coding neurons to CS1 would be the strongest, though it would not appear until after a new learning trial once the drug elevated δ. In short, a prediction error coding model (e.g., Redish 2004) specifies increments in the neural code for δ, most prominently for CS1. That is in stark contrast to the specification (by our incentive salience model below) of CS2 as the stimulus

7 Computational Models of Incentive-Sensitization in Addiction

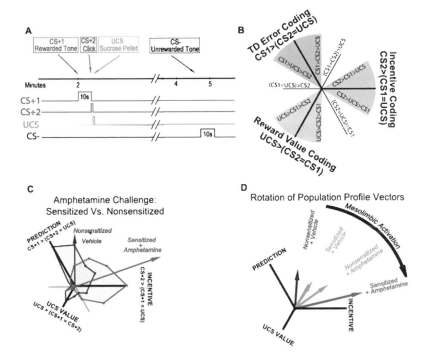

Fig. 7.2 Selective amplification of CS incentive salience (not CS prediction or UCS hedonic impact) by transient amphetamine intoxication and more permanent drug sensitization. Experimental design of the serial CS1/CS2/UCS procedure, and effects of sensitization and amphetamine on neuronal firing profiles in ventral pallidum (**A**). The relative rank-ordering of neuronal responses to CS1/CS2/UCS is defined as the "profile" of a neuron; it can be represented mathematically as the angle of a vector in a two-dimensional space, where the two axes represent two orthogonal contrasts formed from the three responses (**B**). The computation is such that this angular value indexing a response profile exists in a continuum which (1) exhausts all possible firing patterns (i.e., relative orders in firing rates to these three types of stimuli); and (2) guarantees that nearby values represents similar firing patterns. Temporal difference error-coding implies maximal response to CS1 which has the greatest prediction, whereas value-coding implies maximal firing to UCS which has the highest hedonic value. By contrast, incentive-coding implies maximal firing to CS2 that has the greatest motivational impact as it immediately precedes the actual reward. The data panel shows firing in control condition contrasted to the combination of amphetamine plus sensitization (**C**). The summary *arrow panel* shows the averaged neuronal response for each group of rats, illustrating the additive increments produced by sensitization, amphetamine and combination of both (**D**). From Zhang et al. (2009) and modified from Tindell et al. (2005)

most enhanced in incentive salience by a neurobiological activation of mesolimbic systems.

By contrast, our gain-control κ model of incentive salience predicts that mesolimbic activation, even if occurring after learning (e.g., by psychostimulant sensitization or by acute amphetamine administration), may immediately modulate the neuronal computation of CS incentive salience (especially for CS2). In the serial conditioning paradigm, $r_1 = 0$ and $r_2 = r$, so the motivationally controlled incentive salience value is $\tilde{V}_1 = \gamma r$, $\tilde{V}_2 = \kappa r$. Pan et al. (2005) showed that on certain trials when the CS2 is omitted after CS1 there is a "dip" in the firing of dopamine neurons in substantial nigra and VTA. This suggests that following CS1, the animal is indeed expecting CS2 rather than UCS. (Because the temporal discount factor $\gamma < 1$ and sensitization manipulation $\kappa > 1$, $\tilde{V}_1 < \tilde{V}_2$.) In short, with respect to ordering of magnitudes, our model of incentive salience anticipates that CS2 should receive greater motivational impact than CS1, because it is closer in time to UCS, facilitating transfer of incentive properties directly from UCS to CS2 (see Tindell et al. 2005 for supporting evidence regarding relative roles of CS1 and CS2). The reward-proximal CS2 should be most potently enhanced by either persisting neural sensitization induced after learning, or acute amphetamine administration on the day of test, because both manipulations activate mesolimbic dopamine systems.

7.4.1 VP Neuronal Coding for Incentive Salience

These models were tested in a study that focused on mesolimbic output signals that were intercepted in the ventral pallidum (Tindell et al. 2005). The ventral pallidum (VP) lies at the base of the brain behind the nucleus accumbens and in front of ventral tegmental area (VTA) and lateral hypothalamus, and serves as a final common path for mesocorticolimbic circuits. The VP processes compressed representations of mesocorticolimbic reward signals before relaying them back up into corticolimbic and mesocorticolimbic loops, and downward to motor structures (Zahm 2000; Zahm 2006). Regarding mesolimbic dopamine inputs, VP also receives direct dopamine projections from VTA that has been implicated in drug reward (Gong et al. 1996; McFarland et al. 2004), as well as most efferent projections from the nucleus accumbens. Its outputs project upward to mediodorsal thalamus and thence back to corticolimbic loops involving prefrontal, cingulate, and insular cortical areas, and downwards to subcortical sites. VP neurons fire to Pavlovian conditioned stimuli (CS+) that predict rewards as well as to reward UCSs themselves (Tindell et al. 2004). Hence, VP is a prime candidate to study mesolimbic modulations of CS incentive coding.

The Tindell et al. (2005) study tested whether mesolimbic activation by sensitization or pharmacological dopamine release enhances VP firing that codes incentive salience as a motivational transform of CS+ in a manner specified by our model of incentive salience computation, as opposed to pure learning-based models. Rats were trained on a serial Pavlovian conditioning paradigm, CS1/CS2/UCS, where a Pavlovian CS1 (a tone) followed after a 10 second delay by a CS2 (a click) predicted immediate (i.e., within 1 second) reward in the form of a sucrose pellet (UCS); they were also trained on a negative CS− (another tone) that did not lead to any reward. After two weeks of Pavlovian training, physiological activation of dopamine-related mesolimbic brain systems was induced in either or both of the following two ways: by neural sensitization caused by repeated, pulsed psychostimulant drug administration followed by a prolonged drug-free incubation period ("sensitization condition"), or later during testing sessions by acute amphetamine administration that immediately causes dopamine to be released into synapses ("amphetamine challenge condition"). Recordings were done during testing sessions. An offline review of the monitoring video ruled out that VP neuron activity in response to CS2 and UCS presentation were due simply to pure movements.

Each neuronal unit's responses to the three stimuli (CS1, CS2, UCS) were analyzed by a novel data analysis and presentation method, called Profile Analysis (Tindell et al. 2005; Zhang et al. 2009), to derive a profile direction (an angle within 360°). A total of 524 recorded units show that VP neuronal response profiles were broadly dispersed, covering the entire spectrum of value-coding, predictive error-coding, and incentive-coding regions of the profile space. Histograms (polar plot) were constructed by plotting the number of units at each directional angle versus the angular values themselves. Population averages of such profile vectors (i.e., vector sum called a Population Profile Vector), were also plotted for VP neurons under various conditions including normal control, sensitization and/or acute amphetamine conditions. Normally, VP neurons signaled prediction-error preferentially, responding maximally to CS1, secondarily to CS2, and least to UCS. But mesolimbic dopamine activations enhanced incentive salience computations on the fly. Sensitization (Fig. 7.2) and acute amphetamine (Fig. 7.2) both shifted the distributions of response profiles away from predictive error coding (CS1-maximal response) and toward incentive coding (CS2-maximal response). The greatest shift occurred when rats were both pre-sensitized and exposed to acute amphetamine challenge on the same test day (Fig. 7.2).

The effects of mesolimbic dopaminergic activation can be visualized as the rotation of the Population Profile Vectors away from a CS1-maximal/prediction-coding axis (CS1) and towards the CS2-maximal/incentive-coding axis (CS2) (Fig. 7.2). Thus, it can be concluded that while VP neurons in control animals (after training) tend to follow a TD error coding profile, mesolimbic dopaminergic activation causes the neuronal response profiles to shift towards encoding incentive salience. Mesolimbic activation by sensitization, amphetamine administration, or both, specifically and progressively caused VP neurons to increase their firing rates predominantly to CS2, compared with CS1 or UCS. Such results are anticipated by our motivational-based model of incentive salience (with $\kappa > 1$).

7.5 Discussion

Reward cues trigger motivation "wanting", as well as hedonic affects, cognitive expectations and procedural habits. Incentive salience theory posits the motivational value triggered by a CS+ to be based on two separate but integrated inputs: (a) current physiological/neurobiological state; and (b) previously learned associative values. This integration of physiological signals allows drug states, sensitization states, or natural hunger, thirst and other appetitive states to immediately enhance the incentive salience attributed to a previously learned CS+ for relevant reward, without necessarily requiring additional re-learning trials.

To summarize, our analysis supports a computational account of incentive salience as a motivational gain control mechanism that dynamically responds to post-learning shifts in physiological states when computing 'wanting' triggered by a relevant CS for reward. This gain control mechanism modulates motivation on a moment-by-moment basis as brain states vary, gauging the relative importance (tradeoff in values) between primary reward versus expected future reward. Finally, VP circuits, as a crucial node in mesocorticolimbic circuits, may be an important stage in computing the motivational transforms of CS and UCS values alike.

7.5.1 Multiple Motivation-Learning Systems

We stress that other types of learning and motivation exist aside from Pavlovian incentive salience: in particular, cognitive incentives and reward-oriented habits. For example, evidence described elsewhere indicates that 'wanting' (with quotation marks: incentive salience) exists alongside ordinary wanting (without quotation marks: cognitive predictions), which may plausibly be based on full look-ahead cognitive representations of expected goal values and their related act-outcome strategies to obtain those goals (Dayan and Balleine 2002; Dickinson and Balleine 2002; Berridge and Robinson 2003). Ordinarily, wanting and 'wanting' act together to guide behavior toward the same goals, with incentive salience serving to add motivation 'oomph' to cognitive representations. But under some important conditions cognitive and Pavlovian motivation mechanisms may diverge. For example, divergence can lead to 'irrational wanting' in addiction for a target that the individual does not cognitively want, nor predicatively expect to be of high value. Our current model may help to computationally capture the Pavlovian incentive salience limb of that divergence (Berridge and Aldridge 2008).

7.5.2 Contrasting Dynamic Incentive Salience to Cognitive Tree Goals

Loosely speaking, our model could be considered similar to one-step look-ahead in a model-based (tree-search) approach. However, there are important differences between our model and most tree-search models. A full tree-model is usually thought to have an advantage of providing a stable cognitive map of declarative goals and available actions within the tree representation of the world. Our model nevertheless posits a dynamic synergy between current mesolimbic reactivity and the presence of a cue (with its previously acquired association to reward). For example, cue-triggered 'wanting' shoots up upon presentation of a CS, but importantly, also goes down again nearly as soon as the CS is taken away—even when the brain remains in a mesolimbic-activated state (e.g., after amphetamine administration; after sensitization; or after combination of both). Coupling of incentive salience to CS is evident in behavioral cue-triggered 'wanting' experiments (Pavlovian instrumental transfer), where lever-pressing peaks fade away as soon as the CS is removed—even though the dopamine drug or sensitization state that enhanced the cue's motivation-eliciting power persist.

This type of transience is quite typical of motivational states. In particular, the incentive salience mechanism is especially compatible with transient peaks in 'wanting' being tied to CS presence because the rules that underlie Pavlovian controls of incentive salience specify that a synergy exists between CS presence and current mesolimbic state (Robinson and Berridge 1993; Berridge 2007). The physical presence of a Pavlovian CS is a crucial factor in generating incentive salience, and a sporadic CS can lead to up-and-down changes in 'wanting'. This synergy feature is precisely why a drug CS triggers relapse in an addict as a phasic peak of temptation—at least if that CS is encountered in a mesolimbic-activated state.

Our model for the computation of incentive salience implies the motivational magnet property of a drug reward cue is dynamically recomputed based on current physiological states of sensitization and drug intoxication. This dynamic amplification of motivation in addicts may maladaptively pull the addict like a magnet towards compulsively 'wanted' drugs, and so make it harder to escape from the addiction.

Acknowledgements Collection of experimental data that gave rise to this computational model was supported by NIH grants DA017752, DA015188 and MH63649. The writing of this book chapter was also supported by AFOSR grant FA9550-06-1-0298. We thank Dr. Michael F.R. Robinson for helpful comments on an earlier version of the manuscript.

References

Berridge KC (2001) Reward learning: Reinforcement, incentives, and expectations. In: Medin DL (ed) The psychology of learning and motivation, vol 40. Academic Press, New York, pp 223–278
Berridge KC (2004) Motivational concepts in behavioral neuroscience. Physiol Behav 81:179–209
Berridge KC (2007) The debate over dopamine in reward: the case for incentive salience. Psychopharmacology 191:391–431 (2007)
Berridge KC, Aldridge JW (2008) Decision utility, the brain, and pursuit of hedonic goals. Social Cogn 26:621–646
Berridge KC, Robinson TE (1998) What is the role of dopamine in reward: hedonic impact, reward learning, and incentive salience? Brains Res Rev 28:309–369
Berridge KC, Robinson TE (2003) Parsing reward. Trends Neurosci 26:507–513
Berridge KC, Valenstein ES (1991) What psychological process mediates feeding evoked by electrical stimulation of the lateral hypothalamus? Behav Neurosci 105:3–14
Bindra D (1978) How adaptive behavior is produced: a perceptual-motivation alternative to response reinforcement. Behav Brain Sci 1:41–91
Daw ND, Niv Y, Dayan P (2005a) Uncertainty-based competition between prefrontal and dorsal striatal systems of behavioral control. Nat Neurosci 8:1704–1711
Daw ND, Niv Y, Dayan P (2005b) Actions, policies, values, and the basal ganglia. In: Bezard (ed) Recent breakthroughs in basal ganglia research. Nova Publ, New York, pp 91–106
Dayan P (2009) Dopamine, reinforcement learning, and addiction. Pharmacopsychiatry 42(S 01):S56–S65

Dayan P, Balleine BW (2002) Reward, motivation and reinforcement learning. Neuron 36:285–298

Dickinson A, Balleine B (2002) The role of learning in the operation of motivational systems. In: Gallistel CR (ed) Stevens' handbook of experimental psychology: learning, motivation, and emotion, vol 3, 3rd edn. Wiley, New York, pp 497–534

Frederick S, Loewenstein G, O'Donoghue T (2002) Time discounting and time preference: A critical review. J Econ Lit 40:351–401

Giordano LA, Bickel WK, Loewenstein G, Jacobs EA, Marsch L, Badger GJ (2002) Mild opioid deprivation increases the degree that opioid-dependent outpatients discount delayed heroin and money. Psychopharmacology 163:174–182

Glimcher PW, Kable O (2005) Neural mechanisms of temporal discounting in humans. Abstract for 2005 annual meeting of the society for neuroeconomics

Gong W, Neill D, Justice JB Jr (1996) Conditioned place preference and locomotor activation produced by injection of psychostimulants into ventral pallidum. Brain Res 707(1):64–74

Gutkin BS, Dehaene S, Changeux JP (2006) A neurocomputational hypothesis for nicotine addiction. Proc Natl Acad Sci USA 103(4):1106–1111

Koob GF, Le Moal M (2006) Neurobiology of addiction. Academic Press, New York

Laibson D (1997) Golden eggs and hyperbolic discounting. Q J Econ 112(2):443–477

Louie K, Glimcher PW (2005) Intertemporal choice behavior in monkeys: interaction between delay to reward, subjective value, and area LP. Abstract for 2005 annual meeting of the society for neuroeconomics

McClure SM, Daw ND, Montague PR (2003) A computational substrate for incentive salience. Trends Neurosci 26:423–428

McClure SM, Laibson DI, Loewenstein G, Cohen JD (2004) Separate neural systems value immediate and delayed monetary rewards. Science 306:503–507

McFarland K, Davidge SB, Lapish CC, Kalivas PW (2004) Limbic and motor circuitry underlying footshock-induced reinstatement of cocaine-seeking behavior. J Neurosci 24(7):1551–1560

Miller EK, Cohen JD (2001) An integrative theory of prefrontal cortex function. Annu Rev Neurosci 24:167–202

Montague PR, Hyman SE, Cohen JD (2004) Computational roles for dopamine in behavioral control. Nature 760–767

Niv Y, Joel D, Dayan P (2006) A normative perspective on motivation. Trends Cogn Sci 10:375–381

O'Doherty J, Dayan P, Schultz J, Deichmann R, Friston K, Dolan RJ (2004) Dissociable roles of ventral and dorsal striatum in instrumental conditioning. Science 304:452–454

O'Reilly RC (2006) Biologically based computational models of high-level cognition. Science 314:91–94

Pan W-X, Schmidt R, Wickens JR, Hyland BI (2005) Dopamine cells respond to predicted events during classical conditioning: evidence for eligibility traces in the reward-learning network. J Neurosci 25:6235–6242

Puterman ML (1994) Markov decision processes. Wiley, New York

Redgrave P, Gurney K (2006) The short-latency dopamine signal: a role in discovering novel actions? Nat Rev, Neurosci 7:967–975

Redgrave P, Prescott TJ, Gurney K (1999) Is the short-latency dopamine response too short to signal reward error? Trends Neurosci 22:146–151

Redish AD (2004) Addiction as a computational process gone awry. Nature 306:1944–1947

Redish AD, Jensen S, Johnson A (2008) A unified framework for addiction: Vulnerabilities in the decision process. Behav Brain Sci 31(4):415–437; discussion 437–487

Robinson TE, Berridge KC (1993) The neural basis of drug craving: an incentive-sensitization theory of addiction. Brains Res Rev 18:247–291

Robinson TE, Berridge KC (2003) Addiction. Annu Rev Psychol 54:25–53

Robinson TE, Berridge KC (2008) The incentive sensitization theory of addiction: some current issues. Philos Trans R Soc Lond B Biol Sci 363(1507):3137–3146

Schultz W (1998) Predictive reward signal of dopamine neurons. J Neurophysiol 80:1–27

Schultz W (2002) Getting formal with dopamine and reward. Neuron 36:241–263

Schultz W, Dayan P, Montague PR (1997) A neural substrate of prediction and reward. Science 275:1593–1599

Sutton RS, Barto AG (1981) Toward a modern theory of adaptive networks: expectation and prediction. Psychol Rev 88(2):135–170

Tindell AJ, Berridge KC, Aldridge JW (2004) Ventral pallidal representation of pavlovian cues and reward: population and rate codes. J Neurosci 24:1058–1069

Tindell AJ, Berridge KC, Zhang J, Peciña S, Aldridge JW (2005) Ventral pallidal neurons code incentive motivation: amplification by mesolimbic sensitization and amphetamine. Eur J Neurosci 22:2617–2634

Tindell AJ, Smith KS, Pecina S, Berridge KC, Aldridge JW (2006) Ventral pallidum firing codes hedonic reward: when a bad taste turns good. J Neurophysiol 96(5):2399–2409

Tindell AJ, Smith KS, Berridge KC, Aldridge JW (2009) Dynamic computation of incentive salience: "wanting" what was never "liked". J Neurosci 29(39):12220–12228

Toates F (1986) Motivational systems. Cambridge University Press, Cambridge

Zahm DS (2000) An integrative neuroanatomical perspective on some subcortical substrates of adaptive responding with emphasis on the nucleus accumbens. Neurosci Biobehav Rev 24:85–105

Zahm DS (2006) The evolving theory of basal forebrain functional-anatomical 'macrosystems'. Neurosci Biobehav Rev 30:148–172

Zhang J, Berridge KC, Tindell AJ, Smith KS, Aldridge JW (2009) Modeling the neural computation of incentive salience. PLoS Comput Biol 5:1–14

Chapter 8
Understanding Addiction as a Pathological State of Multiple Decision Making Processes: A Neurocomputational Perspective

Mehdi Keramati, Amir Dezfouli, and Payam Piray

Abstract Theories of addiction in neuropsychology increasingly define addiction as a progressive subversion, by drugs, of the learning processes by which animals are equipped with, to adapt their behaviors to the ever-changing environment surrounding them. These normal learning processes, known as Pavlovian, habitual and goal-directed, are shown to rely on parallel and segregated cortico-striatal loops, and several computational models have been proposed in the reinforcement learning framework to explain the different and sometimes overlapping components of this network. In this chapter, we review some neurocomputational models of addiction originating from reinforcement learning theory, each of which explain addiction as a usurpation of one of the well-known models under the effect of addictive drugs. We try to show how each of these partially complete models can explain some behavioral and neurobiological aspects of addiction, and why it is necessary to integrate these models in order to have a more complete computational account for addiction.

8.1 Introduction

Addiction, including addiction to drugs of abuse, is defined as a compulsive orientation toward some certain behaviors, despite the heavy costs that might be followed (Koob and Le Moal 2005b). In the case of drug addiction, addicts are usually portrayed as people who seek and take drugs, even at the cost of adverse social, occupational and health consequences. Although a wide range of effects of drugs on different body and nervous system regions has been shown, it is progressively becoming accepted that the above definition of drug addiction arises from the pharmacological effects of drugs on the brain learning system, that is, the brain circuits involved in adaptively guiding animals' behaviors toward satisfying their needs (Everitt and Robbins 2005; Redish et al. 2008;

M. Keramati
Group for Neural Theory, Ecole Normale Superieure, Paris, France

A. Dezfouli (✉) · P. Piray
Neurocognitive Laboratory, Iranian National Center for Addiction Studies, Tehran University of Medical Sciences, Tehran, Iran
e-mail: a.dezfouli@ut.ac.ir

Belin et al. 2009). In fact, drugs of abuse are notorious for usurpation of the natural learning processes and consequently, understanding normal learning mechanisms has proven to be a prerequisite for understanding addiction as a pathological state of those underlying systems.

Conditioning literature in behavioral psychology has long studied animal behavior and has developed a rich and coherent framework for understanding associative learning by defining several components involved in decision making, most notably Pavlovian, habitual, and goal-directed systems (Dickinson and Balleine 2002). The neural underpinnings of these components and their competitive and collaborative interactions have also been well studied during the last 50 years (Balleine and O'Doherty 2010; Rangel et al. 2008), although there is still a long way to go. This psychological and neurobiological knowledge has paved the way for computational models of decision making to emerge. These models rephrase in a formal language, the developed concepts in the neuropsychology of decision making and thus, guarantee the coherency and self-consistency of the proposed computational theories, as well as quantitatively examining their validity using experimental data. The computational theory of reinforcement learning (RL) (Sutton and Barto 1998), which is the origin of all computational models reviewed in this chapter, is a putative formal framework that has captured many aspects of the psychological and neurobiological knowledge gathered around animal decision making. Within this framework, the "Q-learning" model explains the behavioral characteristics of the habitual process (Sutton and Barto 1998), which is believed to be neurally implemented in the sensorimotor cortico-striatal loop (Yin et al. 2004, 2008). The "actor-critic" models, on the other hand, explain collaboration between Pavlovian and habitual systems and are based on the integrity of limbic and sensorimotor loops (Joel et al. 2002). Finally, "dual-process" models, capture the interplay between habitual and goal-directed processes, and are based on the interaction between sensorimotor and associative loops, respectively (Daw et al. 2005; Keramati et al. 2011).

As addictive drugs are known to usurp the normal learning mechanisms, many of the computational models proposed to date for explaining addiction-like behaviors are based on the RL framework. In fact, each of the five computational models reviewed in this chapter (Redish 2004; Dezfouli et al. 2009; Dayan 2009; Piray et al. 2010; Keramati et al. 2011) explains addiction as a malfunction, due to the effect of drugs, of one of the variants of the RL theory mentioned above. As each model takes into account different, and sometimes overlapping components of the whole learning system, each of them can explain some limited, and sometimes overlapping, behavioral aspects of addiction.

In the following sections, we first briefly discuss some key concepts of the conditioning literature and its neural substrates. The main focus of the first section is on introducing Pavlovian and instrumental forms of associative learning and the multiple kinds of interaction between them, as well as the anatomically parallel and segregated closed loops in the cortico-basal ganglia system that underlie those different associative structures. Based on this literature, potential impairments in these systems induced by pharmacological effects of drugs, and their related behavioral manifestations are explored in the next section. We then review five computational models of addiction, each of which has incorporated the pharmacological effects of drugs into a version of the computational theory of reinforcement learning. The first two models (Redish 2004; Dezfouli et al. 2009) are based on the Q-learning algorithm, which models the habitual decision making process. The second group of models (Dayan 2009; Piray et al. 2010) study the drug-induced pathological state of the actor-critic model, representing the interaction between Pavlovian and habitual process. And the last model (Keramati et al. 2011) relies on the dual process theory of decision making. Finally, we discuss some open avenues for future theoretical efforts for explaining more behavioral and biological evidence on addiction in the RL framework.

8.2 Normal Decision Making Mechanism

Conditioning is an associative learning process by which animals learn to adapt their predictions and behaviors to the occurrence of different stimuli in the environment (Dickinson and Balleine 2002). This learning is made possible by representing the contingencies between different stimuli, responses, and outcomes, in brain associative structures. Psychologists have long made a distinction between Pavlovian and instrumental forms of conditioning. Pavlovian (or classical) conditioning is a form of associative learning where the animal learns that presentation of a neutral stimulus, known as conditioned stimulus (CS), predicts the occurrence of a salient event, known as unconditioned stimuli (US). For this reason, Pavlovian conditioning is also known as stimulus-stimulus (S-S) conditioning. Appearance of the US might evoke an innate, reflexive response called unconditioned response (UR). When this reflexive response is evoked by presenting the CS (which itself predicts the US), it is called a conditioned response (CR). Salivation in response to presentation or prediction of food is a famous example of conditioned or unconditioned responses, respectively. It is important to note that in Pavlovian conditioning, the animal has no control over the occurrence of events in the environment, but only observes. A computational model for learning these S-S associations is presented in Sect. 8.4.2.

In instrumental conditioning, in contrast, the animal learns to choose a sequence of actions so as to attain appetitive stimuli or to avoid aversive ones. At the early stages of exploring a new environment, the animal starts discovering the causal relations between specific actions and their consequent biologically significant outcomes. Based on this instrumental knowledge, at each state like s, the animal deliberates the consequences of different behavioral strategies and then, takes an action like a by which it reaches a desirable outcome like o. Regarding that this kind of instrumental behavior is aimed at gaining access to a certain outcome or goal, it is called goal-directed or stimulus-action-outcome (S-A-O) responding. A formal representation for this system is presented in Sect. 8.4.3. After the animal is extensively trained in the environment, it learns to habitually make a certain response, say a, whenever it finds itself in a certain state, like s, without considering the potential consequences that action might have. Not surprisingly, this type of instrumental behavior is called habitual or stimulus-response (S-R) responding. A computational model representing this type of learning is introduced in Sect. 8.4.1.

Although the three types of learning mechanisms (S-S, S-A-O, S-R) are defined operationally independent from each other, they both collaborate and compete to produce appropriate behavior. The S-S system mainly interacts with the S-R system (Yin and Knowlton 2006; Holland 2004). Conditioned reinforcement phenomenon and Pavlovian-to-instrumental transfer (PIT) are two demonstrations of this interaction, both playing a critical role in addiction to drugs. Conditioned reinforcement refers to the ability of a CS (e.g., a light associated with food) in gaining rewarding properties in order to support the acquisition of a new instrumental response (pressing a lever in order to turn the light on) (Mackintosh 1974). Actor-critic models, explained in Sect. 8.4.2 are proposed to model such an interaction between the two systems (but see Dayan and Balleine 2002). PIT, on the other hand, is a behavioral phenomenon in which non-contingent presentation of a CS markedly elevates responding for an outcome (Lovibond 1983; Estes 1948). Although PIT is suggested to play an important role in addictive behaviors, the computational accounts for the role of this phenomenon in addiction are still not well developed, and thus we do not discuss them in this chapter (see Dayan and Balleine 2002; Niv 2007 for computational models of PIT).

The so far studied interactions between the S-R and S-A-O systems, on the other hand, mainly focus on the competition between these two systems; i.e. these two systems compete for taking the control of behavior. As noted earlier, it has been demonstrated that at the early stages of learning, the behavior is governed by the S-A-O system, whereas extensive learning results in the S-R system winning the competition. The dual process models introduced in Sect. 8.4.3 are developed to model this interaction, and explain how drug-induced imbalance in the interaction between S-R and S-A-O systems can contribute to addictive behaviors.

The three different decision making processes discussed above are demonstrated to depend on topographically segregated, parallel cortico-striato-pallido-thalamo-cortical closed loops (Alexander et al. 1986, 1990; Alexander and Crutcher 1990). These loops include limbic, associative and sensory-motor loops, which are shown to mediate Pavlovian, goal-directed and habitual processes, respectively. Striatum is a central structure in this system, though it should be viewed as only a part of a bigger network. It receives glutamatergic projections from cortex, as well as dopaminergic inputs from Ventral Tegmental Area (VTA) and Substantia Nigra Pars Compacts (SNc). The striatum can be divided into anatomically and functionally heterogeneous subregions. Classically, the ventral subregion is shown to mediate Pavlovian conditioning, whereas the dorsal region is involved in instrumental conditioning (O'Doherty et al. 2004; Yin et al. 2008). Within the dorsal striatum, dorsolateral part and dorsomedial are demonstrated to mediate habitual and goal-directed processes, respectively (Yin et al. 2004, 2005, 2008).

8.3 Aspects of Addictive Behavior

In the general system-level framework within which the computational models of addiction are discussed in this chapter, three criteria for evaluating each model can be proposed. Each criterion is, in fact, a set of theories on which a system-level model of addiction is expected to be based on. Satisfying each of these criteria can improve either the behavioral explanatory power or relevancy to neurobiological reality of the corresponding model. These three criteria are: (1) being based on a model for the normal decision-making system, at both neurobiological and behavioral levels; (2) incorporating the pharmacological effects of drugs on neural systems into the structure of the computational model; and (3) explaining a set of well-known behavioral syndromes of drug addiction. In the previous section, we provided a conceptual framework for the normal decision-making system (basis 1), which will be later used as a basis for the addiction models introduced in this chapter. In this section, we focus on the third basis, and discuss some important behavioral aspects of drug addiction. Discussing the second basis is postponed until the description of computational models in Sect. 8.4.

8.3.1 Compulsive Drug Seeking and Taking

According to the current Diagnostic and Statistical Manual of Mental Disorders (American Psychiatric Association 2000, p. 198) *"The essential feature of substance abuse is a maladaptive pattern of substance use manifested by recurrent and significant adverse consequences related to the repeated use of substances. ... There may be repeated failure to fulfill major role obligations, repeated use in situations in which it is physically hazardous, multiple legal problems, and recurrent social and interpersonal problems."* In other words, the fundamental characteristic of drug addiction is that the consumption of drug doesn't decrease proportionally when its costs (health costs, social costs, financial costs, etc.) increase. In behavioral economic terms, this type of behavior is referred to as inelastic consumption, as opposed to elastic consumption where decreases in demand are significant when price increases. In accordance with this feature, studies looking at the sensitivity of drug consumption to its price, demonstrate that the consumption of cigarettes and heroin among dependent individuals is less elastic (or sensitive) to price, compared to other reinforcers (Petry and Bickel 1998; Bickel and Madden 1999; Jacobs and Bickel 1999). Figure 8.1 presents a simplified environment for computationally investigating the sensitivity of the consumption of drugs to the associated costs (e.g., price). A decision maker (model) has two options: (a) to do nothing (C_1), which leads to the delivery of no reinforcer, and (b) to pay the cost of the drug (C_2), and then receive the drug reinforcer. The relative inelasticity of demand for drugs implies that the probability of selecting the second option (punishment-then-drug) by the model should be insensitive to the cost, as compared to a situation where the

Fig. 8.1 The model has to choose between C_1 which brings no reward, and C_2. Choosing C_2 is followed by a cost, and then a drug reward

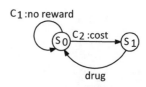

model receives a natural reinforcer instead of the drug (punishment-then-natural-reinforcer). This procedure is used to study the behavior of the model proposed in Sect. 8.4.1.

The compulsive nature of drug-seeking behavior in addicts is tried to be captured in animal models of addiction in various ways. In a variation of such experiments, rats are trained to respond on a seeking lever in order to get access to a taking lever, on which responding leads to the drug. Here, drug seeking and taking are separate actions. In the test phase, seeking responses are measured in the presence of a punishment-paired CS. In fact, during the test phase, the animal doesn't receive punishment nor drugs. Thus, its behavior is measured when no new training is provided and the animal should choose whether to continue going for the drug in the new condition or not (Vanderschuren and Everitt 2004). The formal representation of the procedure is similar to the one in Fig. 8.1: the animal can attenuate aversiveness of the expected electric shock by freezing (C_1), or alternatively, it can press the seeking lever in order to get access to the drug (C_2).

As another attempt to capture compulsivity in animal models, a CS is paired with an electric shock (electric shock plays the role of the cost associated with the drug) during the training phase. In the test phase, if the rat chooses to press the lever while the CS is present, it will receive the electric shock, which is then followed by the delivery of the drug (Deroche-Gamonet et al. 2004; Belin et al. 2008). This procedure is used to examine the behavior of the model proposed in Sect. 8.4.2.2 (a formal representation of the schedule is also provided there). In another experiment (Pelloux et al. 2007), half of the responses (i.e., lever presses) are followed by punishment (and not drug delivery), whereas the other half are followed by drug delivery (and not punishment). From an animal learning point of view, the benefit of this paradigm is that unlike the previous one, the assertiveness of the punishment will not attenuate through its association with the reward (see Pelloux et al. 2007 for more explanation). However, the exact difference between this paradigm and the previous ones from a modeling and behavioral economic point of view needs further investigation.

Intuitively, all the mentioned experiments are to investigate the degree to which the consumption of drugs is sensitive to the associated costs. However, the question of what degree of insensitivity to costs should be regarded as compulsive behavior is still unanswered. At least three types of criteria are used to distinguish between compulsive and non-compulsive drug seeking behavior: (1) Comparing the sensitivity of drug consumption to costs, with the sensitivity of the consumption of natural reinforcers (e.g., sucrose) to costs. Here, the experiments indicate that compared to natural rewards, drug consumption is less sensitive to punishments (Pelloux et al. 2007; Vanderschuren and Everitt 2004); (2) Comparing the behavior of different subpopulations of drug-exposed animals. In such experiments, animals are first divided into groups based on a criterion like the degree of impulsivity, the degree of reactivity to novelty (Belin et al. 2008), or based on results of a test for reinstatement (Deroche-Gamonet et al. 2004). Next, the sensitivity of responses to a punishment is measured and compared between groups, and the group with the lowest sensitivity is considered to be compulsive. In this paradigm, individuals that exhibit compulsive behavior are considered as vulnerable individuals; and (3) Comparing the behavior of animals exposed to drug in different conditions and schedules of drug reinforcement. Here, the main finding is that the inelasticity in drug consumption progressively increases as the history of drug consumption increases. In fact, drug consumption becomes compulsive after a long-term drug exposure (Deroche-Gamonet et al. 2004; Pelloux et al. 2007; Vanderschuren and Everitt 2004).

In conclusion, appearance of compulsive behavior is a function of two independent factors: the degree of drug exposure (criterion 3) and the degree of vulnerability of the individual exposed to drug (criterion 2). This implies that, the more vulnerable the animal is, or the longer the period of exposure to the drug is, the insensitivity of drug consumption to punishments must increase, compared to a natural reward (criterion 1).

8.3.2 Impulsivity

Impulsivity is one of the behavioral traits that is closely related to addiction (Dalley et al. 2011). Addicts are generally characterized as impulsive individuals. They usually exhibit deficiency in response inhibition when it is necessary for reward acquisition, even in non-drug-related tasks. Impulsivity is a multidimensional construct, though two aspects of it seem to be more important: impulsive choice and impaired inhibition. Formal modeling and simulation of situations measuring impulsive choice is rather straight-forward (see below). However, modeling an environment for assessment of impaired inhibition (i.e., inability to inhibit maladaptive behaviors) is hard to achieve, and to our knowledge, there is no computational study on impaired inhibition. Thus, hereafter, we focus on the impulsive choice aspect and refer to it as impulsivity.

Impulsivity is defined as the selection of a small immediate reward over a delayed larger one (Dalley et al. 2011). Figure 8.2 illustrates an environment for the delay discounting task, which is commonly used for the assessment of impulsive behavior. As the figure shows, the model has two choices: one (C_1) leads to an immediate small reward, R_s, and the other (C_2) leads to a delayed (k time steps), but larger reward, R_l. In this environment, the impulsive individuals are those that have more tendency to small rewards, compared to other individuals. A wealth of evidence in human subjects suggests that drug-dependent individuals have more tendency to the small-reward choice, compared to non-dependent individuals (see Bickel and Marsch 2001; Reynolds 2006 for a review). In the same line, animal models report that chronic drug intake causes impulsive choice in rats, as they show less ability to delay gratification compared to control rats (Simon et al. 2007; Paine et al. 2003).

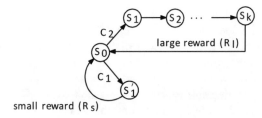

Fig. 8.2 Delay discounting task. The model has two choices, C_1 and C_2. Selection of C_1 leads to a small reward, R_s, only after one time step, whereas by choosing C_2, the model should wait for k time steps, and a large reward, R_l, will be delivered afterwards

It is still unclear whether this increased impulsivity in drug-dependent individuals is a determinant or only a consequence of drug use. However, in human, using self-report measures of impulsivity, it has been reported that youth with impulsive traits are more likely to initiate drug use (see de Wit 2009 for a review). In animals, high impulsivity measured by lack of behavioral inhibition predicts transition to compulsive behavior (Belin et al. 2008). Accordingly, it can be expected from a model of addiction that the more impulsive the model is (as measured by impaired inhibition), the more vulnerable it should be to develop compulsive drug-seeking. However, for concluding that choice impulsivity is also an indicator of vulnerability to develop compulsivity, there should at least be a strong correlation between these two measures of impulsivity. Although it is reported in some studies that impaired inhibition is significantly correlated with impulsivity measured in the delay discounting task (Robinson et al. 2009), other evidence suggest that these two behavioral constructs are not necessarily overlapping (de Wit 2009). Thus, for establishment of links between choice impulsivity and compulsivity, further computational works are needed on modeling impaired inhibition forms of impulsivity.

8.3.3 Relapse

Although compulsive drug taking is an important defining feature of addiction, the most challenging clinical feature of addicts is that they remain vulnerable to relapse, even after long periods of withdrawal (Stewart 2008). Clinical and experimental studies have shown that non-contingent injections of drugs, re-exposure to drug-paired cues, and stress are three factors reinstating drug taking and seeking behavior (Shaham et al. 2003). In a typical reinstatement model of relapse, animals are first trained to acquire responses that lead to the drug (e.g., lever press in order to gain access to the drug). Next, they undergo "extinction training" in which, responses no longer result in the drug outcome. Once the behavior has extinguished, in a subsequent test phase, the effect of different factors triggering relapse (stress, drug priming, drug cues) on the extinguished behavior is determined.

According to these experimental procedures, developing formal representations of the tasks is straightforward. The challenging point, however, is the effect of pharmacological and environmental stimuli on the internal processes of a model, that is, how the effect of drug priming or stress on the brain neurocircuitry can be represented in computational models. In Sect. 8.4.3, we return to these questions and suggest a potential way for modeling these manipulations.

8.4 Computational Accounts

8.4.1 S-R Models

Habit or S-R learning is the ability to learn adaptively to make appropriate responses when some certain stimuli are observed. According to this theory, given a situation or stimulus, if making a certain response produces a reward (a pleasant, biologically salient outcome), then the corresponding S-R association will be potentiated (reinforced) and thus, the probability of taking that response in similar circumstances in the future will increase. Inversely, a behavior will occur less frequently in the future, if it is followed by a punishment (an aversive outcome). In this manner, animals can be viewed as organisms that acquire appropriate behavioral strategies in order to maximize rewards and minimize punishments. This problem, faced by the animals, is analogous to the problem addressed in the machine learning theory of reinforcement learning (RL), which studies how an artificial agent can learn, by trial and error, to make actions to maximize rewards and minimize punishments. Indeed, in recent years, strong links have been forged between a method of RL, called Temporal Difference Reinforcement Learning (TDRL), animal conditioning literature and the potential underlying neural circuits of decision making. The developed neuro-computational models in this interdisciplinary field has provided as an appropriate basis for modeling drug addiction.

In the RL framework, stimulus and response are referred to as "state" and "action", respectively. At each time-step, t, the agent is in a certain state, say s_t, and among the several possible choices, it takes an action, say a_t, on the basis of subjective values that it has assigned to those alternatives through its past experiences in the environment. These assigned values are called Q-values. The more Q-value does an action have, the more likely that action is to be selected for performance. Denoting the probability of taking action a_t at state s_t by $\pi(a_t|s_t)$, the below equation known as the *Softmax* rule reflects this feature:

$$\pi(a_t|s_t) = e^{\beta Q(s_t,a_t)} \Big/ \sum_{b \in \mathbb{A}_{s_t}} e^{\beta Q(s_t,b)} \qquad (8.1)$$

where $\mathbb{A}_{s_t}$ is the set of all available choices at state s_t. β is a free parameter determining the degree of dependence of the policy π on Q-values. In other words, this parameters adjust the exploration/exploitation trade-off.

For making optimal decisions, Q-values are aimed to be proportional to the discounted total rewards that are expected to be received after taking the action onward:

$$Q(s_t, a_t) = E\big[r_t + \gamma r_{t+1} + \gamma^2 r_{t+2} + \cdots | s_t, a_t\big] = E\left[\sum_{i=t}^{\infty} \gamma^{i-t} r_i | s_t, a_t\right] \qquad (8.2)$$

Achieving this objective requires the animal to sufficiently explore the task environment. In the previous equation, $0 < \gamma < 1$ is the discount factor, which indicates the relative incentive value of immediate rewards compared to delayed ones.

To update the prior Q-values, a variant of RL known as TDRL calculates a prediction error signal each time the agent takes an action and receives a reward (as a feedback) from the environment. This prediction error is calculated by comparing the prior expected value of taking that action, $Q(s_t, a_t)$, with its realized value after receiving the reward r_t:

$$\delta_t = \gamma(r_{t+1} + V(s_{t+1})) - Q(s_t, a_t) \tag{8.3}$$

In this equation, $V(s_{t+1})$ is the maximum value of all feasible actions available at the state that comes after taking the action a_t. This prediction error is then utilized to update the estimated value for that action:

$$Q(s_t, a_t) \leftarrow Q(s_t, a_t) + \alpha \delta_t \tag{8.4}$$

where $0 < \alpha < 1$ is the learning rate, determining the degree to which a new experience affects the Q-values. As a critical observation, the phasic activity of midbrain dopamine (DA) neurons is demonstrated to be significantly correlated with the prediction error signal that the TDRL model predicts (Schultz et al. 1997). In fact, dopamine neurons projecting to associative learning structures of the cortico-basal ganglia circuit are believed to carry a teaching signal that modulates the strength of S-R associations and thus, will increase the probability of taking an action in the future, if an unexpected reward has come as a consequence of that action.

TDRL provides a framework for the better understating of the S-R habit formation. In this framework, reinforcement of an association between stimulus s and response a after receiving a reward is equivalent to an increase in $Q(s_t, a_t)$. By utilizing the *softmax* action-selection rule, this will result in increasing the probability of taking that action in the future. By interpreting the TDRL model from another point of view, since only previously learned values accumulated through time determine which action the model takes in a certain state, the behavior of a TDRL model is not sensitive to sudden environmental changes. In other words, it takes several learning trials for the value of actions to be updated according to the new conditions. On the basis of this feature, the TDRL framework is behaviorally consistent with habitual (S-R) responding.

8.4.1.1 Redish's Model

If phasic dopamine activity corresponds to the reward prediction error signal, then after sufficient learning when predictions converge to their true values, the prediction error and thus phasic DA activity should converge to zero. In fact, this happens in the case of natural rewards: after adequate learning trials, the phasic activity of DA neurons vanishes. However, this is not true in the case of drugs such as cocaine and amphetamine. These drugs, through their neuropharmacological mechanisms, increase dopamine concentration within the striatum (Ito et al. 2002;

Stuber et al. 2005). This artificial build up of dopamine readily means that the error signal can not converge to zero in the course of learning and as a consequence, the experienced value of drug-related behaviors will grow more than expected. Based on this argument, a modified version of the TDRL algorithm is proposed Redish (2004) that can explain some behavioral aspects of addiction. Assuming that the pharmacological effect of drugs induces a bias with the magnitude of D on dopaminergic signalling, the error signal equation (8.3) can be rewritten as below when drug is available (Redish 2004):

$$\delta_t^c = \max(\gamma(r_t + V(s_{t+1})) - Q(s_t, a_t) + D(s_t), D(s_t)) \qquad (8.5)$$

This implies that the prediction error signal will always be higher than D, as long as the drug's effect is available:

$$\delta_t^c \geq D(s_t) \qquad (8.6)$$

Hence, by each drug consumption session the value that a decision-maker predicts for drug-seeking and -taking increases. This leads to the over-valuation of this behavior and explains why drug-associated behaviors become more and more insensitive to their harmful consequences through the course of addiction, as measured by the behavior of the model in the environment shown in Fig. 8.1. In fact, as drug-related S-R associations become more and more reinforced, only a more intense adverse event can cancel out the high estimated value of drug-seeking. This model, thus, explains how compulsive drug-seeking habits develop as a result of repeated drug abuse.

Thus, the model proposed in Redish (2004) provides an elegant explanation for progressive inelasticity of drug consumption as a function of the drug exposure history. However, this account does not propose explanations for other addictive behaviors such as impulsivity and relapse. Besides, some predictions of the model have proven inconsistent with some studies that have explicitly investigated the validity of the way in which the effect of drugs are modeled on the error signal.

Firstly, the model predicts that the true value of drug can never be predicted by environmental cues, because it is always better than expected. A behavioral implication of this property is that the "blocking" effect (Kamin 1969) should not occur for the case of drugs (Redish 2004). In fact, the "blocking" phenomenon occurs when a stimulus, as a result of sufficient training, can correctly predict the value of the upcoming outcome. In this case, if a new stimulus is paired with the old one after the training period, since the old stimulus can correctly predict the value of the outcome, no prediction error (teaching signal) should be generated and thus, no new learning will occur. Therefore, it is said that the old highly-trained stimulus blocks other stimuli to be associated with the outcome. However, as the model proposed in Redish (2004) assumes that drugs always induce non-compensable dopamine signalling, it predicts that the blocking effect should not be observed for stimuli that predict drugs. However, experimental results have shown that the "blocking" effect does occur in the case of drugs (Panlilio et al. 2007) and thus, the always-better-than-expected value formation for the drug is not a correct formulation. Secondly, the validity of this method of value learning is investigated even more explicitly. In

Marks et al. (2010), rats were first trained to press two levers in order to receive a large dose of cocaine. Then, the dose associated with one of the levers was decreased. Here, the theory predicts that the value associated with the low-dose lever will not decrease, because drug consumption always increases the value irrespective of the experienced dose (see Eq. (8.6)). At odds with this prediction, the result showed that the lever press performance for the reduced-dose lever has decreased, which indicates that the value of the drug has decreased.

8.4.1.2 Dezfouli et al.'s Model

Borrowing from the model proposed by Redish (2004) (the idea that drugs increase the error signal), we proposed another computational model for drug addiction (Dezfouli et al. 2009) that is based on the supplementary assumption that long-term exposure to drugs causes a long lasting dysregulation in the reward processing system (Koob and Le Moal 2005a). Consistent with behavioral findings, this persistent dysregulation causes less motivation in addicts toward natural rewards like sexually evocative visual stimuli, as well as secondary rewards like money (Garavan et al. 2000; Goldstein et al. 2007).

This dysregulation of the reward system can be modeled in a variant of the TDRL algorithm called "average-reward" TDRL (Mahadevan 1996). In this computational framework, before affecting the current strength of associations, rewards are measured against a level called "basal reward level" (Denoted by ρ_t). As a result, an outcome will have reinforcing effect only if the reward value is higher than the basal reward level. Otherwise, the reinforcing value of the outcome will be negative. The basal reward level, according to this framework, is equal to the average reward per step, which can be computed by an exponentially weighted moving average over experienced rewards (σ is the weight given to the most recent received reward):

$$\rho_t \leftarrow (1-\sigma)\rho_t + \sigma r_t \tag{8.7}$$

In fact, an outcome will reinforce the corresponding association only if it has a rewarding value higher than what the animal receives on average. In this formulation, the value of a state-action is the undiscounted sum of all future rewards measured against ρ_t:

$$Q(s_t, a_t) = E\left[\sum_{i=t}^{\infty}(r_i - \rho_i)|s_t, a_t\right] \tag{8.8}$$

These state-action values can be learned using the following error signal:

$$\delta_t = \gamma(r_{t+1} + V(s_{t+1})) - Q(s_t, a_t) - \rho_t \tag{8.9}$$

Using this error signal, Q-values are updated by the same rule of Eq. (8.4). The definition of the error signal in the average reward RL algorithm does not imply that the value of a state is insensitive to the arrival time of future rewards. In contrast, in Eq. (8.9), the average reward (ρ_t) is subtracted from $V(s_{t+1})$, meaning that by waiting in state s for one time step, the agent loses an opportunity to gain potential future rewards. This opportunity cost is, in average, equal to ρ_t, and is subtracted from the value of the next state. This learning method guides action selection to a policy that maximizes the expected reward per step, rather than maximizing the sum of discounted rewards. As in the simple TDRL framework, the error signal computed by Eq. (8.9) corresponds to the phasic activity of DA neurons. The term ρ_t, on the other hand, is suggested to be coded by the tonic activity of DA neurons (Niv et al. 2007).

Roughly, long-term exposure to drugs causes two, perhaps causally related, effects on the dopamine-dependent reward circuitry. Firstly, chronic exposure to drugs affects the dopamine receptors availability within the striatum. Human subjects and non-human primates with a wide range of drug addictions have shown significant reductions in D2 receptor density within the striatum (Nader et al. 2002; Porrino et al. 2004a; Volkow et al. 2004b). This effect reduces the impact of normal dopamine release that carries the error signal and thus, results in a reduction in the magnitude of the error signal, compared to its normal value (Smith et al. 2006). Secondly, it is proposed that chronic drug abuse causes an abnormal increase in the tonic activity of dopamine neurons (Ahmed and Koob 2005). As the tonic DA activity is hypothesized to encode the ρ_t signal, this second effect of drugs can be modeled by abnormal elevation of the basal reward level. Thirdly, as mentioned earlier, chronic drug exposure causes decreased sensitivity of the reward system to natural rewards. This effect can be interpreted as an abnormal elevation of the level against which reward is measured. In other words, long-term drug abuse elevates the basal reward level to a level that is higher than that of normal subjects. This drug-induced elevation of the basal reward level, ρ_t, can be formally captured by adding a bias to it:

$$\rho_t^c = \rho_t + \kappa_t \qquad (8.10)$$

Normally, κ_t is zero and therefore, rewards are measured against their average level (ρ_t). However, with drug use, κ_t grows and consequently, the basal reward level elevates abnormally to ρ_t^c. This modification covers all the three long-term effects of drugs discussed above. As adding a positive bias to ρ_t leads to a decrease in the error signal (see Eq. (8.9)), it is somehow reflecting the reduced availability of dopamine receptors. Alternatively, if ρ_t is related to the tonic activity of DA neurons, adding a bias to it corresponds to an increase in the tonic activity of these neurons.

According to the above modification to the average reward TDRL algorithm, we rewrite the error signal equation for the case of drugs as follows:

$$\delta_t^c = \max(\gamma(r_t + V(s_{t+1})) - Q(s_t, a_t) + D(s_t), D(s_t)) - \rho_t^c \qquad (8.11)$$

Similar to the model proposed in Redish (2004), the maximization operator reflects the drugs' neuropharmacological effects, but unlike that model, the error signal is not always greater than zero. In this model, although drugs produce extra dopamine through direct pharmacological mechanisms, due to the increase in the basal reward level, the error signal will eventually converge to zero. This property ensures that the estimated value of the drug does not grow unboundedly, which makes the model more biologically plausible. Furthermore, as the prediction error signal in this model can converge to zero after sufficient experience with drugs, no further learning will occur after extensive training. This will result the drug-predicting stimuli to block forming new associations. This is consistent with the report that the blocking effect is observed for the case of drugs (Panlilio et al. 2007).

It should be noted that because abnormal elevation of the basal reward level is a slow process, the error signal under the effect of drugs will be above zero for a relatively long time and thus, drug-seeking habits will be abnormally reinforced. This leads to insensitivity of drug consumption to drug associated punishment, as indicated by the tendency of the model toward C_2 in the environment shown in Fig. 8.1.

As the decision-making system is common for natural and drug reinforcers, deviation of the basal reward level from its normal value can also have adverse effects on decision making in the case of natural rewards. Within the framework proposed above, ρ_t^c determines the cost of waiting. Hence, high values of ρ_t^c in an environment indicate that waiting is costly and thus, guide the decision maker to options with a relatively faster reward delivery. In contrast, low values indicate that the delayed interval before reward delivery is not costly and it is worth waiting for a delayed but large reward. If chronic drug exposure leads to high values of ρ_t^c, then the model's behavioral strategy will shift abnormally toward more immediate rewards, even if their rewarding value is less than that of distant rewards. In other words, in the environment show in Fig. 8.2, preference of the model toward C_1 increases as the degree of prior exposure to drug increases. This is because the cost of waiting is relatively high and the decision-maker prefers to have immediate rewards. This explains why addicts become impulsive after chronic drug abuse (Logue et al. 1992; Paine et al. 2003; Simon et al. 2007).

As another deficit in the decision-making mechanism, since the basal reward level abnormally elevates in addicts, the model predicts that the motivation for natural reinforcers will decrease after long-term drug exposure. This prediction is consistent with behavioral evidence in human addicts (Garavan et al. 2000; Goldstein et al. 2007).

8.4.2 S-S and S-R Interaction: Actor-Critic Models

Actor-critic is a popular reinforcement learning model that subdivides the process of decision making into two subtasks: learning and action-selection (Sutton and Barto 1998). These two tasks are conducted by the "critic" and the "actor" components, respectively.

The critic component is responsible for adaptively predicting the value of states, $V(s_t)$, by utilizing the prediction error signal. Assuming that the agent leaves state s_t, enters state s_{t+1} and receives reward r_t at time t, the critic will compute the prediction error signal based on the received reward and the prior expectation of the agent:

$$\delta_t = \gamma(r_t + V(s_{t+1})) - V(s_t) \qquad (8.12)$$

This prediction error is then used for updating predictions of the critic:

$$V(s_t) \leftarrow V(s_t) + \alpha \delta_t \qquad (8.13)$$

where, as before, α is the learning rate. As the critic only predicts the value of a state ($V(s_t)$), without caring about what action or external cause has led to it, it is suggested to be a model for S-S (Pavlovian) learning.

The actor component, on the other hand, is involved in making decisions about what action to perform at each state, based on its stored preferences for different actions, $P(s_t, a_t)$: the higher the preference toward an action, the higher the probability of taking that action by the actor. The preferences in the actor are learned based on the values learned by the critic: if taking an action by the actor in a state results in an increase in the value of the state (computed by the critic), the preference toward that action will also increase. The converse is also true: if taking an action leads to a decrease in the critic's value of the state, the probability that the actor takes the action again also decreases by decreasing the preference for that action.

For achieving this harmony, the critical feature of the actor-critic model is that the preferences in the actor are updated using the same prediction error signal that is produced and utilized by the critic component:

$$P(s_t, a_t) \leftarrow P(s_t, a_t) + \alpha \delta_t \qquad (8.14)$$

The fact that the actor uses the error signal generated by the critic can be viewed as an interaction between the S-S (critic) and the S-R (actor) systems. Behaviorally, conditioned reinforcement phenomenon implies that a CS which is associated with a reinforcer (e.g., a light associated with food) supports the acquisition of a new instrumental response (pressing a lever in order to turn the light on). Here, the association between the CS and the reinforcer can be learned by the critic component, that is, the value of the state in which the CS is presented (s_{CS}) increases as the reward in the subsequent state (reward delivery state) is experienced. Next, when several actions are available in a state (s_A), the action that leads to s_{CS} obtains a higher preference (learned by the actor), because taking that action leads to an increase in the value of s_A, as predicted by the critic.

In this respect, dissociating the functions of prediction and action-selection in the actor-critic model is reminiscent of the behavioral psychologist dissociation between Pavlovian and instrumental processes (Niv 2007; Joel et al. 2002). Consistently, a relatively rich body of experiments has shown the dissociable role of striatal subdivisions in prediction and action-selection (O'Doherty et al. 2004; Roesch et al. 2009). Based on these observations, critic and actor components can be thought to be neurally implemented by limbic and sensorimotor cortico-striatal loops, respectively.

Dopamine neurons are hypothesized to integrate information across parallel loops in the cortico-basal ganglia circuit (Haber et al. 2000; Haber 2003), by propagating the prediction error signal made by more limbic (ventral) regions toward associative (dorsomedial) and then motor (dorsolateral) areas of the striatum, via the spiral organization of dopamine neurons. By these spiral connections between the striatum and the VTA/SNc, the output of the accumbens shell can affect the functioning of the core region and in the same way, the output of the accumbens core can influence more dorsal domains of the striatum, via SNc. These dopamine spirals that travel from the ventral to the dorsal regions of the striatum can account for the assumption of the model that the prediction error signal used for updating the actor's preferences is the same signal generated and used by the critic (Joel et al. 2002). These behavioral and neurobiological supports of the actor-critic model has made it a popular model for decision making analysis, and the central role that dopamine plays in it, has allowed addiction-modelers to employ it as a basis for their models.

8.4.2.1 Dayan's Model

Recently, inspired by the model proposed in Redish (2004), Dayan proposed an actor-critic model for addiction (Dayan 2009). The model is based on a variant of the actor-critic model called "advantage learning" (Dayan and Balleine 2002) in which, the critic module has the same algorithm as the classical actor-critic model explained above. Thus, the critic module produces a prediction error signal (δ_V) and uses it for both updating its own value predictions (as in Eq. (8.13)) and also feeding it into the actor component. Rather than learning the preference toward actions, the actor component learns the advantage of taking that action over all other actions that has been previously taken in that state. This "advantage" is denoted by $A(s_t, a_t)$. To learn this "advantage", the actor uses a transformed error signal δ_A:

$$\delta_A = \delta_V - A(s_t, a_t) \tag{8.15}$$

This signal is then used to update the expected advantages:

$$A(s_t, a_t) \leftarrow A(s_t, a_t) + \alpha \delta_A \tag{8.16}$$

The actor utilizes advantages instead of classic preferences to choose among different possible actions. After sufficient learning, as the best action will be the action that the agent takes frequently, its advantage over previously taken actions will tend to zero and the advantage of other alternatives will become negative in their steady levels.

The basis of this model is a hypothesis suggested in Everitt and Robbins (2005) that explains addiction, at a behavioral level, as a transition from voluntary control over drug consumption at the early stages to rigid habitual and compulsive behavior in later stages. Specifically, the hypothesis indicates that this behavioral shift is based on a transition of control over drug-seeking behavior from limbic structures, such as prefrontal cortex (PFC) and nucleus accumbens (NAc), to more motor structures, particularly dorsal striatum. Neurobiological evidence has suggested that this shift is mainly mediated by striatal-midbrain spiraling network that connects the ventral regions of the striatum to more the dorsal parts (Belin and Everitt 2008). According to the Dayan's model, the pharmacological effect of drugs on the dopamine spirals will not only affect the actor indirectly through its effect on the critic's error signal, δ_V, but will also directly affect the actor's updating mechanism due to its effect on δ_A.

In fact, if the pharmacological effect of drugs is assumed to be equal to D, then it will be augmented to the critic's error signal and thus, the critic's value for a drug state will converge to $\gamma(r_D + V(s_{t+1})) + D$. Similarly, due to the effect of drugs on δ_A, the advantage of drug-related actions will increase by D units. This abnormality has been interpreted as a reason to explain why drug-seeking behavior becomes compulsive. The model in Dayan (2009) can also explain how addictive drugs can induce abnormal drug-seeking behavior without abnormally affecting the addict's expectations stored in the critic.

8.4.2.2 Piray et al.'s Model

So far, we have described models that have explained addiction as a disease that is pervasively augmented by drug experience. However, like other diseases, addiction requires a suitable host, that is, a susceptible individual, to spread (Nader et al. 2008). Indeed, overwhelming evidence has shown that only a subpopulation of humans, as well as animals, that have experienced drugs, show symptoms of addiction (compulsive drug seeking and taking) (O'Brien et al. 1986). Some behavioral traits and neural vulnerabilities have been hypothesized to predispose addiction (Koob and Le Moal 2005a; Everitt et al. 2008; Nader et al. 2008). Importantly, a large body of literature suggests a crucial role for dopamine receptors in predisposition to exhibit addiction-like behavior. For example, Dalley and colleagues have shown that lower density of D2 receptors in NAc, but not dorsal striatum, of rats, predicts higher tendency to cocaine self-administration and also addiction-like behavior (Dalley et al. 2007; Belin et al. 2008). Similar results have been reported in non-human primates' neuroimaging studies (Nader et al. 2008), as well as in human studies (Volkow et al. 2008). Moreover, it has been reported recently that low D1 receptor availability within NAc predisposes tendency to cocaine self-administration (Martinez et al. 2009).

In a similar line, a wealth of evidence has shown the important role of dopamine receptors in the development of obesity (Johnson and Kenny 2010) and pathological gambling (Steeves et al. 2009). This is computationally important because a common framework for these diseases and drug addiction, as suggested by Volkow et al. (2008) and Potenza (2008), cannot be constructed only by focusing on the direct pharmacological effects of drugs (Ahmed 2004), but instead, there should be a model that some elements of it bootstrap abnormal and compulsive tendency to rewarding stimuli.

Recently, we proposed a simple actor-critic like model to capture this feature of addiction (Piray et al. 2010). The model relies on three assumptions motivated by neurobiological evidence: (1) VTA dopamine neurons encode action-dependent prediction error (Roesch et al. 2007; Morris et al. 2006) and ventral striatal neurons encode action-dependent values (Roesch et al. 2009; Nicola 2007; Ito and Doya 2009), (2) lower co-availability of D1 and D2 receptors, that is, lower availability of either D1 or D2, in NAc is a necessary condition for addiction to both drug and food to develop (Hopf et al. 2003; Ikemoto et al. 1997; Dalley et al. 2007;

Johnson and Kenny 2010; Martinez et al. 2009), and (3) the first leg of the spiral, that is, posteromedial VTA to NAc shell, is involved in appetitive but not aversive learning (Ford et al. 2006; Ikemoto 2007).

The translation of these assumptions to actor-critic components is straightforward. The first assumption can be interpreted as an action-dependent value representation in the critic, $V(s_t, a_t)$, and also action-dependent prediction error, instead of action-independent ones (see Eq. (8.12):

$$\delta_t = \gamma(r_t + V(s_{t+1})) - V(s_t, a_t) \tag{8.17}$$

$V(s_{t+1})$ is again the value of the best available choice at state s_{t+1} (see Piray et al. 2010 for further discussion).

To model the second assumption, we need to suppose a role for dopamine receptors in terms of the actor-critic model. In line with previous studies (Rutledge et al. 2009; Frank et al. 2007), we have assumed that the availability of dopamine receptors modulates the learning rate (see Smith et al. 2006; Dezfouli et al. 2009 for other ways of modeling the function of dopamine receptors in RL models). Thus, a slight modification in the critic's learning rule, Eq. (8.13), is required:

$$V(s_t, a_t) \leftarrow V(s_t, a_t) + \kappa_c \alpha \delta_t \quad \text{if } r > 0 \tag{8.18}$$

where κ_c corresponds to the availability of dopamine receptors in the NAc. In this formulation, the second assumption can be realized by normalizing the parameter κ_c to one for a healthy subject, and setting it to a value less than one ($\kappa_c < 1$) for individuals who are susceptible to addiction. Finally, the third assumption implies that only appetitive, but not aversive, learning is modulated by the availability of dopamine receptors. Thus, Eq. (8.18) should only be used for appetitive learning; and for learning the value of aversive outcomes ($r < 0$), the prediction error computed by Eq. (8.17) will be used directly.

The behavior of the model can be examined in the task introduced in Deroche-Gamonet et al. (2004). In this experiment, animals learn to self-administer drugs by performing a lever-press action firstly. In the next phase, the lever-press action gets paired with an acute shock punishment. It has been reported that only a proportion of rats, almost 20 percent, that had prolonged experience with drugs, show compulsive behavior.

Figure 8.3 illustrates the behavior of the model in an environment that models the mentioned experiment. As the figure shows, the simulated individual selects the drug-related lever, even after removing the drug reward and instead, giving an acute punishment (phase 2). Since the critic's value is updated with $\kappa_c \alpha \delta$, but the actor's preference is updated by $\alpha \delta$, when $\kappa_c < 1$, the preference toward action a in phase 1 increases abnormally, whereas it increases in a normal way in the critic. In phase 2, however, both the value and the preference are updated by an equal amount and thus, as the figure shows, the amount of drop in both the critic's value and the actor's preference is equal. This drop is sufficient for the critic's value, $V(s, a)$, to converge to r_{sh}, but is not enough to make the preference, $P(a, s)$, negative. For action b, as the reward associated with it is zero, its value and preference remain zero. Hence, in phase 2, while the value of action a falls below the value of action

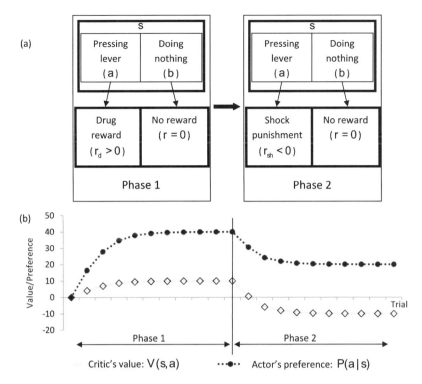

Fig. 8.3 (a) A model with vulnerability to addiction ($\kappa_c = 0.25$) performs the task illustrated in the figure. In state s, the model chooses between two actions. Action a results in a drug reward (drug taking action, $r_d = 10$) and action b results in no reward. After sufficient learning in this phase, the drug reward will be removed and action a is paired with a shock punishment, r_{sh} (phase 2). (b) The performance of the model proposed in Piray et al. (2010) in the mentioned task. While the optimal behavior in phase 2 is choosing b, the vulnerable model chooses a. This is because the preference toward a in phase 1 is abnormally exaggerated in the actor, while its value is normal in the critic. Moreover, since learning from punishment is required in this phase, both value and preference will be updated equally and thus, the amount of drop in both the critic's value and the actor's preference is equal. After a while, the critic's value converges to r_{sh} and thus, the prediction error by performing a converges to zero. As a result, no more changes in the value and the preference associated with a occurs. This effect will decrease the critic's value for a to a level below that of b (zero), but is not sufficient to make the preference toward a negative. The origin of this behavior is the abnormal increase in the actor's preference (habit) toward a in phase 1

b, the preference toward action a is still above that of action b. In fact, when the critic's value converges to r_{sh}, the prediction error of performing a converges to zero and so, no change in the value and preference associated with a occurs.

Notably, if we assume that by chronic administration of drugs, the availability of receptors will further decrease, which is supported by neurobiological data (Nader et al. 2002; Porrino et al. 2004a; Volkow et al. 2004b), the discrepancy between the values and the preferences in the appetitive system will further increase through learning (see Piray et al. 2010 for details). Hence, the insensitivity of addicts to the negative consequences of drug-taking will increase after a prolonged experience with drugs.

This model has two major behavioral implications. First, the compulsivity only appears in vulnerable individuals. Second, compulsivity does not depend on the pharmacological effect of drugs and thus, the model can explain compulsive tendency to natural rewards, such as palatable foods and gambling, in a common framework with compulsive drug taking. The important neurobiological implication of the model is that compulsivity depends on abnormally strong actor's preferences toward drugs; however, it is the critic's deficit that is the origin of this abnormal behavior. Thus, the model accounts for the progressive shift of behavior control during drug consumption from ventral to dorsal striatum, which is initiated by the ventral striatal vulnerabilities and mediated by the dopaminergic spiralling network (Everitt et al. 2008; Porrino et al. 2004b).

8.4.3 S-R and S-A-O Interaction: Dual-Process Models

Whereas actor-critic models have tried to model some properties of the limbic and sensorimotor loops as well as their interaction, dual-process models are focused on the sensorimotor and associative loops, responsible for making habitual and goal-directed decisions, respectively, as well as competitive and collaborative interactions between them (Daw et al. 2005; Keramati et al. 2011). In this section we explain a dual-process model that we have proposed recently (Keramati et al. 2011), which we believe has important implications for explaining some aspects of addiction.

In this model, similar to the seminal dual-process model (Daw et al. 2005), the fundamental nature of the habitual system is the same as a simple TDRL model discussed in previous sections. This system is capable of enforcing or weakening an association between a state and an action (denoted by $Q^H(s_t, a_t)$, hereafter), based on the prediction error signal, which is hypothesize to be carried by the phasic activity of dopamine neurons (Schultz et al. 1997). At the time of decision making, the established associations can be exploited, and as all the information needed for making a choice between several alternatives is accumulated in S-R associations from previous experiences, the habitual responses can be made within a short interval after the stimulus is presented. However, this speed in action selection doesn't come without cost: because many learning trials are required for the outcomes of an action to affect an association, the strength of associations are low-elastic to the outcomes, making the habitual responses inaccurate, particularly under changing motivational or environmental conditions.

In contrast, the goal-directed system is hypothesized to learn through experience the causal relationship between actions and outcomes, so that it has access to a decision tree at the time of decision making and can deliberate the consequences of different alternatives. Denoting the learned dynamics of the environment by $\hat{p}_T(s \xrightarrow{a} s')$ (indicating the probability of traveling from state s to s' by taking action a), and the reward function by $\hat{p}_R(r|s,a)$ (indicating the probability of receiving reward r by taking action a at state s), the estimated value of a state-action pair can be calculated by the below recursive equation. This algorithm is intuitively equivalent to a full-depth search in a decision tree for finding the maximum attainable reward by taking each of the available choices:

$$\hat{Q}^G(s_t, a_t) = E[\hat{p}_R(r|s_t, a_t)] + \gamma \sum_{s'} \hat{p}_T\left(s_t \xrightarrow{a_t} s'\right).\hat{V}(s') \quad (8.19)$$

Although this system can estimate the value of actions more accurately and more optimally, it is not as fast as the habitual system because of the cognitive load (tree search) required for value estimation. Thus, the animals' decision making machinery is always confronted with a trade-off between speed and accuracy; that is, whether to make a fast, but inaccurate habitual response, or to wait for the goal-directed system to make a more optimal decision. This trade-off is hypothesized to be based on a cost-benefit analysis. Assuming that the time needed for the goal-directed system to accurately calculate the estimated value of each available action is τ, the cost of deliberating for each response will be $\bar{R}\tau$, where $\bar{R}$ is the amount of reward that the animal is expected to receive at each unit of time. This variable can be simply computed by taking an average over the rewards obtained through time in the past, as in Eq. (8.7). As discussed before, this average reward signal is hypothesized to be carried by the tonic activity of dopamine neurons (Niv et al. 2007). Thus, if for whatever reason the tonic firing rate of dopamine neurons elevates, the model predicts that the cost of goal-directed responding will increase and consequently, decisions will be made more habitually.

The benefit of deliberation for a certain action, on the other hand, is equal to how much the animal estimates that having the exact value of that action will help it improve its decision policy. This parameter, called "value of perfect information (VPI)", is computable using the estimated Q-values and their corresponding uncertainties cached in the habitual system. Without going into details of the algorithm, one critical prediction of the model is that if the values of two competing actions, estimated by the habitual system, are very close together, then knowing their exact values would greatly help the animal make the optimal decision between those two choices. In contrast, if at a certain state, the estimated value of one of the feasible choices is markedly greater than other actions, and its uncertainty is low, then it can be inferred by the animal that it is less likely that having perfect information about the value of actions will change its initial conjecture about the best choice, made by the habitual system. Thus, under such conditions, the goal-directed system will not contribute to the decision making process.

As the consistency of the model with behavioral and neuronal findings is discussed in the original paper (Keramati et al. 2011), we focus here on the implications of the model for addiction.

All the previous computational theories of addiction discussed in this chapter explain how drug-seeking and drug-taking habits consolidate through the course of addiction as a result of neuroplasticity in different regions of the cortico-basal ganglia circuit, under the effect of dopamine bursts. Although these models have proven fruitful to some degrees, many theories of addiction emphasize on impairment of top-down cognitive control as the essential source of compulsivity. Inability of addicts in breaking habits that have evident adverse consequences is attributed to dysfunctional prefrontal cortical executive control over abnormally strong maladaptive habits. In fact, the evolution of control over behavior from ventral to dorsal striatum, discussed in the previous section, is followed by a shift within the dorsal striatum from action-outcome to stimulus-response mechanisms (Pierce and Vanderschuren 2010; Belin et al. 2009).

Taking into account the effect of drugs on phasic dopamine, the dual-process model discussed above can explain how addictive drugs, by over-reinforcing stimulus-response associations, result in the estimated value of the habitual system for drug-seeking choices becoming maladaptively high. As a consequence, the *VPI* signal (benefit of deliberation) for those actions will be very low after long-term drug consumption and thus, the individual will make habitual and automatic responses, without considering the possible consequences. Consistent with this prediction, it has been reported that short-term drug seeking is a goal-directed behavior, whereas after prolonged drug exposure, drug seeking becomes habitual (Zapata et al. 2010). According to the models introduced in the previous sections, this is equivalent to the insensitivity of drug consumption to harmful consequences.

Beside the direct effect of addictive drugs on reinforcing drug-seeking S-R associations through their pharmacological effect on the dopaminergic circuit, they also pathologically subvert higher level learning mechanisms responsible for suppressing inflexible responses. Protracted exposure to drugs of abuse is widely reported to associate with behavioral deficits in tasks that require cognitive areas of the brain to be involved (Rogers and Robbins 2001; Almeida et al. 2008). Reduction in the activity of the PFC regions in abstinent addicts is also reported in many imaging studies (Goldstein and Volkow 2002; Volkow et al. 2004a). Interestingly, extended access to cocaine has shown to induce long-lasting impairments in working memory-dependent tasks, accompanied with decreased density of neurons in dorsomedial PFC (George et al. 2008). Considering the role of this region in goal-directed decision making, the atrophy of the associative cortex induced by drugs can further disrupt the balance between the goal-directed and habitual systems in favor of the latter. One simple way to model these morphological neuroadaptations in the dual-process framework is to assume that debility of the goal-directed system corresponds to its weakness in searching for the accurate estimated value of actions in the decision tree. Thus, reaching an acceptable level of accuracy (searching deep enough) to obtain "perfect information" will require more time (τ) in addicts, compared to healthy individuals. The assumption that the low performance of addicts in cognitive tasks can be modelled by a higher-than-normal τ can be tested by comparing the addicts' reaction time with that of healthy individuals, at the early stages of learning when responding is still goal-directed. Furthermore, since the deliberation time constitutes the cost of deliberation, another prediction that comes from this assumption is that addicts, because of having higher-than-normal deliberation cost, are less prone to deliberate and thus, more prone to make habitual responses than normal subjects. Thus, habitual responding for natural rewards must appear earlier in addicts, compared to non-addict subjects. In addition, addicts must be more vulnerable than healthy subjects to commit actions with catastrophic consequences, not only in drug-associated cases, but also in other aspects of their daily lives.

The arbitration between the two systems is not only under the effect of long-lasting brain adaptations (like the two mechanisms described above: drug affects *VPI* and τ signals), but some transient changes in the brain decision making variables might also affect the arbitration between the two systems for a short period of time. For example, if for any reason the tonic dopamine, which is assumed to encode the average reward signal, increases for a certain period, the model predicts that the cost of deliberation will increase and thus, the individual will be more susceptible to make habitual responses during that period. This prediction of the model can explain why drug relapse is often precipitated by exposure to drug-associated cues, non-contingent drug injection, or stress (Shaham et al. 2003; Kalivas and McFarland 2003). In fact, the model explains that after prolonged abstinence, these three triggers of relapse revive the habitual system by increasing tonic dopamine and therefore, result in the dormant maladaptive habits to drive the behavior again toward drug consumption. Stress, as a potent trigger of relapse, has shown to increase extracellular concentration of dopamine in cortical and subcortical brain regions in both animal models of addiction (Thierry et al. 1976; Mantz et al. 1989) and humans (Montgomery et al. 2006). Intermittent tail-shock stress, for example, increases extracellular dopamine relative to the baseline by 39% and 95% in nucleus accumbens and medial frontal cortex, respectively (Abercrombie et al. 1989). Interestingly, protracted exposure to stress, similar to the effect of chronic drug consumption, results in the atrophy of the medial prefrontal cortex and the associative striatum, as well as hypertrophy of the sensorimotor striatum. These structural changes are accompanied with progressive behavioral insensitivity to the outcome of responses (Dias-Ferreira et al. 2009). This phenomenon can be explained in a similar argument proposed for explaining the long-lasting effect of drugs on the associative loop. Exposure to drug cues and drug-priming (non-contingent injection of drugs), as other triggers of relapse, are also well-known to increase extracellular dopamine for a considerable period of time (Di Chiara and Imperato 1988; Ito et al. 2002).

In sum, the dual-process model proposed above, explains the story of addiction in a scenario like this: at the early stages of drug self-administration, similar to responding for natural rewards, responding for drugs is controlled by the goal-directed system. After extensive training, as a result of a decrease in the *VPI* signal, as well as an increase in the average reward signal and deliberation time (as described before), the habitual behavior takes control over behavior. At this stage, as no drug is delivered to the animal anymore (extinction period), the average reward signal will drop significantly and thus, the goal-directed system will again take control over behavior. Finally, when a relapse trigger is experienced by the animal, the average reward signal increases again and thus, the habitual system can again come to the scene. Hence, the high values assigned to drug-seeking behavior by the habitual system will make the animal motivated to start responding for the drug again.

As explained above, this scenario is based on the assumption that the goal-directed system doesn't predict maladaptively high values for drug-seeking behaviors. This assumption implies that animals with inactivated brain regions underlying the habitual system should not develop compulsive behavior.

Furthermore, it is assumed that after the extinction period, the habitual system assigns a high value to drug seeking and taking behavior when the animal is exposed to relapse-triggering conditions. This property cannot be explained by the computational models of the S-R system introduced previously. This is because during the extinction phase, drug taking action is not followed by a drug reward and thus, drug seeking and taking actions lose their assigned values. This implies that the habitual system will not exhibit a compulsive behavior after extinction training. To explain the fact that drug-related behaviors regain high values after the animal faces relapse-triggers, it is necessary to incorporate more complicated mechanisms into the habitual system to represent the effect of relapse-triggers on habitual responding (see Redish et al. 2007 for the case of cue-induced relapse).

Finally, the explained scenario predicts that the reinstatement of drug seeking behavior is due to the transition of control from the goal-directed system to the habitual system. However, it is still unclear whether the drug seeking response after reinstatement is under the control of the habitual system or the goal directed system (Root et al. 2009).

8.5 Conclusion

Drug addiction is definitely a much more complicated phenomenon, both behaviorally and neurally, than the simplified image presented in this chapter. Neurally, different drugs have different sites of action and even for a certain drug like cocaine, the dopaminergic system is not the only circuit that is under the pharmacological effect. For example, serotonergic (Dhonnchadha and Cunningham 2008; Bubar and Cunningham 2008) and glutamatergic (Kalivas 2009) systems are also shown to be affected by drugs. However, the computational theory of reinforcement learning has proven to be an appropriate framework to approach this complex phenomenon. The great advantage of this framework is in its ability to bridge between behavioral and neural findings. Moreover, modeling DA receptors' availability within the actor-critic framework (Piray et al. 2010), as an example, shows that the RL framework is also potentially capable of modeling at least some of the detailed neuronal mechanisms.

There are still many steps to be taken in order to improve the current RL-based models of addiction. On important step is to develop an integrated model that can have all the three learning processes (Pavlovian, habitual and goal-directed) at the same time. Such a model would be expected to explain several behavioral aspects of addiction like loss of cognitive control, as well as the influence that Pavlovian predictors of drugs can exert on habitual and/or goal-directed systems (the role of PIT in cue-triggered relapse).

References

Abercrombie ED, Keefe KA, DiFrischia DS, Zigmond MJ (1989) Differential effect of stress on in vivo dopamine release in striatum, nucleus accumbens, and medial frontal cortex. J Neurochem 52:1655–1658

Ahmed SH (2004) Addiction as compulsive reward prediction. Science (New York) 306:1901–1902

Ahmed SH, Koob GF (2005) Transition to drug addiction: a negative reinforcement model based on an allostatic decrease in reward function. Psychopharmacology 180.473–490

Alexander GE, Crutcher MD (1990) Functional architecture of basal ganglia circuits: neural substrates of parallel processing. Trends Neurosci 13:266–271

Alexander GE, DeLong MR, Strick PL (1986) Parallel organization of functionally segregated circuits linking basal ganglia and cortex. Annu Rev Neurosci 9:357–381

Alexander GE, Crutcher MD, De Long MR (1990) Basal ganglia-thalamocortical circuits: parallel substrates for motor, oculomotor, "prefrontal" and "limbic" functions. Prog Brain Res 85:119–146

Almeida PP, Novaes MAFP, Bressan RA, de Lacerda ALT (2008) Executive functioning and cannabis use. Rev Bras Psiquiatr (São Paulo, 1999) 30:69–76

American Psychiatric Association (2000) Diagnostic and statistical manual of mental disorders: DSM-IV-TR, 4th edn, Washington, DC

Balleine BW, O'Doherty JP (2010) Human and rodent homologies in action control: corticostriatal determinants of goal-directed and habitual action. Neuropsychopharmacology 35:48–69

Belin D, Everitt BJ (2008) Cocaine seeking habits depend upon dopamine-dependent serial connectivity linking the ventral with the dorsal striatum. Neuron 57:432–441

Belin D, Mar AC, Dalley JW, Robbins TW, Everitt BJ (2008) High impulsivity predicts the switch to compulsive cocaine-taking. Science (New York) 320:1352–1355

Belin D, Jonkman S, Dickinson A, Robbins TW, Everitt BJ (2009) Parallel and interactive learning processes within the basal ganglia: relevance for the understanding of addiction. Behav Brain Res 199:89–102

Bickel WK, Madden GJ (1999) A comparison of measures of relative reinforcing efficacy and behavioral economics: cigarettes and money in smokers. Behav Pharmacol 10:627–637

Bickel WK, Marsch LA (2001) Toward a behavioral economic understanding of drug dependence: delay discounting processes. Addiction (Abingdon) 96:73–86

Bubar MJ, Cunningham KA (2008) Prospects for serotonin 5-HT2R pharmacotherapy in psychostimulant abuse. Prog Brain Res 172:319–346

Dalley JW, Fryer TD, Brichard L, Robinson ESJ, Theobald DEH et al (2007) Nucleus accumbens d2/3 receptors predict trait impulsivity and cocaine reinforcement. Science (New York) 315:1267–1270

Dalley JW, Everitt BJ, Robbins TW (2011) Impulsivity, compulsivity, and top-down cognitive control. Neuron 69(4):680–694

Daw ND, Niv Y, Dayan P (2005) Uncertainty-based competition between prefrontal and dorsolateral striatal systems for behavioral control. Nat Neurosci 8:1704–1711

Dayan P (2009) Dopamine, reinforcement learning, and addiction. Pharmacopsychiatry 42(1):S56–S65 Suppl

Dayan P, Balleine BW (2002) Reward, motivation, and reinforcement learning. Neuron 36:285–298

de Wit H (2009) Impulsivity as a determinant and consequence of drug use: a review of underlying processes. Addict Biol 14:22–31

Deroche-Gamonet V, Belin D, Piazza PV (2004) Evidence for addiction-like behavior in the rat. Science (New York) 305:1014–1017

Dezfouli A, Piray P, Keramati MM, Ekhtiari H, Lucas C et al (2009) A neurocomputational model for cocaine addiction. Neural Comput 21:2869–2893

Dhonnchadha BAN, Cunningham KA (2008) Serotonergic mechanisms in addiction-related memories. Behav Brain Res 195:39–53

Di Chiara G, Imperato A (1988) Drugs abused by humans preferentially increase synaptic dopamine concentrations in the mesolimbic system of freely moving rats. Proc Natl Acad Sci USA 85:5274–5278

Dias-Ferreira E, Sousa JC, Melo I, Morgado P, Mesquita AR et al (2009) Chronic stress causes frontostriatal reorganization and affects decision-making. Science (New York) 325:621–625

Dickinson A, Balleine BW (2002) The role of learning in motivation. In: Gallistel CR (ed) Steven's handbook of experimental psychology: learning, motivation, and emotion, 3rd edn, vol 3. Wiley, New York, pp 497–533

Estes WK (1948) Discriminative conditioning; effects of a pavlovian conditioned stimulus upon a subsequently established operant response. J Exp Psychol 38:173–177

Everitt BJ, Robbins TW (2005) Neural systems of reinforcement for drug addiction: from actions to habits to compulsion. Nat Neurosci 8:1481–1489

Everitt BJ, Belin D, Economidou D, Pelloux Y, Dalley JW et al (2008) Neural mechanisms underlying the vulnerability to develop compulsive drug-seeking habits and addiction. Philos Trans R Soc Lond B, Biol Sci 363:3125–3135

Ford CP, Mark GP, Williams JT (2006) Properties and opioid inhibition of mesolimbic dopamine neurons vary according to target location. J Neurosci 26:2788–2797

Frank MJ, Moustafa AA, Haughey HM, Curran T, Hutchison KE (2007) Genetic triple dissociation reveals multiple roles for dopamine in reinforcement learning. Proc Natl Acad Sci USA 104:16311–16316

Garavan H, Pankiewicz J, Bloom A, Cho JK, Sperry L et al (2000) Cue-induced cocaine craving: neuroanatomical specificity for drug users and drug stimuli. Am J Psychiatry 157:1789–1798

George O, Mandyam CD, Wee S, Koob GF (2008) Extended access to cocaine self-administration produces long-lasting prefrontal cortex-dependent working memory impairments. Neuropsychopharmacology 33:2474–2482

Goldstein RZ, Volkow ND (2002) Drug addiction and its underlying neurobiological basis: neuroimaging evidence for the involvement of the frontal cortex. Am J Psychiatry 159:1642–1652

Goldstein RZ, Alia-Klein N, Tomasi D, Zhang L, Cottone LA et al (2007) Is decreased prefrontal cortical sensitivity to monetary reward associated with impaired motivation and self-control in cocaine addiction? Am J Psychiatry 164:43–51

Haber SN (2003) The primate basal ganglia: parallel and integrative networks. J Chem Neuroanat 26:317–330

Haber SN, Fudge JL, McFarland NR (2000) Striatonigrostriatal pathways in primates form an ascending spiral from the shell to the dorsolateral striatum. J Neurosci 20:2369–2382

Holland PC (2004) Relations between pavlovian-instrumental transfer and reinforcer devaluation. J Exp Psychol, Anim Behav Processes 30:104–117

Hopf FW, Cascini MG, Gordon AS, Diamond I, Bonci A (2003) Cooperative activation of dopamine d1 and d2 receptors increases spike firing of nucleus accumbens neurons via g-protein betagamma subunits. J Neurosci 23:5079–5087

Ikemoto S (2007) Dopamine reward circuitry: two projection systems from the ventral midbrain to the nucleus accumbens-olfactory tubercle complex. Brains Res Rev 56:27–78

Ikemoto S, Glazier BS, Murphy JM, McBride WJ (1997) Role of dopamine d1 and d2 receptors in the nucleus accumbens in mediating reward. J Neurosci 17:8580–8587

Ito M, Doya K (2009) Validation of decision-making models and analysis of decision variables in the rat basal ganglia. J Neurosci 29:9861–9874

Ito R, Dalley JW, Robbins TW, Everitt BJ (2002) Dopamine release in the dorsal striatum during cocaine-seeking behavior under the control of a drug-associated cue. J Neurosci 22:6247–6253

Jacobs EA, Bickel WK (1999) Modeling drug consumption in the clinic using simulation procedures: demand for heroin and cigarettes in opioid-dependent outpatients. Exp Clin Psychopharmacol 7:412–426

Joel D, Niv Y, Ruppin E (2002) Actor-critic models of the basal ganglia: new anatomical and computational perspectives. Neural Netw 15:535–547

Johnson PM, Kenny PJ (2010) Dopamine d2 receptors in addiction-like reward dysfunction and compulsive eating in obese rats. Nat Neurosci 13:635–641

Kalivas PW (2009) The glutamate homeostasis hypothesis of addiction. Nat Rev, Neurosci 10:561–572
Kalivas PW, McFarland K (2003) Brain circuitry and the reinstatement of cocaine-seeking behavior. Psychopharmacology 168:44–56
Kamin L (1969) Predictability, surprise, attention, and conditioning. In: Campbell BA, Church RM (eds) Punishment and aversive behavior. Appleton-Century-Crofts, New York, pp 279–296
Keramati M, Dezfouli A, Piray P (2011) Speed-accuracy trade-off between the habitual and the goal-directed processes. PLoS Comput Biol 7(5):1–25
Koob GF, Le Moal M (2005a) Plasticity of reward neurocircuitry and the 'dark side' of drug addiction. Nat Neurosci 8:1442–1444
Koob GF, Le Moal M (2005b) Neurobiology of addiction. Academic Press, San Diego
Logue A, Tobin H, Chelonis J, Wang R, Geary N et al (1992) Cocaine decreases self-control in rats: a preliminary report. Psychopharmacology 109:245–247
Lovibond PF (1983) Facilitation of instrumental behavior by a pavlovian appetitive conditioned stimulus. J Exp Psychol, Anim Behav Processes 9:225–247
Mackintosh NJ (1974) The psychology of animal learning. Academic Press, London
Mahadevan S (1996) Average reward reinforcement learning: foundations, algorithms, and empirical results. Mach Learn 22:159–195
Mantz J, Thierry AM, Glowinski J (1989) Effect of noxious tail pinch on the discharge rate of mesocortical and mesolimbic dopamine neurons: selective activation of the mesocortical system. Brain Res 476:377–381
Marks KR, Kearns DN, Christensen CJ, Silberberg A, Weiss SJ (2010) Learning that a cocaine reward is smaller than expected: a test of redish's computational model of addiction. Behav Brain Res 212:204–207
Martinez D, Slifstein M, Narendran R, Foltin RW, Broft A et al (2009) Dopamine d1 receptors in cocaine dependence measured with PET and the choice to self-administer cocaine. Neuropsychopharmacology 34:1774–1782
Montgomery AJ, Mehta MA, Grasby PM (2006) Is psychological stress in man associated with increased striatal dopamine levels?: a [11C]raclopride PET study. Synapse (New York) 60:124–131
Morris G, Nevet A, Arkadir D, Vaadia E, Bergman H (2006) Midbrain dopamine neurons encode decisions for future action. Nat Neurosci 9:1057–1063
Nader MA, Daunais JB, Moore T, Nader SH, Moore RJ et al (2002) Effects of cocaine self-administration on striatal dopamine systems in rhesus monkeys: initial and chronic exposure. Neuropsychopharmacology 27:35–46
Nader MA, Czoty PW, Gould RW, Riddick NV (2008) Positron emission tomography imaging studies of dopamine receptors in primate models of addiction. Philos Trans R Soc Lond B, Biol Sci 363:3223–3232
Nicola SM (2007) The nucleus accumbens as part of a basal ganglia action selection circuit. Psychopharmacology 191:521–550
Niv Y (2007) The effects of motivation on habitual instrumental behavior. PhD thesis, The Hebrew University of Jerusalem, Interdisciplinary Center for Neural Computation
Niv Y, Daw ND, Joel D, Dayan P (2007) Tonic dopamine: opportunity costs and the control of response vigor. Psychopharmacology 191:507–520
O'Brien CP, Ehrman R, Ternes J (1986) Classical conditioning in human opioid dependence. In: Goldberg SR, Stolerman IP (eds) Behavioral analysis of drug dependence, 1st edn. Academic Press, London, pp 329–356
O'Doherty J, Dayan P, Schultz J, Deichmann R, Friston K et al (2004) Dissociable roles of ventral and dorsal striatum in instrumental conditioning. Science (New York) 304:452–454
Paine TA, Dringenberg HC, Olmstead MC (2003) Effects of chronic cocaine on impulsivity: relation to cortical serotonin mechanisms. Behav Brain Res 147:135–147
Panlilio LV, Thorndike EB, Schindler CW (2007) Blocking of conditioning to a cocaine-paired stimulus: testing the hypothesis that cocaine perpetually produces a signal of larger-than-expected reward. Pharmacol Biochem Behav 86:774–777

Pelloux Y, Everitt BJ, Dickinson A (2007) Compulsive drug seeking by rats under punishment: effects of drug taking history. Psychopharmacology 194:127–137

Petry NM, Bickel WK (1998) Polydrug abuse in heroin addicts: a behavioral economic analysis. Addiction (Abingdon) 93:321–335

Pierce RC, Vanderschuren LJMJ (2010) Kicking the habit: the neural basis of ingrained behaviors in cocaine addiction. Neurosci Biobehav Rev 35:212–219

Piray P, Keramati MM, Dezfouli A, Lucas C, Mokri A (2010) Individual differences in nucleus accumbens dopamine receptors predict development of addiction-like behavior: a computational approach. Neural Comput 22:2334–2368

Porrino LJ, Daunais JB, Smith HR, Nader MA (2004a) The expanding effects of cocaine: studies in a nonhuman primate model of cocaine self-administration. Neurosci Biobehav Rev 27:813–820

Porrino LJ, Lyons D, Smith HR, Daunais JB, Nader MA (2004b) Cocaine self-administration produces a progressive involvement of limbic, association, and sensorimotor striatal domains. J Neurosci 24:3554–3562

Potenza MN (2008) The neurobiology of pathological gambling and drug addiction: an overview and new findings. Philos Trans R Soc Lond B, Biol Sci 363:3181–3189

Rangel A, Camerer C, Montague PR (2008) A framework for studying the neurobiology of value-based decision making. Nat Rev, Neurosci 9:545–556

Redish AD (2004) Addiction as a computational process gone awry. Science (New York) 306:1944–1947

Redish AD, Jensen S, Johnson A, Kurth-Nelson Z (2007) Reconciling reinforcement learning models with behavioral extinction and renewal: implications for addiction, relapse, and problem gambling. Psychol Rev 114:784–805

Redish AD, Jensen S, Johnson A (2008) A unified framework for addiction: vulnerabilities in the decision process. Behav Brain Sci 31:415–437; discussion 437–487

Reynolds B (2006) A review of delay-discounting research with humans: relations to drug use and gambling. Behav Pharmacol 17:651–667

Robinson ESJ, Eagle DM, Economidou D, Theobald DEH, Mar AC et al (2009) Behavioural characterisation of high impulsivity on the 5-choice serial reaction time task: specific deficits in 'waiting' versus 'stopping'. Behav Brain Res 196:310–316

Roesch MR, Calu DJ, Schoenbaum G (2007) Dopamine neurons encode the better option in rats deciding between differently delayed or sized rewards. Nat Neurosci 10:1615–1624

Roesch MR, Singh T, Brown PL, Mullins SE, Schoenbaum G (2009) Ventral striatal neurons encode the value of the chosen action in rats deciding between differently delayed or sized rewards. J Neurosci 29:13365–13376

Rogers RD, Robbins TW (2001) Investigating the neurocognitive deficits associated with chronic drug misuse. Curr Opin Neurobiol 11:250–257

Root DH, Fabbricatore AT, Barker DJ, Ma S, Pawlak AP et al (2009) Evidence for habitual and goal-directed behavior following devaluation of cocaine: a multifaceted interpretation of relapse. PLoS ONE 4:e7170

Rutledge RB, Lazzaro SC, Lau B, Myers CE, Gluck MA et al (2009) Dopaminergic drugs modulate learning rates and perseveration in parkinson's patients in a dynamic foraging task. J Neurosci 29:15104–15114

Schultz W, Dayan P, Montague PR (1997) A neural substrate of prediction and reward. Science (New York) 275:1593–1599

Shaham Y, Shalev U, Lu L, Wit HD, Stewart J (2003) The reinstatement model of drug relapse: history, methodology and major findings. Psychopharmacology 168:3–20

Simon NW, Mendez IA, Setlow B (2007) Cocaine exposure causes long-term increases in impulsive choice. Behav Neurosci 121:543–549

Smith AJ, Li M, Becker S, Kapur S (2006) Linking animal models of psychosis to computational models of dopamine function. Neuropsychopharmacology 32:54–66

Steeves TDL, Miyasaki J, Zurowski M, Lang AE, Pellecchia G et al (2009) Increased striatal dopamine release in parkinsonian patients with pathological gambling: a [11C] raclopride PET study. Brain 132:1376–1385

Stewart J (2008) Psychological and neural mechanisms of relapse. Philos Trans R Soc Lond B, Biol Sci 363:3147–3158

Stuber GD, Wightman RM, Carelli RM (2005) Extinction of cocaine self-administration reveals functionally and temporally distinct dopaminergic signals in the nucleus accumbens. Neuron 46:661–669

Sutton RS, Barto AG (1998) Reinforcement learning: an introduction. MIT Press, Cambridge

Thierry AM, Tassin JP, Blanc G, Glowinski J (1976) Selective activation of mesocortical DA system by stress. Nature 263:242–244

Vanderschuren LJMJ, Everitt BJ (2004) Drug seeking becomes compulsive after prolonged cocaine self-administration. Science (New York) 305:1017–1019

Volkow ND, Fowler JS, Wang G (2004a) The addicted human brain viewed in the light of imaging studies: brain circuits and treatment strategies. Neuropharmacology 47(1):3–13 Suppl

Volkow ND, Fowler JS, Wang G, Swanson JM (2004b) Dopamine in drug abuse and addiction: results from imaging studies and treatment implications. Mol Psychiatry 9:557–569

Volkow ND, Wang G, Fowler JS, Telang F (2008) Overlapping neuronal circuits in addiction and obesity: evidence of systems pathology. Philos Trans R Soc Lond B, Biol Sci 363:3191–3200

Yin HH, Knowlton BJ (2006) The role of the basal ganglia in habit formation. Nat Rev, Neurosci 7:464–476

Yin HH, Knowlton BJ, Balleine BW (2004) Lesions of dorsolateral striatum preserve outcome expectancy but disrupt habit formation in instrumental learning. Eur J Neurosci 19:181–189

Yin HH, Ostlund SB, Knowlton BJ, Balleine BW (2005) The role of the dorsomedial striatum in instrumental conditioning. Eur J Neurosci 22:513–523

Yin HH, Ostlund SB, Balleine BW (2008) Reward-guided learning beyond dopamine in the nucleus accumbens: the integrative functions of cortico-basal ganglia networks. Eur J Neurosci 28:1437–1448

Zapata A, Minney VL, Shippenberg TS (2010) Shift from Goal-Directed to habitual cocaine seeking after prolonged experience in rats. J Neurosci 30:15457–15463

Part III
Economic-Based Models of Addiction

Chapter 9
Policies and Priors

Karl Friston

Abstract This chapter considers addiction from a purely theoretical point of view. It tries to substantiate the idea that addictive behaviour is a natural consequence of abnormal perceptual learning. In short, addictive behaviours emerge when behaviour confounds its own acquisition. Specifically, we consider what would happen if behaviour interfered with the neurotransmitter systems responsible for optimising the conditional certainty or precision of inferences about causal structure in the world. We will pursue this within a rather abstract framework provided by free-energy formulations of action and perception. Although this treatment does not touch upon many of the neurobiological or psychosocial issues in addiction research, it provides a principled framework within which to understand exchanges with the environment and how they can be disturbed. Our focus will be on behaviour as active inference and the key role of prior expectations. These priors play the role of policies in reinforcement learning and place crucial constraints on perceptual inference and subsequent action. A dynamical treatment of these policies suggests a fundamental distinction between *fixed-point policies* that lead to a single attractive state and *itinerant policies* that support wandering behavioural orbits among sets of attractive states. Itinerant policies may provide a useful metaphor for many forms of behaviour and, in particular, addiction. Under these sorts of policies, neuromodulatory (e.g., dopaminergic) perturbations can lead to false inference and consequent learning, which produce addictive and preservative behaviour.

9.1 Introduction

This chapter provides a somewhat theoretical account of behaviour and how addiction can be seen in terms of aberrant perception. Its contribution is not to provide a detailed model of addictive behaviour (see Ahmed et al. 2009 for a nice review of current models) but rather to describe a principled framework that places existing ideas in a larger context. This exercise highlights the archi-

K. Friston (✉)
The Wellcome Trust Centre for Neuroimaging, University College London, Queen Square,
London WC1N 3BG, UK
e-mail: k.friston@ucl.ac.uk

tecture of adaptive behaviour, in relation to perception, and the ways in which things can go wrong. Its main conclusion is that addictive behaviour may be an unfortunate and rather unique consequence of a pathological coupling between behaviour (e.g., drug taking) and the perceptual learning (e.g., abnormal modulation of synaptic plasticity) that supports behaviour (cf., Alcaro et al. 2007; Zack and Poulos 2009). This coupling can be particularly disruptive because learning is fundamental for making predictions about exchanges with the world and these predictions prescribe behaviour. In what follows, we will spend some time developing a normative framework for perception and action, with a special emphasis on behavioural policies as prior expectations about how the world unfolds. Having established the basic structure of the problem faced by adaptive agents, we will consider how pathologies of learning manifest behaviourally and show that addictive behaviour is almost impossible to avoid, unless perceptual inference and learning are optimal

This chapter comprises three sections. In Sect. 9.2, we review a free-energy principle for the brain. In Sect. 9.3, we focus on a key element of this formulation; namely, prior expectations that reflect innate or epigenetic constraints. In Sect. 9.4, we use the policies from Sect. 9.3 to illustrate failures in learning and behaviour using simulations.

9.2 The Free-Energy Formulation

This section considers the fundaments of normal behaviour using a free-energy account of action and perception (Friston et al. 2006). Its agenda is to establish an intimate relationship between action and perception and to sketch their neurobiological substrates. In brief, we will see that an imperative for all adaptive (biological) agents is to resist a natural tendency to disorder (Evans 2003) by minimising the surprise (unexpectedness) of sensory exchanges with the world. This imperative can be captured succinctly by requiring agents to minimise their free-energy, where free-energy is an upper bound on surprise. When one unpacks this mathematically, minimisation of surprise entails two things. First, it requires an optimisation of perceptual representations of sensory input of the sort implied by the Bayesian brain hypothesis. Second, it requires an active sampling of the sensorium to select sensory inputs that are predicted and predictable. These two facets of free-energy minimisation correspond to perception and action respectively. Basically, we will see that perceptual predictions enslave action to ensure they come true. We will start with a heuristic overview of the free-energy principle and then reprise the basic ideas more formally. By the end of this section we will have expressed perceptual inference, learning and action in terms of ordinary differential equations that describe putative neuronal dynamics underlying active inference. These dynamics can be regarded as a form of evidence accumulation, because free-energy is a bound approximation to log model-evidence. The ensuing scheme rests on internal models of the world used by agents to make predictions. In the subsequent section, we will look at the basic forms that these models can take and the prior expectations about state-transitions (i.e., policies) they entail.

9.2.1 Free-Energy and Self-organisation: Overview

Free-energy is a quantity from information theory that bounds the evidence for a model of data (Feynman 1972; Hinton and van Camp 1993; MacKay 1995). Here, the data are sensory inputs and the model is encoded by the brain. More precisely, free-energy is greater than the negative log-evidence or 'surprise' inherent in sensory data, given a model of how they were generated. Critically, unlike surprise itself, free-energy can be evaluated because it is a function of sensory data and brain states. In fact, under simplifying assumptions (see below), it is just the amount of prediction error.

The motivation for the free-energy principle is simple but fundamental. It rests upon the fact that self-organising biological agents resist a tendency to disorder and therefore minimise the entropy of their sensory states. Under ergodic assumptions, minimising entropy corresponds to suppressing surprise over time. In brief, for a well-defined agent to exist it must occupy a limited repertoire of states (e.g., a fish in water). This means the equilibrium density of an ensemble of agents, describing the probability of finding an agent in a particular state, must have low entropy: A distribution with low entropy just means a small number of states are occupied most of the time. Because entropy is the long-term average of surprise, agents must avoid surprising states (e.g., a fish out of water). But there is a problem; agents cannot evaluate surprise directly because this would require access to all the hidden states in the world causing sensory input. However, an agent can avoid surprising exchanges with the world if it minimises its free-energy, because free-energy is always bigger than surprise.

Mathematically, the difference between free-energy and surprise is the divergence between a probabilistic representation (recognition density) encoded by the agent and the true conditional distribution of causes of sensory input. This enables the brain to reduce free-energy by changing its representation, which makes the recognition density an approximate conditional density. This corresponds to Bayesian inference on unknown states of the world causing sensory data (Knill and Pouget 2004; Kersten et al. 2004). In short, the free-energy principle subsumes the Bayesian brain hypothesis; or the notion that the brain is an inference machine (von Helmholtz 1866; MacKay 1956; Neisser 1967; Gregory 1968, 1980; Ballard et al. 1983; Dayan et al. 1995; Lee and Mumford 2003; Friston 2005). In other words, biological agents must engage in some form of Bayesian perception to avoid surprises. However, perception is only half the story; it makes free-energy a good proxy for surprise but it does not change the sensations themselves or their surprise.

To reduce surprise, we have to change sensory input. This is where the free-energy principle comes into its own: it says that action should also minimise free-energy (Friston et al. 2009, 2010). We are open systems in exchange with the environment; the environment acts on us to produce sensory impressions and we act on the environment to change its states. This exchange rests upon sensory and effector organs (like photoreceptors and oculomotor muscles). If we change the environment or our relationship to it, sensory input changes. Therefore, action can reduce free-energy (i.e., prediction errors) by changing sensory input, while perception reduces free-energy by changing predictions. In short, we sample the world to ensure our predictions become a self-fulfilling prophecy and that surprises are avoided. In this view, perception enables action by providing veridical predictions (more formally, by making the free-energy a tight bound on surprise) that guide active sampling of the sensorium. This is active inference.

In summary, (i) agents resist a natural tendency to disorder by minimising a free-energy bound on surprise; (ii) this entails acting on the environment to avoid surprises, which (iii) rests on making Bayesian inferences about the world. In this view, the Bayesian brain is mandated by the free-energy principle. Free-energy is not used to finesse perception, perceptual inference is necessary to minimise free-energy. This provides a principled explanation for action and perception that serve jointly to suppress surprise or prediction error; but it does not explain how the brain does this or how it encodes the representations that are optimised. In what follows, we look more formally at what minimising free-energy means for the brain.

9.2.2 Free-Energy and Self-Organisation: Active Inference from Basic Principles

Our objective is to minimise the average uncertainty (entropy) about generalised sensory states $\tilde{s} = s \oplus s' \oplus s'' \ldots \in S$, sampled by a brain or model or the world m ($\oplus$ means concatenation). Generalised states comprise the state itself, its velocity, acceleration, jerk, etc. The average uncertainty is

$$H(S|m) = -\int p(\tilde{s}|m) \ln p(\tilde{s}|m) \, d\tilde{s} \tag{9.1}$$

Under ergodic assumptions, this is proportional to the long-term average of surprise, also known as negative log-evidence $-\ln p(\tilde{s}(t)|m)$

$$H(S|m) \propto -\int_0^T dt \ln p(\tilde{s}(t)|m) \tag{9.2}$$

It can be seen that sensory entropy accumulates negative log-evidence over time. Minimising sensory entropy therefore corresponds to maximising the accumulated log-evidence for an agent's model of the world. Although sensory entropy cannot be minimised directly, we can induce an upper bound $\mathcal{S}(\tilde{s}, q) \geq H(S)$ that can be evaluated using a recognition density $q(t) := q(\vartheta)$ on the generalised causes (i.e., environmental states and parameters) of sensory signals. We will see later that these causes comprise time-varying states $u(t) \subset \vartheta$ and slowly varying parameters $\varphi(t) \subset \vartheta$. This bound is the path-integral of free-energy $\mathcal{F}(t)$, which is created by simply adding a non-negative function of the recognition density to surprise:

$$\mathcal{S} = \int dt \, \mathcal{F}(t)$$

$$\mathcal{F}(t) = D_{KL}(q(\vartheta)\|p(\vartheta|\tilde{s},m)) - \ln p(\tilde{s}(a)|m)$$
$$= D_{KL}(q(\vartheta)\|p(\vartheta|m)) - \langle \ln p(\tilde{s}(a)|\vartheta,m)\rangle_q$$
$$= \langle \ln q(\vartheta)\rangle_q - \langle \ln p(\tilde{s}(a),\vartheta|m)\rangle_q \qquad (9.3)$$

This non-negative function is a Kullback-Leibler divergence $D_{KL}(q(\vartheta)\|p(\vartheta|\tilde{s},m))$, which is only zero when $q(\vartheta) = p(\vartheta|\tilde{s},m)$ is the true conditional density. This means that minimising free-energy, by optimising $q(\vartheta)$, makes the recognition density an approximate conditional density on sensory causes. The free-energy can be evaluated easily because it is a function of $q(\vartheta)$ and a generative model $p(\tilde{s},u|m)$ entailed by m. One can see this by rewriting the last equality in Eq. (9.3) in terms of $\mathcal{H}(t)$, the neg-entropy of $q(t)$ and an energy $\mathcal{L}(t)$ expected under $q(t)$.

$$\mathcal{F}(t) = \langle \mathcal{L}(t)\rangle_q - \mathcal{H}(t)$$
$$\mathcal{L}(t) = -\ln p(\tilde{s}(a),\vartheta|m)$$
$$\mathcal{H}(t) = -\langle \ln q(\vartheta)\rangle_q \qquad (9.4)$$

In physics, $\mathcal{L}(t)$ is called Gibb's energy and reports the joint surprise about sensations and their causes. If we assume that the recognition density $q(\vartheta) = \mathcal{N}(\mu,\mathcal{C})$ is Gaussian (the Laplace assumption), then we can express free-energy in terms of the mean and covariance of the recognition density

$$\mathcal{F} = \mathcal{L}(\mu) + \frac{1}{2}\mathrm{tr}(\mathcal{C}\mathcal{L}_{\mu\mu}) - \frac{1}{2}\ln|\mathcal{C}| - \frac{n}{2}\ln 2\pi e \qquad (9.5)$$

Where $n = \dim(\mu)$. Here and throughout, subscripts denote derivatives. We can now minimise free-energy with respect to the conditional precision $\mathcal{P} = \mathcal{C}^{-1}$ (inverse covariance) by solving $\partial_\Sigma \mathcal{F} = 0 \Rightarrow \delta_\Sigma \mathcal{S} = 0$ to give

$$\mathcal{F}_\Sigma = \frac{1}{2}\mathcal{L}_{\mu\mu} - \frac{1}{2}\mathcal{P} = 0 \Rightarrow \mathcal{P} = \mathcal{L}_{\mu\mu} \qquad (9.6)$$

This allows one to simplify the expression for free-energy by eliminating $\mathcal{C}$ to give

$$\mathcal{F} = \mathcal{L}(\mu) + \frac{1}{2}\ln|\mathcal{L}_{\mu\mu}| - \frac{n}{2}\ln 2\pi \qquad (9.7)$$

Crucially, Eq. (9.7) shows that free-energy is a function of the conditional mean, which means all we have worry about is optimising the means or (approximate) conditional expectations. Their optimal values are the solution to the following differential equations. For the generalised states $\tilde{u}(t) \subset \vartheta$

$$\dot{\mu}^{(u)} = \mu'^{(u)} - \mathcal{F}_u$$
$$\dot{\mu}'^{(u)} = \mu''^{(u)} - \mathcal{F}_{u'}$$
$$\vdots$$
$$\dot{\tilde{\mu}}^{(u)} = \mathcal{D}\tilde{\mu}^{(u)} - \mathcal{F}_{\tilde{u}} \qquad (9.8)$$

Where $\mathcal{D}$ is a derivative matrix operator with identity matrices above the leading diagonal, such that $\mathcal{D}\tilde{u} = [u', u'', \ldots]^T$. Here and throughout, we assume all gradients are evaluated at the mean; here $\tilde{u} = \tilde{\mu}^{(u)}$. The stationary solution of Eq. (9.8), in a frame of reference that moves with the generalised motion of the mean, minimises free-energy and its path integral. This can be seen by noting $\dot{\tilde{\mu}}^{(u)} - \mathcal{D}\tilde{\mu}^{(u)} = 0 \Rightarrow \mathcal{F}_{\tilde{u}} = 0 \Rightarrow \delta_{\tilde{u}} S = 0$. This ensures that when free-energy is minimised the mean of the motion is the motion of the mean: i.e., $\dot{\tilde{\mu}}^{(u)} = \mathcal{D}\tilde{\mu}^{(u)}$. For slowly varying parameters $\varphi(t) \subset \vartheta$, we can use the a formally related scheme, which ensures their motion disappears

$$\dot{\mu}^{(\varphi)} = \mu'^{(\varphi)}$$
$$\dot{\mu}'^{(\varphi)} = -\mathcal{F}_\varphi - \kappa \mu'^{(\varphi)} \qquad (9.9)$$

Here, the solution $\dot{\tilde{\mu}}^{(\varphi)} = 0$ minimises free-energy, under constraint that the motion of the expected parameters is small: i.e., $\mu'^{(\varphi)} \to 0$. One can see this by noting that when $\dot{\mu}^{(\varphi)} = \dot{\mu}'^{(\varphi)} = 0 \Rightarrow \mathcal{F}_\varphi = 0 \Rightarrow \delta_\varphi S = 0$. Equations (9.8) and (9.9) prescribe recognition dynamics for the expected states and parameters respectively. The dynamics for states can be thought of as a gradient descent in a frame of reference that moves with the expected motion of the world (cf., a moving target). Conversely, the dynamics for the parameters can be thought of as a gradient descent that resists transient fluctuations with the damping term $\mathcal{F}_{\varphi'} = \kappa \mu'^{(\varphi)}$ (see Appendix A for a perspective from conventional decent schemes). It is this damping that instantiates prior knowledge that fluctuations in the parameters are small. These recognition dynamics minimise free-energy with respect to the conditional expectations underlying perception but what about action?

9.2.2.1 Action and Perception

The second equality in Eq. (9.3) equality shows that free-energy can also be suppressed by action, through its effects on hidden states and ensuing sensory signals. The key term here is the accuracy term, $\langle \ln p(\tilde{s}(a)|\vartheta, m) \rangle_q$ which, under Gaussian assumptions, this is just the amount of prediction error. This means action should change the motion of sensory states so that they conform to conditional expectations. This minimises surprise, provided perception makes free-energy a tight bound on surprise. In short, the free-energy principle prescribes optimal perception and action

$$\mu(t)^* = \arg\min_\mu \mathcal{F}(\tilde{s}(a), \mu)$$
$$a(t)^* = \arg\min_a \mathcal{F}(\tilde{s}(a), \mu) \qquad (9.10)$$

Action reduces to sampling input that is expected under the recognition density (i.e., sampling selectively what one expects to experience). In other words, agents must necessarily (if implicitly) make inferences about the causes of their sensory signals and sample signals that are consistent with those inferences. In summary, the free-energy principle requires the internal states of an agent and its action to suppress free-energy. This corresponds to optimising a probabilistic model of how sensations are caused, so that the resulting predictions can guide active sampling of sensory data. The requisite interplay between action and perception (i.e., active inference) ensures the agent's sensory states have low entropy. This recapitulates the notion that "perception and behaviour can interact synergistically, via the environment" to optimise behaviour (Verschure et al. 2003). Active inference is an example of *self-referenced* learning (Maturana and Varela 1980; Porr and Wörgötter 2003) in which "the actions of the learner influence its own learning without any valuation process" (Porr and Wörgötter 2003).

9.2.2.2 Summary

In conclusion, we have derived recognition dynamics for expected states (in generalised coordinates of motion) and parameters, which cause sensory samples. The solution to these equations minimise free-energy and therefore minimise a bound on sensory surprise or (negative) log-evidence. Optimisation of the expected states and parameters corresponds to perceptual inference and learning respectively. The precise form of the recognition dynamics depends on the energy $\mathcal{L} = -\ln p(\tilde{s}, \vartheta | m)$ associated with a particular generative model. In what follows, we consider dynamic models of the world.

9.2.3 Dynamic Generative Models

We now look at hierarchal dynamic models (discussed in Friston 2008) and assume that any sensory data can be modelled with a special case of these models. Consider the state-space model

$$\begin{aligned} s &= f^{(v)}(x, v, \theta) + \omega^{(v)} : \omega^{(v)} \sim \mathcal{N}\big(0, \Sigma^{(v)}(x, v, \gamma)\big) \\ \dot{x} &= f^{(x)}(x, v, \theta) + \omega^{(x)} : \omega^{(x)} \sim \mathcal{N}\big(0, \Sigma^{(x)}(x, v, \gamma)\big) \end{aligned} \quad (9.11)$$

The nonlinear functions $f^{(u)} : u = v, x$ represent a sensory mapping and equations of motion respectively and are parameterised by $\theta \subset \varphi$. The states $v \subset u$ are referred to as sources or causes, while hidden states $x \subset u$ mediate the influence of the causes on sensory data and endow the system with memory. We assume the random fluctuations $\omega^{(u)} \in \Omega$ are analytic, such that the covariance of $\tilde{\omega}^{(u)}$ is well defined. This model allows for state-dependent changes in the amplitude of random fluctuations, which speaks to a key distinction between the effect of states on first and second-order sensory dynamics. These effects are mediated by the vector and matrix functions $f^{(u)} \in \Re^{\dim(u)}$ and $\Sigma^{(u)} \in \Re^{\dim(u) \times \dim(u)}$ respectively, which are parameterised by first and second-order parameters $\theta, \gamma \subset \varphi$. Under local linearity assumptions, the generalised motion of the sensory response and hidden states can be expressed compactly as

$$\tilde{s} = \tilde{f}^{(v)} + \tilde{\omega}^{(v)}$$
$$\mathcal{D}\tilde{x} = \tilde{f}^{(x)} + \tilde{\omega}^{(x)} \quad (9.12)$$

Where the generalised predictions are

$$\tilde{f}^{(u)} = \begin{bmatrix} f^{(u)} = f^{(u)} \\ f'^{(u)} = f_x^{(u)} x' + f_v^{(u)} v' \\ f''^{(u)} = f_x^{(u)} x'' + f_v^{(u)} v'' \\ \vdots \end{bmatrix} \quad (9.13)$$

Equation (9.12) means that Gaussian assumptions about the random fluctuations specify a generative model in terms of a likelihood and empirical priors on the motion of hidden states

$$p(\tilde{s}|\tilde{x}, \tilde{v}, \theta, m) = \mathcal{N}(\tilde{f}^{(v)}, \tilde{\Sigma}^{(v)})$$
$$p(\mathcal{D}\tilde{x}|x, \tilde{v}, \theta, m) = \mathcal{N}(\tilde{f}^{(x)}, \tilde{\Sigma}^{(x)}) \quad (9.14)$$

These probability densities are encoded by their covariances $\tilde{\Sigma}^{(u)}$ or precisions $\tilde{\Pi}^{(u)} := \tilde{\Pi}^{(u)}(x, v, \gamma)$ with precision parameters $\gamma \subset \varphi$ that control the amplitude and smoothness of the random fluctuations. Generally, the covariances factorise; $\tilde{\Sigma}^{(u)} = V^{(u)} \otimes \Sigma^{(u)}$ into a covariance proper and a matrix of correlations $V^{(u)}$ among generalised fluctuations that encodes their smoothness. Given this generative model, we can now write down the energy as a function of the conditional means, which has a simple quadratic form (ignoring constants)

$$\mathcal{L} = \frac{1}{2}\tilde{\varepsilon}^{(v)T}\tilde{\Pi}^{(v)}\tilde{\varepsilon}^{(v)} - \frac{1}{2}\ln|\tilde{\Pi}^{(v)}|$$
$$+ \frac{1}{2}\tilde{\varepsilon}^{(x)T}\tilde{\Pi}^{(x)}\tilde{\varepsilon}^{(x)} - \frac{1}{2}\ln|\tilde{\Pi}^{(x)}|$$
$$+ \frac{1}{2}\tilde{\varepsilon}^{(\varphi)T}\tilde{\Pi}^{(\varphi)}\tilde{\varepsilon}^{(\varphi)} - \frac{1}{2}\ln|\tilde{\Pi}^{(\varphi)}| \quad (9.15)$$
$$\tilde{\varepsilon}^{(v)} = \tilde{s} - \tilde{f}^{(v)}$$
$$\tilde{\varepsilon}^{(x)} = \mathcal{D}\tilde{\mu}^{(x)} - \tilde{f}^{(x)}$$
$$\tilde{\varepsilon}^{(\varphi)} = \tilde{\mu}^{(\varphi)} - \tilde{\eta}^{(\varphi)}$$

Here, the auxiliary variables $\tilde{\varepsilon}^{(j)} : j = v, x, \varphi$ are prediction errors for sensory data, the motion of hidden states and parameters respectively. The predictions for the states are $\tilde{f}^{(u)}(\mu)$ and the predictions for the parameters are the prior expectations $\tilde{\eta}^{(\varphi)}$. Equation (9.16) assumes flat priors on the states and that priors $p(\varphi|m) = \mathcal{N}(\tilde{\eta}^{(\varphi)}, \tilde{\Sigma}^{(\varphi)})$ on the parameters are Gaussian, where κ is the precision on the motion of the parameter (see Eq. (9.9)).

9.2.3.1 Perceptual Inference and Predictive Coding

Usually, these models are cast in hierarchical form to make certain conditional independences explicit. Hierarchical forms may look more complicated but they are simpler than the general form above. They are useful because they provide an empirical Bayesian perspective on inference and learning that may be exploited by the brain. Hierarchical dynamic models have the following form

$$s = f^{(1,v)}(x^{(1)}, v^{(1)}, \theta) + \omega^{(1,v)}$$
$$\dot{x}^{(1)} = f^{(1,x)}(x^{(1)}, v^{(1)}, \theta) + \omega^{(1,x)}$$
$$\vdots$$
$$v^{(i-1)} = f^{(i,v)}(x^{(i)}, v^{(i)}, \theta) + \omega^{(i,v)}$$
$$\dot{x}^{(i)} = f^{(i,x)}(x^{(i)}, v^{(i)}, \theta) + \omega^{(i,x)}$$
$$\vdots$$
(9.16)

The random terms $\omega^{(i,u)}$ are conditionally independent and enter each level of the hierarchy. They play the role of observation error or noise at the first level and induce random fluctuations in the states at higher levels. The causes $v = v^{(1)} \oplus v^{(2)} \oplus \cdots$ link levels, whereas the hidden states $x = x^{(1)} \oplus x^{(2)} \oplus \cdots$ link dynamics over time. In hierarchical form, the output of one level acts as an input to the next. This input can enter nonlinearly to produce quite complicated generalised convolutions with deep (hierarchical) structure. If we substitute Eq. (9.16) into the recognition dynamics of Eq. (9.8) (ignoring the derivatives of curvatures and state-dependent noise), we get the following hierarchical message passing scheme

$$\dot{\tilde{\mu}}^{(i,v)} = \mathcal{D}\tilde{\mu}^{(i,v)} + \tilde{f}_{\tilde{v}}^{(i,v)T}\xi^{(i,v)} + \tilde{f}_{\tilde{v}}^{(i,x)T}\xi^{(i,x)} - \xi^{(i+1,v)}$$
$$\dot{\tilde{\mu}}^{(i,x)} = \mathcal{D}\tilde{\mu}^{(i,x)} + \tilde{f}_{\tilde{x}}^{(i,v)T}\xi^{(i,v)} + \tilde{f}_{\tilde{x}}^{(i,x)T}\xi^{(i,x)} - \mathcal{D}^T\xi^{(i,x)}$$
$$\xi^{(i,v)} = \tilde{\Pi}^{(i,v)}\tilde{\varepsilon}^{(i,v)}$$
$$\xi^{(i,x)} = \tilde{\Pi}^{(i,x)}\tilde{\varepsilon}^{(i,x)}$$
$$\tilde{\varepsilon}^{(i,v)} = \tilde{\mu}^{(i-1,v)} - \tilde{f}^{(i,v)}$$
$$\tilde{\varepsilon}^{(i,x)} = \mathcal{D}\tilde{\mu}^{(i,x)} - \tilde{f}^{(i,x)}$$
(9.17)

In neural network terms, Eq. (9.17) suggests that error-units receive messages from the states in the same level and the level above. Conversely, state-units are driven by error-units in the same level and the level below, were $\tilde{f}_{\tilde{w}}^{(i,u)} : u = v, x$ are the forward connection strengths to the state unit representing $w \in \tilde{v}, \tilde{x}$. Critically, recognition requires only the (precision-weighted) prediction error from the lower level $\xi^{(i,v)}$ and the level in question, $\xi^{(i,x)}$ and $\xi^{(i+1,v)}$ (see Fig. 9.1 and Mumford 1992). These constitute bottom-up and lateral messages that drive conditional expectations $\tilde{\mu}^{(i,u)}$ towards a better prediction, which reduces the prediction error in the level below. These top-down and lateral predictions correspond to $\tilde{f}^{(i,u)}$. This is the essence of recurrent message passing between hierarchical levels to optimise free-energy or suppress prediction error (see Friston 2008 for a more detailed discussion). This scheme can be regarded as generalisation of linear predictive coding (Rao and Ballard 1999).

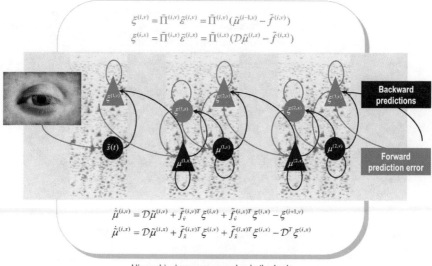

Fig. 9.1 Schematic detailing the neuronal architectures that could encode conditional expectations about the states and parameters of (three levels of) a hierarchical model of the world. This schematic shows the speculative cells of origin of forward driving connections that convey prediction error from a lower area to a higher area and nonlinear backward connections that are used to construct predictions. These predictions try to explain input from lower areas by suppressing prediction error. In this scheme, the sources of forward connections are superficial pyramidal cells and the sources of backward connections are deep pyramidal cells. The differential equations relate to the optimisation scheme detailed in the main text. The state-units and their efferents are in *black* and the error-units in *red*; with causal states on the right and hidden states on the left. For simplicity, we have assumed the output of each level is a function of, and only of, hidden states. This induces a hierarchy over levels and, within each level, a hierarchical relationship between states, where causes predict the motion of hidden states

Equation (9.17) shows that precision effectively sets the synaptic gain of error-units to their top-down and lateral inputs. Therefore, changes in precision $\tilde{\Pi}^{(i,u)}$ correspond to neuromodulation of error-units encoding precision-weighted prediction error $\xi^{(i,u)}$. This translates as an optimisation of synaptic gain of principal (superficial pyramidal) cells that elaborate prediction error (see Mumford 1992; Friston 2008) and fits comfortably with (among other things) the modulatory effects of dopaminergic and cholinergic neurotransmission. We will exploit this interpretation in the final section. We next consider learning.

9.2.3.2 Perceptual Learning and Associative Plasticity

Perceptual learning corresponds to optimising the first-order parameters $\theta \subset \varphi$. Equation (9.9) describes a process that is remarkably similar to models of associative plasticity based on correlated pre and post-synaptic activity. This can be seen most easily by assuming an explicit form for the generating functions; for example (for a single parameter and ignoring high-order derivatives)

$$\begin{aligned} f_j^{(i,x)} &= \theta x_k^{(i)} \quad \Rightarrow \\ \dot{\mu}^{(\theta)} &= \mu'^{(\theta)} \\ \dot{\mu}'^{(\theta)} &= -\tilde{\mu}_k^{(i,x)T} \xi_j^{(i,x)} - \Pi^{(\theta)} \mu^{(\theta)} - \kappa \mu'^{(\theta)} \end{aligned} \quad (9.18)$$

Here $\mu^{(\theta)}$ is the connection strength mediating the influence of the k-th hidden state on the motion of the j-th, at hierarchical level $i = 1, 2, \ldots$. This strength changes in proportion to a 'synaptic tag' $\mu'^{(\theta)}$ that accumulates in proportion to the product of the k-th pre-synaptic input $\tilde{\mu}_k^{(i,x)}$ and post-synaptic response $\xi_j^{(i,x)}$ of the j-th error unit (first term of Eq. (9.18)). The tag is auto-regulated by the synaptic strength and decays with first-order kinetics (second and third terms respectively). Crucially, this activity-dependent plasticity rests on (precise) prediction errors that are accumulated by the 'tag'. This highlights the fact that learning (optimising synaptic efficacy) depends on an optimal level of precision encoded by the synaptic gain of error units. Similar equations can be derived for the optimisation of the gain or precision parameters $\gamma \subset \varphi$. However, in this work we will use fixed values and change them to simulate pathology. We conclude this section by examining the dynamics prescribing optimal action.

9.2.3.3 Action

Because action can only affect the free-energy through the sensory data, it can only affect sensory prediction error. If we assume that action performs a gradient descent on free-energy, it is prescribed by:

$$\begin{aligned} \dot{a} &= -\mathcal{F}_a \\ &= -\tilde{\varepsilon}_a^{(v)T} \xi^{(v)} \\ \tilde{\varepsilon}_a^{(v)} &= f_{\tilde{x}}^{(v)} \sum_i \mathcal{D}^{-i} \left(f_{\tilde{x}}^{(x)} \right)^{i-1} f_a^{(x)} \end{aligned} \quad (9.19)$$

The partial derivative of the error with respect to action is the partial derivative of the sensory samples with respect to action. In biologically plausible instances of this scheme, this partial derivative would have to be computed on the basis of a mapping from action to sensory consequences, which are usually quite simple; for example, activating an intrafusal muscle fibre elicits stretch receptor activity in the corresponding spindle (see Friston et al. 2010 for discussion).

9.2.3.4 Summary

In conclusion, we have established some simple dynamics for active inference that implement recognition or perceptual inference, learning and behaviour. However, we have said nothing about the form of models biological agents might call upon. In the next section, we turn to some fundamental questions about the nature of generative models underlying active inference and, in particular, the role of $f^{(x)}(x, v, \theta)$ in furnishing formal priors on the motion of hidden states in the world.

9.3 Priors and Policies

In this section, we focus on the equations of motion that constitute an agent's generative model of its world. In the previous section, we saw that every agent or phenotype can be regarded as a model of its environment (econiche and internal milieu). Mathematically, this model corresponds to the form of the equations of motion describing hidden states. If these forms are subject to selective pressure, we can regard evolution as optimising formal priors on the environmental dynamics to which each phenotype is exposed. Because these dynamics describe a flow through different states (i.e., state-transitions), they correspond to policies. This section tries to establish the different sorts of priors or policies that might have emerged at an evolutionary scale. It also tries to relate existing formulations (such as optimal control theory, dynamic programming and reinforcement learning) to the dynamical framework that ensues. Briefly, we will see that there are two fundamentally different sorts of policies one could entertain. The first class of (fixed-point) policies can be derived from vector calculus and equilibrium arguments about ensemble densities on the states agents occupy (Birkhoff 1931; Moore 1966; McKelvey and Palfrey 1995; Haile et al. 2008; see Eldredge and Gould 1972 for an evolutionary take on equilibria). These equilibria arguments suggest that the states that are most likely to be occupied (peaks of the ensemble density) require the local policy (flow) to have negative divergence. We will refer to this as the *divergence-constraint*. Mathematically, divergence measures the rate at which flow disperses or dispels a density at any particular point in state-space. This somewhat abstract treatment (and in particular the divergence-constraint) leads to putative policies that ensure attractive states are occupied with the greatest probability. Important examples of these value-based policies are considered in optimal control (Bellman 1952; Sutton and Barto 1981; Todorov 2006) and reinforcement learning (Rescorla and Wagner 1972; Watkins and Dayan 1992; Friston et al. 1994; Montague et al. 1995; Daw and Doya 2006; Daw et al. 2006; Dayan and Daw 2008; Niv and Schoenbaum 2008). This class of policies rests on assuming that all hidden states are equipped with a particular cost, which has the important implication that the optimal flow (prior or policy) has fixed-point attractors. These attract states to low-cost invariant sets; more formally global random attractors $\mathcal{A}(\omega)$, when considering random fluctuations $\omega \in \Omega$ on the states (Matheron 1975; Crauel and Flandoli 1994; Crauel 1999). One of the main purposes of this section is to suggest that although fixed-point policies may provide useful heuristics, they are not necessarily optimal or indeed tenable in a general (dynamical) setting. This is because the external and internal milieu is changing constantly and does not support fixed-point attractors in the state-space of any phenotype. Put simply, any agent that aspires to a fixed state is doomed, both ethologically and physiologically. To accommodate this, we introduce the notion of itinerant policies, whose implicit attractors are space filling and support wondering (possibly chaotic) trajectories or orbits (e.g., Maturana and Varela 1980; Haken 1983; Freeman 1994; Tsuda 2001; Tyukin et al. 2003; Tschacher and Haken 2007; Tyukin et al. 2009; Rabinovich et al. 2008). Put simply, this means an agent will move through its state-space, sampling different weakly attracting states (attractors in the Milnor sense; Tyukin et al. 2009; Colliaux et al. 2009 or attractor ruins; Rabinovich et al. 2008; Gros 2009) in an itinerant fashion.

The basic idea behind the construction of these itinerant policies (priors) rests on the destruction or vitiation of (weakly) attracting sets. We will focus on attractors that destroy themselves (autovitiate), when they have been occupied too long or other imperatives come into play. This sort of policy will be illustrated with a simple simulation of active inference that leads to exploration and exploitation, under physiologically plausible constraints. The associated model (agent) will be used in the next section to see what would happen if we confound its ability to infer and learn optimally.

9.3.1 Set-up and Preliminaries

The distinction between fixed-point and itinerant policies arises from the following distinction among different subsets of hidden states: $x \supseteq \{x^{(a)}, x^{(p)}, x^{(q)}\}$. This partition acknowledges the fact that, from the agent's perspective, there are two proper disjoint subsets of states. The first comprises those states that can be affected by action $x^{(a)} \subset x$; namely states that support the motion of effectors (e.g., motor plant) and causal (e.g., Newtonian) mechanics in the external milieu. We will call these *physical states*. The other subset $x^{(p)} \subset x \backslash x^{(a)}$ represents states in the internal milieu, which must be maintained within certain bounds (e.g., physiological states that determine interoceptive signals; Davidson 1993). To help remember what these refer to, we will call them *physiological states* and represent the bounds with an indicator or cost-function $c(x^{(p)}) = 0 : x^{(p)} \in \mathcal{A}^{(p)}$ that is zero on the interior of some bounded (attractive or low cost) set $\mathcal{A}^{(p)}$ and one otherwise. Note that the cost-function is defined only on the physiological states. Indeed, one could define the physiological states as the domain of the cost-function. We will use the notion of an indicator or cost-function extensively below for two reasons. First, it is the sort of constraint that can be specified epigenetically and is therefore consistent with the evolutionary perspective above (cf., Traulsen et al. 2006; Maynard Smith 1992). For example, it is not inconceivable that natural selection has equipped us with indicator functions that register when (inferred) blood sugar falls outside the normal 3.6 and 5.8 mM range. Second, utility, loss, or cost-functions are an integral part of optimal control in reinforcement learning and optimal decision (game) theory in economics (e.g., Shreve and Soner 1994; Camerer 2003; Coricelli et al. 2007; Johnson et al. 2007). The remaining hidden states will be called *manifold-states* $x^{(q)} = x \backslash \{x^{(a)}, x^{(p)}\} \subset x$ for reasons that will become clear later.

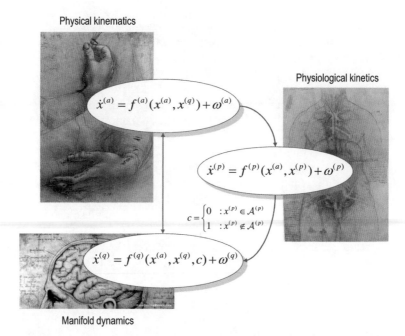

Fig. 9.2 Schematic showing the partition of hidden states into physical states, physiological states, and manifold-states. Physical states correspond, heuristically, to mechanics of the physical world, such as the movement of the motor plant and physical objects. The physiological states pertain to the internal milieu and exhibit kinetics that depend upon physical states. The manifold-states represent the remaining hidden states that govern causal dynamics in the sensorium. These affect (and can be affected by) the physical states but are only affected by the physiological states through indicator or cost-functions reporting whether the physiological states occupy a particular subset: $\mathcal{A}^{(p)}$. The stochastic differential equations describing each partition are a probabilistic summary of their dynamics. The *arrows* represent conditional dependencies and the schematic can be regarded as a Bayesian dependency graph

With this partition in place, we can now consider the conditional dependencies among the subsets. We will assume that physiological states depend on and only on themselves and physical states (e.g., changes in blood sugar after ingestion). The physical states depend upon themselves and manifold-states that shape the manifold that contains the flow of physical states (e.g., forces on manipulanda in the immediate environment). Finally, the manifold-states per se can be influenced by the physical states and physiological states, where the latter influence is mediated by a cost-function. The partition into physical and physiological states means that action cannot affect physiological states directly. This is important and respects the constraints biological agents evolve under. For example, no amount of voluntary (striatal) muscle activity can directly increase blood sugar, it can only do so vicariously by changing physical states that affect physiology. We can summarise these dependencies mathematically with the following equations of motion, which are shown as a dependency graph in Fig. 9.2.

9 Policies and Priors   251

$$f^{(x)} = \begin{bmatrix} f^{(a)}(x^{(a)}, x^{(q)}) \\ f^{(q)}(x^{(a)}, x^{(q)}, c) \\ f^{(p)}(x^{(a)}, x^{(p)}) \end{bmatrix} \tag{9.20}$$

$$c = \begin{cases} 0 & :x^{(p)} \in \mathcal{A}^{(p)} \\ 1 & :x^{(p)} \notin \mathcal{A}^{(p)} \end{cases}$$

These equations of motion are part of the agent's generative model and induce formal priors on state-transitions (i.e., a policy). Our objective now is to find constraints on their form that disclose the nature of implicit policies. Clearly, the only explicit constraint we have is the indicator or cost-function on physiological states. This defines the physiological states the agent expects to be in a priori. In what follows, we will use this cost-function in two distinct ways. First, we will use it to define low-cost attractors in state-space using equilibrium arguments. This requires a rather abstract formulation of the problem, which ignores the distinction between physical and physiological states and leads to conventional (fixed-point) policies. We then reinstate the partition and use indicator or cost-functions to engender flow in the physical space that destroys costly fixed-points in the physiological space. This leads to itinerant policies, which we will use to examine pathological policies in the last section.

9.3.2 Fixed-Point Policies: The Equilibrium Perspective

In this subsection, we will consider policies as prior expectations on flow that lead to low-cost equilibrium densities. This perspective provides a fundamental (divergence) constraint on local flow that can be exploited directly (or is met implicitly) in schemes based upon value; the path-integral of cost. However, to pursue this analysis we need to make a rather severe and implausible assumption. Namely, that we can ignore the conditional dependencies implicit in the partition above and assume that all states can be treated equally. This means the policy reduces to $f := f^{(x)}(x, v, \theta)$. With this simplifying assumption, one can appeal to standard results in vector calculus that describe the evolution of the probability density on the states the agent could occupy as a function of time. This is the ensemble density of the previous section. It can be regarded as either the probability distribution of an infinite number of copies of the agent, observed simultaneously. Alternatively, under ergodic assumptions, this is the same as the probability that an agent will be found in a particular state when observed at different times. This probability is also called the sojourn time and reflects the relative amount of time each state is occupied. The evolution of the ensemble density over time is described by the Fokker-Planck equation

$$\begin{aligned} \dot{p}(x|m) &:= \Lambda p \\ &= \nabla \cdot (\Gamma \nabla - f) p \\ &= \nabla \cdot \Gamma \nabla p - \nabla \cdot (pf) \\ &= \nabla \cdot \Gamma \nabla p - p \nabla \cdot f - f \cdot \nabla p \end{aligned} \tag{9.21}$$

Here Γ is half the amplitude (variance) of random fluctuations on the states. At equilibrium, $\dot{p}(\tilde{x}|m) = 0$ and

$$p(x|m) := p = \frac{\nabla \cdot \Gamma \nabla p - f \cdot \nabla p}{\nabla \cdot f} \quad (9.22)$$

Notice that as the divergence $\nabla \cdot f$ increases, the sojourn time (i.e., the proportion of time a state is occupied) falls. Crucially, at the peaks of the ensemble density, the gradient is zero and its curvature is negative, which means the divergence must be negative (from Eq. (9.22))

$$\left.\begin{array}{r} p > 0 \\ \nabla p = 0 \\ \nabla \cdot \nabla p < 0 \end{array}\right\} \Rightarrow \nabla \cdot f < 0 \quad (9.23)$$

This divergence-constraint simply says that any policy or flow must have negative divergence at (low cost) maxima of the equilibrium density. One can exploit this constraint by ensuring that all costly fixed-points have positive divergence. Essentially, this destroys any fixed-points in the environment by making them unstable. These policies are easy to construct. For example, the following (Newtonian) policy can be made to satisfy the divergence-constraint very simply by ensuring $\chi(c) \leq 0$, where

$$f = \begin{bmatrix} x' \\ -c\varphi_x(x) + \chi(c)x' \end{bmatrix} \Rightarrow \nabla \cdot f = \chi(c) \quad (9.24)$$

This flow (policy) describes the Newtonian motion of a unit mass in a potential energy well $\varphi(x, \theta)$, where cost plays the role of negative dissipation or friction (and vitiates fixed points in costly regions). Crucially, under this policy, divergence is a function of, and only of, cost. This means the associated ensemble density can only have maxima in regions, where $\chi(c) \leq 0$. Put simply, this ensures that agents are expelled from high-cost regions of state-space and get 'stuck' in attractive (flat) regions. We can illustrate this sort of policy by revisiting a benchmark problem in optimal control:

9.3.2.1 The Mountain-Car Problem

The mountain-car problem can be envisaged as follows: one has to move a car from the bottom of valley and keep it there. However, the car is too heavy to simply drive up the hill. This means that the target can only be accessed by starting on the opposite side of the valley to gain enough momentum to carry it up the other side. This represents an interesting problem, when considered in the state-space of position and velocity, $x, x' \in \tilde{x}$; the agent has to move *away* from the target location ($x = 1$) to attain its goal and execute a very circuitous movement (cf., avoiding obstacles). This problem can be specified with the following equations

9 Policies and Priors

$$g = \begin{bmatrix} \dot{\mathbf{x}} \\ \dot{\mathbf{x}}' \end{bmatrix}$$

$$\mathbf{f} = \begin{bmatrix} \mathbf{x}' \\ -\varphi_{\mathbf{x}}(\mathbf{x}) - \frac{1}{4}\mathbf{x}' + \sigma(a) \end{bmatrix} \qquad (9.25)$$

$$\varphi_{\mathbf{x}} = \begin{cases} 2\mathbf{x}+1 & :\mathbf{x} \leq 0 \\ \mathbf{x}^2(1+5\mathbf{x}^2)^{-3/2} + \mathbf{x}^4/16 & :\mathbf{x} > 0 \end{cases}$$

We have used bold to highlight the fact that the states and functions are the true values generating sensory data (as distinct from any hidden states assumed by a generative model of these data). Crucially, at $\mathbf{x} = 0$ the force on the car cannot be overcome by the agent, because a squashing function $-1 \leq \sigma(a) \leq 1$ is applied to action to prevent it being greater than one. Divergence-based policies provide a remarkably simple and effective solution to problems of this sort and can be implemented under active inference using policies with the form of Eq. (9.24) (see Friston et al. 2010 for more details). These policies are entailed by the agent's generative model of its sensory inputs. For example,

$$f^{(v)} = \begin{bmatrix} x \\ x' \end{bmatrix}$$

$$f^{(x)} = \begin{bmatrix} x' \\ -c\varphi_x(x) + \chi(c)x' \end{bmatrix}$$

$$\varphi_x = \theta_1(x - \theta_2) \qquad (9.26)$$

$$\chi = \frac{1}{4} - 32(1-c)$$

$$c = \begin{cases} 0 & :|x-1| \leq \Delta \\ 1 & :|x-1| > \Delta \end{cases}$$

Figure 9.3 shows how paradoxical but adaptive behaviour (e.g. moving away from a target to ensure it is secured later) emerges from these simple priors on the motion of hidden states. This example used $\Delta = \frac{1}{16}$, $\theta_1 \approx 0.6$ and $\theta_2 \approx -0.2$. These simulations of active inference involve integrating the states in the environments (e.g., Eq. (9.25)) and the agent (Eqs. (9.17) and (9.19)) simultaneously as described in Appendix B.

Clearly, the construction of policies that use divergence to vitiate costly fixed-points rests on knowing the form of the policy. In principle, this is no problem, because we are talking about the agent's prior expectations or model of its environment. At no point do we assume that any of the states in the generative model actually exist. For example, the true landscape that exerts forces on a mountain car (Eq. (9.25) and Fig. 9.3) is much more complicated than the agent's model of this landscape, which is a simple quadratic approximation (Eq. (9.26)). This highlights the fact that our expectations about the world and its actual causal structure do not have to be formally equivalent to support adaptive policies. However, it is clearly important that there is a sufficient homology between modelled and experienced causal structure, otherwise the agent will be perpetually surprised by 'obstructions' to its path. This begs the question as to whether there is any universal form of policy

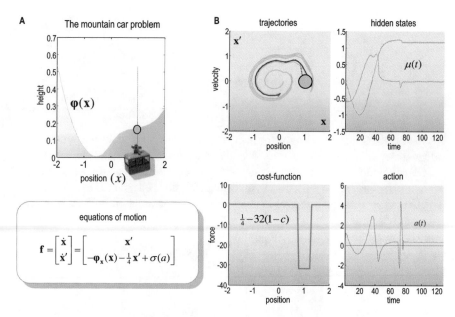

Fig. 9.3 This figure shows how paradoxical but adaptive behaviour (e.g., moving away from a target to ensure it is secured later) emerges from simple priors on the (Newtonian) motion of hidden states in the world. **A**: The *upper panel* shows the landscape or potential energy function (with a minimum at position $x = -0.5$) that exerts forces on a mountain car. The car is shown at the target position on the hill at $x = 1$, indicated by the *cyan ball*. The equations of motion of the car are shown below the figure. Crucially, at $x = 0$ the agent cannot overcome the force on the car because a squashing function $-1 \leq \sigma(a) \leq 1$ is applied to action to prevent it being greater than one. This means that the agent can only access the target by starting halfway up the left hill to gain enough momentum to carry it up the other side. **B**: The results of active inference under priors that destabilise fixed-points outside the target domain. The priors are encoded in a cost-function $c(x)$ (*lower left*), which acts like negative friction. When 'friction' is negative the car expects to go faster. The inferred hidden states (*upper right*: position in *blue* and velocity in *green*) show that the car explores its landscape until it encounters the target. At this point, friction increases (i.e., cost decreases) dramatically to prevent the car from escaping the target (by falling down the hill). The ensuing trajectory is shown in *blue* (*upper left*) in the phase-space of position and velocity. The *paler lines* provide exemplar trajectories from other trials with different starting positions. In the real world, friction is constant. However, the car 'expects' friction to change with its position, enforcing exploration or exploitation. These expectations are fulfilled by action (*lower right*)

that would comply with the divergence-constraint. An example of a universal form is afforded by policies based upon value.

9.3.2.2 Value-Based Policies

In what follows, we consider the key notion of value $V(x)$ as a function of state-space that reports the relative probability or sojourn time a state is occupied at equilibrium. Let flow be decomposed into the gradient of value and an orthogonal component $f = \nabla V + \zeta$, such that $\nabla V \cdot \zeta = 0$, where the value of a state is proportional to its log-sojourn time or density at equilibrium

$$V = \Gamma \ln p \quad \Rightarrow \quad p\nabla V = \Gamma \nabla p$$
$$p := p(x|m) = \exp(V/\Gamma) \tag{9.27}$$

Equation (9.27) implies (intuitively) that if ζ is orthogonal to value (log-density) gradients, it must also be orthogonal to the density gradients per se: $\nabla V \cdot \zeta = 0 \Rightarrow \nabla p \cdot \zeta = 0$. If we now substitute Eq. (9.27) into the Fokker-Planck equation (9.21) and solve for the equilibrium density that satisfies $\Lambda p = 0$, we obtain (using standard results from vector calculus)

$$\Lambda p = \nabla \cdot (p\nabla V) - \nabla \cdot (p\nabla V) - \nabla \cdot p\zeta = 0 \quad \Rightarrow$$
$$\nabla \cdot p\zeta = p\nabla \cdot \zeta - \zeta \cdot \nabla p = 0 \quad \Rightarrow \quad \nabla \cdot \zeta = 0 \tag{9.28}$$

This means that the orthogonal flow $\zeta = \nabla \times W$ is divergence-free and can be expressed in terms of a vector-potential $W(x)$. This is just an example of the Helmholtz decomposition (also known as the fundamental theorem of vector calculus). It means we can express any policy as the sum of irrotational (curl-free) ∇V and solenoidal (divergence-free) $\nabla \times W$ components. If the two components are orthogonal, then the scalar-potential $V(x)$ defines the equilibrium density and its attracting states; that is, the scalar-potential is value. This equivalence rests on the orthogonality condition $\nabla V \cdot \zeta = 0$, which we will call the *curl-constraint*. Under this constraint, curl-free flow prescribed by value counters the change in the equilibrium density due to random fluctuations. Conversely, divergence-free flow follows isoprobability contours and does not change the equilibrium density. Finally, it is easy to show that value is a Lyapunov function for policies that conform to the curl-constraint

$$f = \nabla V + \zeta : \nabla V \cdot \zeta = 0$$
$$= \nabla V + \nabla \times W \tag{9.29}$$
$$\dot{V}(x(t)) = \nabla V \cdot f = \nabla V \cdot \nabla V + \nabla V \cdot \zeta = \nabla V \cdot \nabla V \geq 0$$

Lyapunov functions increase (or decrease) with time and are used to prove the stability of fixed-points in dynamical systems. This means every policy that satisfies the curl-constraint increases its value as a function of time. The notion of a Lyapunov function is introduced here, because of its relationship to value or attraction in optimal control and decision (game) theory, respectively:

9.3.2.3 Optimal Control and Reinforcement Learning

In optimal control theory and its ethological variants (i.e., reinforcement learning), adaptive behaviour is formulated in terms how agents navigate state-space to access sparse rewards and avoid costly regimes. The aim is to find a (proximal) policy that attains long-term (distal) rewards. In terms of the above, a policy $f = \nabla V + \zeta$ is specified via the scalar-potential or value $V(x)$ also known as (negative) cost-to-go.

In this sense, value is sometimes called a navigation function. The value-function is chosen to minimise expected cost. More formally, the cost-to-go of a state is the cost expected over future states. In the deterministic limit $\Gamma \to 0$, this is just the path integral of cost

$$V(x) = -\int_t^\infty d\tau c(x(\tau)) \Rightarrow$$
$$\dot{V}(x(t)) = c(x) = \nabla V \cdot f \geq 0 \tag{9.30}$$

This says that cost is the rate of increase in value (the Lyapunov function). Crucially, Eq. (9.30) shows that the maxima of the equilibrium density can only exist where cost is zero, at which point value stops increasing and the divergence-constraint is satisfied

$$\begin{aligned}\nabla V(x) &= 0 \Rightarrow c(x) = 0\\ \nabla \cdot \nabla V(x) &< 0 \Rightarrow \nabla \cdot f < 0\\ \nabla \cdot f &= \nabla \cdot \nabla V + \nabla \cdot \zeta\\ &= \nabla \cdot \nabla V\end{aligned} \tag{9.31}$$

Heuristically, we can regard value as guiding flow towards points where there is no cost (i.e., no gradients). This means that, in principle, we have a way to prescribe equilibria with maxima (attracting fixed-points) that are specified with a cost-function. Equation (9.30) shows that the cost-function can be derived easily, given the policy and implicit value-function. However, to specify a policy with cost, we have to derive the value-function from the cost-function; that is, solve Eq. (9.30) for value. This is the difficult problem optimal control and value-learning deal with:

In the deterministic limit, the equilibrium density becomes a point mass at the maximum of the value function (see Eq. (9.27)). This is the fixed-point to which all trajectories are attracted. Value-based policies represent universal solutions that do not require any knowledge about the form of the equations of motion generating sensory contingencies. However, this is also their weakness, because we require the solution of Eq. (9.30) under unknown constraints. This leads to the celebrated Hamilton-Jacobi-Bellman equation in optimal control theory (Bellman 1952), for which there is no general solution. However, there is a vast literature on approximate solutions based upon dynamic programming and stochastic iteration. Variants of these schemes appear as temporal difference models (Sutton and Barto 1981) and Q-learning (Watkins and Dayan 1992) in machine learning, and as heuristics in psychological studies of reinforcement learning (Rescorla and Wagner 1972). Almost invariably, these approximate solutions rest on updating explicit representations of the value-function using a prediction error on cost (or reward). This is called a reward prediction error, which we will return to in the discussion. We will not pursue this enormous field here for one simple reason: fixed-point policies are not solutions to real-world problems. This is because there are no valuable fixed-points in dynamical systems: an organism can only occupy a fixed-point when it is frozen or petrified (i.e., dead).

Furthermore, from a technical point of view, value-based (fixed-point) policies are incomplete. This is because real-world (non-abstract) systems do not satisfy the curl-constraint: Although, the Helmholtz decomposition provides a universal form for policies, with curl and divergence-free components, there is no fundamental lemma or requirement for these components to be orthogonal. This means the scalar-potential is not necessarily a Lyapunov function (i.e., a value-function) or a useful navigation function (see Eq. (9.29)). The interactions among states that violate the curl-constraint are implicit in the conditional dependencies in Eq. (9.20) (for nonlinear equations of motion). In the next subsection, we relax the simplifying assumptions necessary for the abstract formulations used in economics and reinforcement learning and turn to itinerant policies.

9.3.3 Itinerant Policies

In this subsection, we look at functional forms for policies using the (non-abstract) set up that distinguishes between physical, physiological and other hidden states. Here, we consider attractive states that are not fixed-points but bounded sets that arise from itinerant (wandering or searching) dynamics. This is sensible, given the nature of the environment, and speaks to optimising space-filling attractors that ensure low cost equilibria.

The importance of itinerancy has been articulated many times in the past (see Nara 2003), particularly from the perspective of computation and autonomy (see van Leeuwen 2008; with a focus on Milnor attractors). It has also been considered formally in relation to cognition (e.g., Gros 2009, with a focus on attractor relics, ghosts or ruins) and implicitly in ethology (e.g., Panksepp et al. 1984). The ethological perspective is useful here because it suggests that some species are equipped with prior expectations that they will engage in exploratory or social play, For example, 'rough and tumble play' may be a fundamental form of play comprising a unique set of behaviours that can be distinguished from aggression and other childhood activities. Indeed, there is growing interest in understanding brain dynamics per se in terms of itinerancy and metastability (e.g., Jirsa et al. 1994; Breakspear and Stam 2005; Bressler and Tognoli 2006). Tani et al. (2004) consider itinerant dynamics in terms of bifurcation parameters that generate multiple goal-directed actions on the behavioural side, and optimisation of the same parameters when recognising actions. They provide a series of elegant robotic simulations to show generalisation by learning with this scheme. See also Herrmann et al. (1999) for interesting simulations of itinerant exploration, using just prediction errors on sensory samples over time.

We will see below that it is fairly easy to construct itinerant policies. Furthermore, they can have constant (negative) divergence at all points in state-space. This means that their equilibria depend on the divergence-free component of flow (i.e., the component that is discounted by the curl-constraint in fixed-point policies). Although there may not be a universal form for itinerant policies, the principles upon which they are based may be universal.

One universal principle (which we exploit here) is the vitiation or destruction of costly attractors. A key difference between general vitiative mechanisms and the divergence-based vitiation above is that the destruction of costly attractors can be state and time-dependent. This idea appears in several guises and has found important applications in a number of domains. For example, it is closely related to the notion of autopoiesis and self-organisation in situated (embodied) cognition (Maturana and Varela 1980). It is formally related to the destruction of gradients in synergetic treatments of intentionality (Tschacher and Haken 2007). Mathematically, it is finding a powerful application to universal optimisation schemes (Tyukin et al. 2003) and, indeed, as models of perceptual categorisation (Tyukin et al. 2009). The dynamical phenomena, upon which these schemes rest, involve an itinerant wandering through state-space along heteroclinic channels (orbits connecting different fixed-points). Crucially, these attracting sets are weak (Milnor) attractors or attractor ruins that expel the state until it finds the next weak attractor or ruin. The result is a sequence of transitions through state-space that, in some instances, can be stable and repeating. The resulting stable heteroclinic channels have already been proposed as a metaphor for neuronal dynamics and underlying cognitive processing (Rabinovich et al. 2008). Furthermore, the notion of Milnor or ruined attractors underlies much of the technical and cognitive literature on itinerant dynamics. For example, Tyukin et al. (2009) can explain "a range of phenomena in biological vision, such as mental rotation, visual search, and the presence of multiple time scales in adaptation" using the concept of weakly attracting sets. It is this sort of policy we exploit in the final part of this section.

9.3.3.1 Itinerant Control and Autovitiation

The basic idea is to construct a policy (equations of motion) in which costly states in the physiological subspace change the manifold on which the physical states are evolving. In principle, the only ergodic solution, under this sort of policy, is one in which an attractor (manifold) in the physical subspace induces a low-cost attractor in the physiological subspace. Clearly, this rests upon the existence of such solutions. The mathematical treatment of the existence of these solutions is not necessarily simple. Indeed, it is only recently that the conditions for the existence of stable heteroclinic channels have been established (Rabinovich et al. 2008). Furthermore, even the existence of weakly attracting (Milnor) sets presents some deep challenges (see Tyukin et al. 2003). Generally, attractors are invariant sets that attract states from their neighbourhood, known as a basin of attraction (like a pudding basin that collects its contents at its base). Milnor attractors generalise this notion so that the basin of attraction is not required to be in the neighbourhood of the attractor (like a pudding basin or sieve 'riddled' with holes). This allows the states to escape the attractor when subject to small random fluctuations (like shaking the pudding basin). Attractor ruins result from changing the manifold to destroy an attractor but preserve its characteristic ability to attract trajectories (like a basin with a hole at the base, from which its contents can escape slowly). A key distinction between different sorts of itinerancy is based on whether the manifold supporting itinerant flow is fixed or changing. Milnor attractors and attractor ruins support itinerant dynamics with Type I complexity (Friston 2000); that is, the manifold is invariant. Conversely, when dynamical systems are coupled to each other, the states of one system can change the manifold (topology or shape of the pudding basin) of another, leading to Type II complexity (Friston 2000). This sort of itinerancy rests on the construction (autopoiesis) and destruction (autovitiation) of attractors in one subspace by changes in the states of another. This is the mechanism we will pursue, given the partition in Eq. (9.20).

We will forego further mathematical discussion and try to illustrate the basic idea with a simple example. This example has been chosen because it embodies autovitiation using intuitive constructs from neurobiology. Consider the following policy

$$\begin{aligned} f^{(x)} &= \begin{bmatrix} f^{(a)} \\ f^{(q)} \end{bmatrix} \\ f^{(a)} &= f^{(a,k)}\left(x^{(a)}\right) : k = \arg\max_i x_i^{(q)} \\ f_i^{(q)} &= h\left(x^{(a)}, x^{(q)}\right) : x^{(a)} \notin \mathcal{A}_i^{(a)} \\ f_i^{(q)} &< h\left(x^{(a)}, x^{(q)}\right) : x^{(a)} \in \mathcal{A}_i^{(a)} \end{aligned} \quad (9.32)$$

This policy describes coupled nonlinear systems in physical $x^{(a)}$ and manifold-subspaces $x^{(q)} = [x_1^{(q)}, \ldots, x_K^{(q)}]$. Physical flow is 'selected' by the (k-th) manifold-state with the highest value, where each alternative flow $f^{(a,k)}(x)$ has a unique attractor $\mathcal{A}_k^{(a)}$. More formally, for all real $t > T$ there exists a time $T \in \mathfrak{R}^+$ for which $x(t)^{(a)} \in \mathcal{A}_i^{(a)}$, under $f^{(a,i)}(x) : i \subset 1, \ldots, K$. For each attractor there is a corresponding manifold-state. These change according to some arbitrary function $h(x^{(a)}, x^{(q)})$. Crucially, all the manifold-states experience the same change unless the physical-state occupies the attractor selected by the manifold-state. In this instance, the manifold states decreases, relative to its competitors. The attractor is vitiated when its manifold-state ceases to be the largest and another physical flow supervenes. This is a simple and fairly universal scheme that ensures all the attractors are visited at some point. The key aspect of these schemes is that attractors are destroyed when occupied.

There are clearly many ways that we could have constructed itinerant schemes to illustrate this sort of policy. We elected to use competition among attractors in the physical state-space for several reasons. First, dynamics of this sort can be cast in the abstract form required for conventional value-based policies. This is because the system will visit a discrete number of attractive states $\mathcal{A}_i^{(a)} : i \in 1, \ldots, K$ with well defined probabilities. This will be pursued in a later communication using model-based reinforcement learning. Second, the, saltatory migration from one attractor (pattern) to the next is a ubiquitous phenomenon in neuronal dynamics; manifest as synfire chains (Abeles et al. 2004), reproducible patterns in neuronal avalanches (Pasquale et al. 2008) and 'loss-less' saltatory transitions observed in local field potentials (Thiagarajan et al. 2010). Functionally, the use of attractors with associated basins of attraction, provides a generic way of 'tiling' any space and bears a formal resemblance to classical receptive fields in vision or, indeed, place-cells in spatial navigation (O'Keefe and Dostrovsky 1971; Sheynikhovich et al. 2009; Robbe and Buzsáki 2009). This means that itinerant policies may furnish a model of saccadic eye movements during exploration of visual scenes (Chen and Zelinsky 2006) or in the context of foraging and spatial exploration. In what follows, we will adopt the second heuristic and associate the attractors $\mathcal{A}_i^{(a)}$ with $i \in 1, \ldots, K$ locations in something like a Morris water-maze (Morris 1984). To emulate conditioned place-preference (e.g., Seip et al. 2008), we have to augment the itinerant scheme above (Eq. (9.32)) with physiological states that can moderate the vitiation of rewarding attractors. For simplicity, we will deal with just four locations and two physiological states.

9.3.3.2 The Generative Model

The particular policy we will focus on for the remainder of this paper is part of the following generative model

$$s = f^{(v)} + \omega^{(v)}$$
$$\dot{x} = f^{(x)} + \omega^{(x)}$$

$$f^{(v)} = \begin{bmatrix} x^{(a)} \\ x'^{(a)} \\ x^{(p)} \end{bmatrix}$$

$$f^{(x)} = \begin{bmatrix} f^{(a)} \\ f'^{(a)} \\ f^{(p)} \\ f^{(q)} \end{bmatrix} \tag{9.33}$$

$$= \begin{bmatrix} x'^{(a)} \\ 8(\alpha_k - x^{(a)}) - 4x'^{(a)} \\ \theta^T \beta(x^{(a)}) - x^{(p)} \\ \theta c(x^{(p)}) - 4\beta(x^{(a)}) - \sum_i x_i^{(q)} \end{bmatrix} \Rightarrow \nabla \cdot f = -4 - 1 - K$$

$$\beta_i = \begin{cases} 0 & :|\alpha_i - x^{(a)}| \geq \Delta \\ 1 & :|\alpha_i - x^{(a)}| < \Delta \end{cases} \quad c_j = \begin{cases} 0 & :x_j^{(p)} \geq \tau \\ 1 & :x_j^{(p)} < \tau \end{cases}$$

$$i \in 1, \ldots, K, \quad j \in 1, \ldots, J, \quad k = \arg\max_i x_i^{(q)}$$

To complete the specification of this model, we will use the following values (unless otherwise stated): A sensory log-precision of eight $\Pi^{(v)} = 8 \Leftrightarrow \omega_i^{(v)} \sim N(0, e^{-8})$, a log-precision of four or six on the motion of hidden states: $\Pi^{(a)} = 4$, $\Pi^{(p)} = 6$, $\Pi^{(q)} = 4$, a spatial threshold of $\Delta = \frac{1}{8}$ and a physiological threshold of $\tau = \frac{1}{8}$.

The sensory mapping $f^{(v)}$ means that the agent has access to its position and velocity and (in this example) two physiological states $x^{(p)} = [x_1^{(p)}, x_2^{(p)}]^T$ (e.g., blood sugar and osmolarity). The second line describes the policy in terms of formal expectations about the generalised motion of hidden states: The agent assumes

9 Policies and Priors

that it pulled to the location, α_k under a degree of friction. This location is the point attractor $\alpha_k \subseteq \mathcal{A}_k^{(a)}$ associated with the highest manifold-state. This means we can regard $x_i^{(q)}$ as the attractiveness of its corresponding location. The manifold-states are subject to three influences, the third is just a non-specific return to zero (mediated by the sum over physiological states). The second mediates itinerancy by vitiating the attractiveness of fixed-points when the agent is in their neighbourhood; i.e., $|\alpha_i - x^{(a)}| < \Delta$. The first makes some locations progressively more attractive, when the cost-function $c(x^{(p)})$ reports that a physiological state has fallen below threshold, $\tau = \frac{1}{8}$. This cost-dependent attractiveness depends on parameters θ_{ij} that encode an association between the j-th physiological-state and the i-th location. These parameters also mediate an increase in the physiological-state—a reward—when the location is occupied, as reported by the indicator function $\beta(x^{(a)})$. In the absence of any reward, the physiological states simply decay with first-order kinetics.

These dynamics mean that when a physiological state falls below threshold this costly state is reported by a (vector) cost-function. This increases the attraction of locations in proportion to a parameterised association between each location and the costly physiological state (cf., Drive Reduction Theory; Hull 1943). The attractiveness of the appropriate location increases until it supervenes over remaining locations, at which point it draws the agent towards it. When the agent is sufficiently close, the physiological state is replenished and the agent is rewarded. This construction of interdependent physical, physiological and manifold dynamics ensures that no physiological state will remain below threshold for long. The ensuing physiological homeostasis depends on physiological imperatives vitiating (non-rewarding) physical attractors. In the absence of any cost (i.e., all physiological states are above some lower bound) all locations will compete with each other, until they are all visited in turn. This is a simple example of a system that shows cost-dependent heteroclinic channels which, in ethological terms includes both exploration and exploitation (e.g., Nowak and Sigmund 1993). Note that the divergence of this policy is a negative constant (see Eq. (9.33)). This means that the self-organising dynamics conform to the divergence constraint but are mediated by changes in divergence-free flow.

9.3.3.3 The Generative Process

Hitherto, we have described the policy as if it were a description of a real environment. However, the policy is just the agent's fantasy about an unknown environment. Crucially, this model is can be much more structured than the environment in which the agent is immersed. The actual generative process we will use can be written as follows.

$$\mathbf{f}^{(v)} = \begin{bmatrix} \mathbf{x}^{(a)} \\ \mathbf{x}'^{(a)} \\ \mathbf{x}^{(p)} \end{bmatrix}$$

$$\mathbf{f}^{(x)} = \begin{bmatrix} \mathbf{x}'^{(a)} \\ a - 2\mathbf{x}^{(a)} - 4\mathbf{x}'^{(a)} \\ \theta\beta(\mathbf{x}^{(a)}) - \mathbf{x}^{(p)} \end{bmatrix} \quad (9.34)$$

Where $\omega_i^{(u)} \sim N(0, e^{-16}) : u = v, x$. Here, the only forces acting upon the agent are those that it generates itself with action. In other words, although the agent has a concept of fixed-points to which it is variously attracted, the environment per se has no such attractors (other than a fixed-point at $\mathbf{x} = 0$). However, a number of the locations do deliver rewards. The mapping between these locations and the rewards is encoded by the (unknown) parameters $\theta_{ij} \in \{0, 1\}$. These play the same role as the parameters of the agent's generative model. If the true parameters and those used by the agent are the same, then the agent will happily navigate its environment alternately visiting rewarding locations to replenish its physiology (e.g., eating and drinking at different locations). However, to achieve this it has to learn the correct parameters. Crucially, this learning is purely perceptual and driven by the prediction errors established by conditional expectations about physiological rewards at every location. This is a key attribute of the current scheme and highlights the critical role of perceptual learning (parameter optimisation) in acquiring and maintaining appropriate policies (cf., conditioned place-preference in animal studies; Seip et al. 2008). We will return to this in the last section.

In summary, we have described an itinerant policy in terms of a generative model that prescribes the motion of physical and physiological states and how they couple to each other. Under active inference, this policy will enslave action to fulfil implicit prior expectations, under the constraints afforded by the real generative process in the environment. To illustrate this, we integrated the differential equations describing active inference from the first section, using the generative process and model above (Eqs. (9.33) and (9.34)). In this example, we used the correct mapping between rewards and locations ($\theta_{ij} = \boldsymbol{\theta}_{ij}$) such that the first location (upper right) replenished the first physiological state and the second location (lower left) replenished the second physiological state. The resulting behaviour is shown in Fig. 9.4. The upper left panel shows the predicted sensory input and its associated prediction errors (dotted red lines). This sensory input corresponds to the position and motion of the agent in two dimensions and the two physiological states. The underlying conditional expectations of these hidden states, which include the manifold-states, are shown on the upper right. The corresponding physical trajectory is shown on the lower left superimposed on the four attractor locations (cyan circles). This trajectory was driven purely by active inference, with the action controlling forces in two dimensions (shown in the lower right panel). The trajectory here shows that the two rewarding locations (upper right and lower left) are visited most frequently, with occasional excursions to the remaining two locations. The numbers by each location represent the percentage of time spent within $\Delta = \frac{1}{8}$ of the location.

Figure 9.5 provides a more detailed description of the conditional expectations about the physiological and manifold (internal) states in the upper panel and the true physiological states in the lower panels. The upper panel shows the expected physiological states (solid lines) and the manifold-states (broken lines). The key thing to take from these time courses is the recurrent build-up and self-destruction of manifold-states, as each attracting fixed-point is visited and consequently rendered less attractive. Crucially, the attractors delivering rewards become more attractive

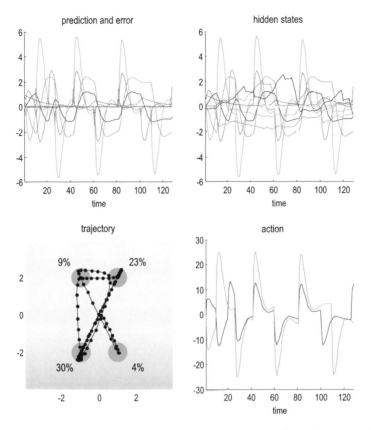

Fig. 9.4 Conditional expectations and behaviour under an itinerant policy. The *upper right panel* shows the conditional expectations of hidden states, while the *upper left panel* shows the corresponding predictions of sensory input (*solid lines*) and prediction errors (*dotted red lines*). Action tries to suppress these prediction errors and is shown on the *lower right*. These action variables exert forces in two orthogonal directions to produce the movements shown on the *lower left*. The ensuing path is shown as a continuous *blue line*, where each dot represents a single time bin in the simulations. The *cyan circles* represent the four attractors used in this itinerant policy. It can be seen that most of the time is spent at the two locations that supply physiological rewards: 23% for the first (*upper right*) and 30% for the second (*lower left*)

after the physiological state falls below some threshold (red dotted line in all panels). This ensures that the physiological states are lower bounded as seen in the lower left panel. This shows the first (blue) and second (green) levels of the physiological variable as a function of time. It can be seen that whenever the level falls below threshold, the values are replenished rapidly by a visit to the appropriate attractor. The same data are shown on the lower right. Here the two physiological states have been plotted against each other to show how they are always (jointly) above or near threshold.

These simulations were integrated as described in Appendix B and (Friston et al. 2010) using log-precisions of eight and four on the sensory input and motion of

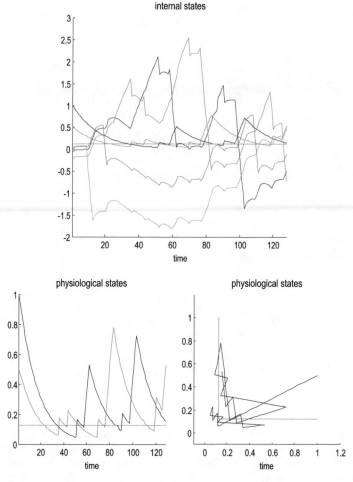

Fig. 9.5 This figure provides a more detailed description of the conditional expectations of the physiological (and manifold) states in the *upper panel* and the true physiological states in the *lower panels*. The *upper panel* shows the expected physiological states (*solid lines*) and the manifold-states (*broken lines*). The key thing to take from these dynamics is the recurrent build up and autovitiation of manifold-states, as each attracting fixed-point is visited and consequently rendered unattractive. Crucially, the attractors delivering rewards become more attractive after the physiological state falls below some threshold (*red dotted lines in all panels*). This ensures that the physiological states are lower bounded, as shown in the *lower left panel*. This shows the levels of first (*blue*) and second (*green*) physiological variables as functions of time. It can be seen that whenever the level falls below threshold, the values are rapidly replenished by a visit to the appropriate attractor. The same data are shown on the *lower right*. Here, the two physiological states have been plotted against each other to show how they are always (jointly) above or near threshold

physical states, respectively. These values are crucial for implementing any policy, as we will see in the next section, where we use low and high precisions to simulate pathological behaviour.

9.3.4 Summary

This section has focused on plausible forms for the motion of hidden states in generative models of the world. These forms correspond to formal priors or policies, which themselves have been optimised (by evolution in a biological setting). We have introduced the distinction between fixed-point and itinerant policies. Fixed-point policies (from optimal control theory and reinforcement learning) can be elaborated using an equilibrium perspective on abstract models of state-space, under constraints on divergence-free flow (the curl-constraint). When this constraint is satisfied, the scalar-potential guiding flow becomes a Lyapunov function and the (log of the) equilibrium density; that is, value. Conversely, itinerant policies are called for when one partitions hidden states into those that can be controlled directly and those which cannot. Both fixed-point and itinerant policies must conform to a divergence-constraint, in that the flow at low-cost points of the equilibrium density must have negative divergence. Furthermore, both sorts of policies rest upon the destruction or vitiation of costly fixed-points (either directly by making divergence or value depend on cost or indirectly using cost-dependent autovitiation). The notion of vitiating attractors to create itinerant dynamics along heteroclinic channels can be exploited in itinerant policies using fairly simple schemes. We have seen an example of one such scheme that will be used in the next section to study some of its key modes of failure.

If you have got this far through the arguments then you must either be very interested, or an editor (or both). Furthermore, you may be thinking "this is all plausible but its just common sense dressed up in the rhetoric of dynamical systems". In one sense this is true; however, it is worth reflecting on what has been achieved: We now have a model of exploratory behaviour and conditioned place-preference that is detailed to the level of forces, friction and physiology, using neurobiologically tenable computations. Furthermore, at no point did we need to invoke any (abstract) reinforcement learning scheme: the only learning required is conventional associative plasticity that is an integral part of perception. In the final section, we will use this model to see how abnormal perceptual inference and learning can have profound effects on behaviour.

9.4 Pathological Policies

In this section, we provide some simple case studies, using simulations to show how behaviour breaks down when perception is suboptimal. Specifically, we will look at the effect of changing the precision of random fluctuations on the hidden states. This may seem a rather arbitrary target for simulated lesions; however, there are some key reasons for starting here. Up until now, we have treated the precisions as known quantities. In more general treatments they are optimised using update or recognition schemes that are not dissimilar to those used for perceptual learning (see Friston 2008). This optimisation of the precisions corresponds to optimising uncertainty about prediction errors and the consequent predictions. As noted in the first section, precision may be encoded in the post-synaptic gain of prediction error units. The most likely candidates for these prediction error units are the principal (superficial pyramidal) cells originating forward connections in the cortex (see Friston 2008). In the present context, an important determinant of post-synaptic gain is classical neuromodulation. For example, changes in post-synaptic sensitivity due to the effect of dopaminergic or cholinergic neurotransmission on slow conductances following depolarisation. This premise is important in terms of clinical neuroscience because the vast majority of neuropsychiatric disorders are associated with abnormalities in neuromodulatory neurotransmission at one level or another (e.g., Liss and Roeper 2008; Goto et al. 2010). Indeed, the very fact that most psychotropic treatments target these systems testifies to this fact. Furthermore, the drugs most commonly associated with addictive behaviour affect dopaminergic and related classical neuromodulatory systems:

The mesocorticolimbic dopamine (DA) system comprises DA producing cells in the ventral tegmental area (VTA) of the midbrain and projects to forebrain structures including the nucleus accumbens (NAcc), medial prefrontal cortex (mPFC) and amygdala. It is generally thought that this system evolved to mediate behaviours essential for survival (Kelley and Berridge 2002; Panksepp et al. 2002) and that it plays an essential role in mediating biological incentives. Acute exposure to all drugs of abuse directly or indirectly increases DA neurotransmission in the NAcc and repeated drug exposure results in enduring changes in mesocorticolimbic brain regions (Berke and Hyman 2000; Henry and White 1995; Nestler 2005; Pierce and Kalivas 1997). These drugs include psychostimulants (e.g., cocaine, amphetamine and its derivatives methamphetamine and methlyenedioxy methamphetamine), opiates (e.g., heroin and morphine) and other common drugs of abuse (e.g., alcohol and nicotine). Psychostimulants act directly on dopaminergic terminals in the NAcc (Khoshbouei et al. 2003), while opiates act indirectly by inhibiting GABAergic neurons in the VTA with disinhibition of DA neurons.

In what follows, we will repeat the simulations of the previous section but using suboptimal low and high levels of precision on the motion of hidden states. This produces two characteristic failures of behaviour and learning that map, roughly, onto the psychomotor poverty and bradykinesia associated with Parkinson's disease on the one hand and stereotyped perseverative behaviours that are reminiscent of addiction on the other. We first consider the affect of reducing precision.

9.4.1 Simulating Parkinsonism

In the first simulations, we will look at the effects of reducing precision on the motion of hidden states. This can be seen as a crude model of neurodegeneration in ascending dopaminergic systems, which would reduce synaptic gain and precision $\tilde{\Pi}^{(i,u)}$ in Eq. (9.17). To simulate this reduction, we repeated the foraging simulations above, using progressively lower levels of precision on the motion of physical states:

$\Pi^{(a)} \in 4, 2, 0$. The results of these simulations are shown in Fig. 9.6, in terms of the trajectories in physical subspace (left panels) and the physiological subspace (right panels). It is immediately obvious that the accuracy and speed of locomotion is impaired, with a progressive failure to hit the targets and pronounced over-shooting. The physiological sequelae of this impaired behaviour are shown in terms of a progressive failure to keep the physiological states above threshold. Indeed, in the lower right panel, the physiological states are sometimes close to zero.

The reason for this loss of control is simple. Action is driven by sensory prediction errors (see Eq. (9.19)). These prediction errors depend upon precise predictions. If the precision or certainty about the inferred motion of hidden states falls, more weight is placed on sensory evidence. Heuristically, a low precision on the empirical priors afforded by the motion of hidden states means that conditional predictions are based upon sensory evidence. Because action tries to reduce prediction errors it now depends more on what is sensed, as opposed to what is predicted. In the absence of precise predictions, the agent will simply stop moving. We can see the beginnings of this motor poverty in Fig. 9.6 (lower panels), where the forces exerted by action are attenuated, resulting in trajectories with a much lower curvature. If we continued reducing the level of precision (cf., dopamine), the agent would ultimately become akinetic. We have illustrated this behaviour in a variety of simulations previously, for example, the same behaviour can be elicited using the mountain car example in Fig. 9.3, as shown in Friston et al. (2010).

Figure 9.7 shows the action and underlying sensory prediction errors associated with the trajectories in Fig. 9.6. The action (in both directions) is shown as a function of time in the left panels. The right panels show the corresponding prediction error on the four physical states (position and velocity in two directions). The key thing to take from these results is the progressive reduction in the amplitude of action due to an underlying fall in the amplitude of sensory prediction errors. This leads to smaller forces on the physical motion of the agent and the bradykinesia seen in Fig. 9.6. The progressive reduction in sensory prediction errors reflects a loss of confidence (precision) in top-down prior expectations about movement, which would normally subtend itinerant behaviour. This example is used to highlight the key role of precision, especially the precision of predictions about the motion of hidden states. If these predictions become less precise, they have less influence, relative to sensory information and consequently exert less influence over action. In this view, pathologies that involve a loss of neuromodulation can be regarded as subverting the potency of empirical prior expectations that maintain adaptive behaviour.

9.4.1.1 Summary

In summary, we have seen how perceptual synthesis plays a crucial role in providing predictions that action can fulfil. However, if these predictions are under confident, they will fail to elicit sufficient sensory prediction errors to engage behaviour. A key mechanism, by which conditional confidence can be undermined, is false inference about the amplitude of random fluctuations on hidden states. This leads to

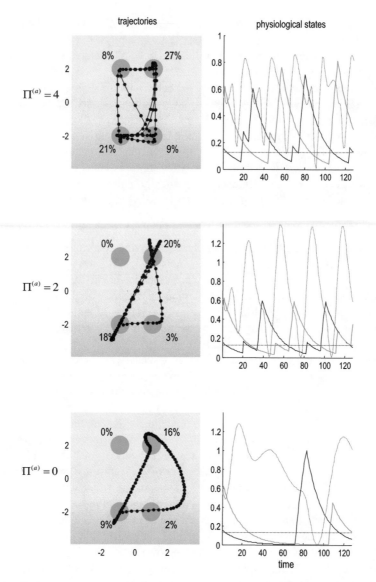

Fig. 9.6 This figure shows the (true) trajectories and resulting physiological states (using the same format as Figs. 9.4 and 9.5) for different levels of precision on the motion of physical states (i.e., position and velocity). The *top row* shows normal behaviour elicited with a log-precision of four. The *remaining two rows* show progressive pathology in behaviour, when using log-precisions of two and zero, respectively. The *left panels* show deterioration of the trajectories, with a generalised slowing of movements and a loss of accuracy, when locating the target (attracting fixed-points). This slowing is reflected in the number of times a target is visited. This is indicated in the *right panels* by the *dotted lines*, which report the distance from the centre. In an extreme case (log-precision of zero), only one definite movement has been emitted in the 128 second simulated exposure. These simulations are meant to reproduce the characteristic psychomotor slowing, bradykinesia and loss of fine movement control associated with Parkinsonism due to neurodegeneration or psycholytic therapy

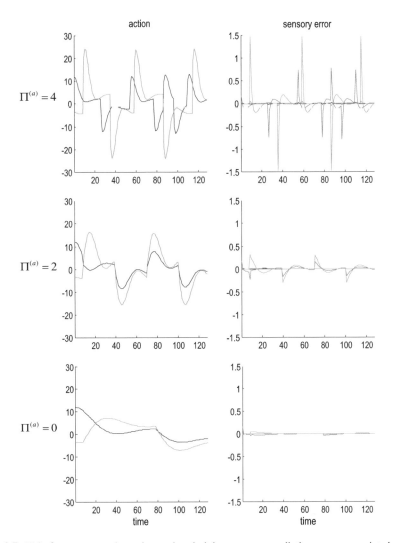

Fig. 9.7 This figure reports the action and underlying sensory prediction errors associated with the trajectories in the previous figure. The action (in both directions) is shown as a function of time in the left column, while the right column shows the corresponding prediction error on the four physical states (position and velocity in two directions). The key thing to take from these results is the progressive reduction in the amplitude of action due to an underlying fall in the amplitude of sensory prediction errors. This leads to smaller forces on the physical motion of the agent and the bradykinesia seen in the previous figure. The reduction in sensory prediction error reflects a loss of confidence (precision) in top–down prior expectations about movements, which would normally subtend itinerant activity

the adoption of pathologically low precision on internal prediction errors that may be associated with the failure of synaptic gain control associated with Parkinsonism (e.g., Zhao et al. 2001). In this context, impaired inference about proprioceptive states translates into a failure of motor intention. This mechanism also sits comfortably with the role of substantia nigra-amygdala connections in surprise-induced enhancement of attention in the perceptual domain: Lesion studies in rats (Lee et al. 2006) show that these connections are "critical to mechanisms by which the coding of prediction error by midbrain dopamine neurons is translated into enhancement of attention and learning modulated by the cholinergic system". Furthermore, low dose apomorphine, which is thought to inhibit DA release by activating pre-synaptic DA autoreceptors, decreases the frequency of itinerant behaviours (e.g., Niesink and Van Ree 1989). Interestingly, increasing precision has relatively little effect on perceptual inference and the attending behaviour; however, it can have a profound effect on perceptual learning. We consider this in the next section, where we ask what would happen if the precision or gain was too high? Here, the consequences are expressed less in terms of locomotion but more in terms of deleterious effects on perceptual learning that determines the organisation of behaviour.

9.4.2 Simulating Addiction

Hitherto, all our simulations have assumed the agent has learned the association between the locations in its environment and the physiological rewards available. These are encoded by the parameters $\theta_{ij} \in \varphi$ in the generative model. In the final simulations, we study how these associations can be acquired and the effects of increasing precision (e.g., dopamine) on this learning.

9.4.2.1 Normal Learning

To study the effects of learning, we changed the reward contingencies by moving the reward usually available at the second location (lower left) to the third location (upper right). This presents an interesting problem under active inference, because action fulfils expectations and the agent expects to be rewarded at the first and second location. It must now undo this association to discover something unexpected, while acting to fulfil its expectations. Itinerant policies meet this challenge easily because, by their construction, they explore all putative reward locations in an itinerant fashion. In brief, the itinerant policy means the agent expects to visit most states at some point and therefore its behaviour will follow suit. This ensures that new associations between the physical and physiological dynamics are encountered and remembered, through optimisation of the parameters encoded by connection strengths (synaptic efficacy). An illustration of perceptual learning under an itinerant policy is shown in Fig. 9.8. This summarises the results of perceptual learning after 128 seconds of exploration, following a switch in the location of the second

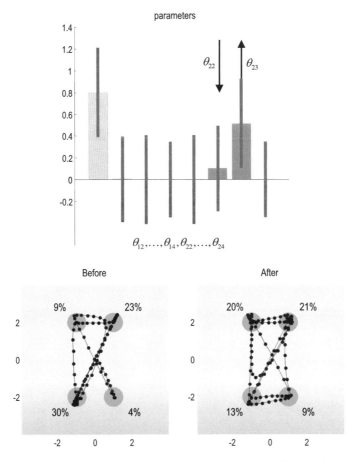

Fig. 9.8 This figure summarises the results of perceptual learning after 128 seconds of exploration, following a switch in the location of the second reward. This location was switched from the lower left to the upper left attractor. The *upper panel* shows the parameter expectations (*grey bars*) and the 90% conditional confidence intervals (*red lines*). The eight parameters constitute the matrix of coefficients θ that associate the two rewards with the four attracting locations. Before learning, rewards were available at the first and second locations (corresponding to parameters one and six). The switch of the location of the second reward corresponds to re-setting the sixth parameter from one to zero $\theta_{22} \to 0$ with a complimentary increase in the seventh parameter from zero to one $\theta_{23} \to 1$. The *top panel* shows that the true values are contained within the 90% confidence intervals and a degree of 'reversal learning' has occurred (*arrows* above the parameters in *dark gray*). The corresponding behaviour (before and after learning) is shown in the *lower panels* (*left* and *right* respectively), using the same format as in previous figures. Before learning, the old and new locations of the second reward were visited 30% and 9% of the time respectively. Conversely, after learning this ratio reversed, such that the newly rewarded location is now visited 20% of the time

reward. This can be regarded as a simulation of reversal learning, in the context of conditioned place-preference (McDonald et al. 2002). The reward location was switched from the lower left to the upper left. The upper panel shows the parameter expectations (grey bars) and the 90% conditional confidence intervals (red bars). It should be noted that these confidence intervals (which are based upon the conditional precisions in Eq. (9.6)), are not represented explicitly by the agent. However, they provide a useful measure of the implicit certainty the agent has in its expectations about causal structure in its world. The eight parameters correspond to the matrix of coefficients $\theta \in \varphi$ that associate the two rewards with the four attracting locations. Before learning, rewards were available at the first and second locations (corresponding to parameters one and six). The switch of the location of the second reward to the third location corresponds to a reduction in the sixth parameter θ_{22} (from one to zero) and a complimentary increase in the seventh parameter θ_{23} (from zero to one). The top panel shows that the true values are contained within the 90% confidence intervals and a degree of reversal learning has occurred. The corresponding behaviour before and after learning is shown in the lower panels (left and right, respectively). Before learning, the old and new locations of the second reward were visited 30% and 9% of the time, respectively. After learning, this ratio has reversed, such that the newly rewarded location is now visited 20% of the time. Note that there is no imperative to spend all the time at a rewarding location; just to emit a sufficient number of visits to ensure the physiological states do not fall to very low levels (data not shown). This learning occurred with a log-precision on the motion of the physiological states of four; $\Pi^{(p)} = 4$. Next, we examine what happens with inappropriately high levels of precision on physiological kinetics.

9.4.2.2 Pathological Learning

We repeated the above simulations but using a pathologically high level of precision that can be thought of (roughly) as a hyper-dopaminergic state. The motivation for this is based on the fact that most addictive behaviours involve taking drugs that cross the blood/brain barrier and augment neuromodulatory transmission. For example, acute exposure to psychostimulants increases extracellular DA levels in the NAcc and this increase is significantly enhanced after repeated exposure; due to increased activity of DA neurons and alterations in DA axon terminals (Pierce and Kalivas 1997). Although a very simplistic interpretation of addiction, we can associate increases in extracellular DA levels with an increase in precision. Intuitively speaking, this means the agent becomes overly confident about its internal predictions, in relation to the sensory evidence encountered. So what effect will this have on learning?

Figure 9.9 reports the results of simulated learning under increasing levels of log-precision on the motion (kinetics) of the physiological states. The left panels show the corresponding behaviour using the same format as in previous figures. The right panels show the conditional expectations and confidence following a 128 second exposure to the environment, after the location of the second reward was switched. The first row reproduces the results of Fig. 9.8 showing veridical, if incomplete, reversal learning (a decrease in parameter six and an increase in parameter seven).

9 Policies and Priors

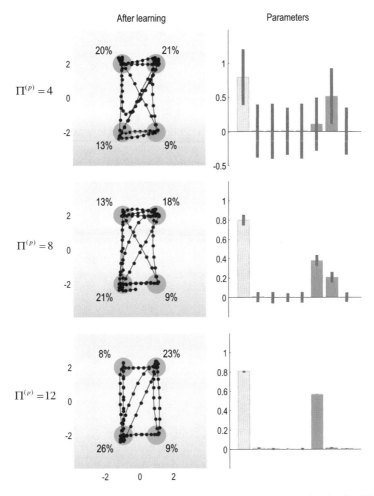

Fig. 9.9 This figure reports the results of simulated learning under increasing levels of log-precision on the motion or dynamics of the two physiological states. The *left column* shows the corresponding trajectories using the same format as in previous figure. The *right column* shows the conditional expectations and confidence intervals following 128 second exposure to the environment, after the location of the second reward had been switched. These use the same format as the *upper panel* of the previous figure. The *first row* reproduces the results of Fig. 9.8 showing veridical, if incomplete, learning of the switched locations (a decrease in parameter six and an increase in parameter seven). This reversal learning is partially (*middle row*) and completely (*lower row*) blocked as log-precision increases from four to eight and from eight to twelve. The failure to learn the change in the association between locations and rewards is reflected in the occupancy of the corresponding locations. For example, the newly rewarding location (*upper left*) is visited on 20%, 13% and 8% of the time as precision increases and learning fails

This learning is partially (middle row) and completely (lower row) blocked as the log-precision increases from four to eight and from eight to twelve. The failure to learn the change in the association between locations and rewards is reflected in the occupancy of the corresponding locations. For example, the newly rewarding location (upper left) is visited on 20%, 13% and 8% of the time, as precision increases and learning fails. There is a concomitant retention of place-preference for the previously rewarded location (lower left). The reason for this failure of reversal learning and consequent failure to adaptively update place-preference is reflected in the conditional confidence intervals on the parameters. These reveal a progressive reduction in conditional uncertainty (increase in conditional precision), which interferes with learning. The mechanism of this interference is quite subtle but illuminating: Recall from Sect. 9.2 (Eq. (9.18)) that learning (associative plasticity) is driven by the appropriate prediction error, here prediction errors about the motion or changes in physiological states. These are extremely sensitive to the assumed precision about fluctuations in these states as shown in the next figure:

Figure 9.10 shows the conditional expectations or predictions about the motion of physiological states and their associated prediction errors (left and right columns, respectively). The upper rows correspond to a roughly optimal log-precision of four, while the middle and lower rows show the results for pathologically high log-precisions (cf. hyper-dopaminergic states) of 8 and 12, respectively. The corresponding increase in precision means that the conditional representations of changes in physiological state (here the second physiological variable) are over confident and, in extreme cases, a fantasy. This is shown in the left panels in terms of the conditional expectations (solid lines) and the true changes (dotted lines). These are in good agreement for appropriate levels of precision but not at high levels of precision (see lower row). When precision is very high, the agent expects to be rewarded when it visits the old location. This expectation is so precise that it completely ignores sensory evidence to the contrary. These false predictions are reflected in a progressive fall in prediction error (see right column); such that, at high levels of precision, there is no prediction error when there should be. For example, look at the prediction error at around 20 seconds, when the second reward is elicited for the first time. In summary, a high precision leads to over confident inference about the states of the world and their motion, which subverts appropriate prediction errors and their ability to drive associative plasticity. This leads to false expectations about exteroceptive and interoceptive signals and a consequent failure of active inference (behaviour). This example highlights the complicated but intuitive interplay between perceptual inference, learning and action.

9.5 Discussion

In summary, we have seen how inappropriately high levels of precision in generalised predictive coding schemes can lead to false, over confident, predictions that do not properly reflect the true state of the world. This leads to an inappropriately low expression of prediction errors signalled, presumably, by (superficial pyramidal) principal cells in the cortex and a concomitant failure of associative plasticity in their synaptic connections. This failure to learn causal contingencies or associations in the environment results in maladaptive 'place-preferences' as reflected

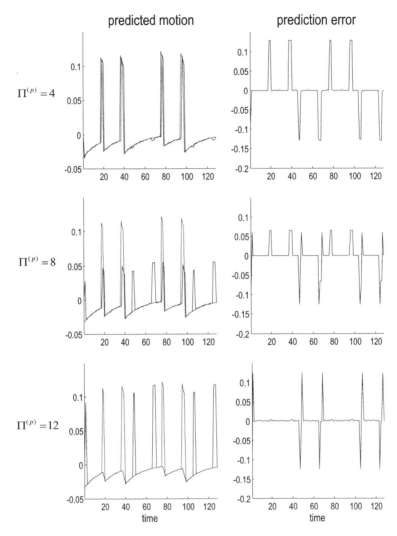

Fig. 9.10 This figure shows the conditional expectations or predictions about the motion of physiological states and their associated prediction errors (*left* and *right columns*, respectively). The *upper rows* correspond to a roughly optimal log-precision of four, while the *middle* and *lower rows* show the results for pathologically high log-precisions (cf., hyper-dopaminergic states) of eight and twelve, respectively. The increase in precision means that the conditional representations of changes in the physiological state (here the second physiological variable) are overconfident and, in extreme cases, illusory. This is shown in the *left panels* in terms of the conditional expectations (*solid lines*) and the true changes (*dotted lines*). These are in good agreement for appropriate levels of precision but represent a 'fantasy' at very high levels of precision (see *lower row*). These overconfident predictions are reflected in a progressive fall in prediction error (see *right column*), such that, at high levels of precision there is no prediction error when there should be. In short, a high precision leads to overconfident inference, which subverts appropriate prediction errors and their ability to drive associative plasticity

in the ensuing perseverative behaviour. This may represent one way in which addictive behaviour could be understood. The implicit explanation for why high levels of precision are maintained in addictive (preservative) behaviour rests upon the assumption that the behaviour per se results in the brain adopting inappropriately high levels of precision. Neurobiologically speaking, this translates into inappropriately high levels of post-synaptic gain in specific neuronal populations. This is consistent with the action of nearly all known drugs of abuse, which affect the mesocorticolimbic dopamine system. Clearly, there can be many ways in which to associate dopaminergic and other neuromodulatory mechanisms with the various parameters and states of predictive coding models. We have chosen to focus on the role of classical neuromodulators in optimising the sensitivity or gain of cells and have equated this with the brain's representation of the precision of random fluctuations in the environment: in other words, a representation of uncertainty. This is certainly consistent with some electrophysiological interpretations of dopaminergic firing, in which phasic dopamine release may represent reward prediction error per se and sustained or tonic firing represents the level of uncertainty (Fiorillo et al. 2003). For example, prediction error on the physiological states could be encoded by phasic discharges in the dopaminergic system, whereas the post-synaptic gain of DA error units may be influenced by (or cause) tonic discharge rates.

Traditionally, midbrain dopamine neurons in the substantia nigra and ventral tegmental area (VTA) are thought to encode reward prediction error (Montague et al. 1996; Schultz et al. 1997; Schultz 1998; Salzman et al. 2005). Activity in these neurons reflects a mismatch between expected and experienced reward that emulates the prediction errors used in (abstract) value-learning theories (Friston et al. 1994; Montague et al. 1996; Sutton and Barto 1981). Indeed, aberrant reward prediction error accounts have proposed for addictive behaviour (Lapish et al. 2006; Redish 2004) and the maintenance of maladaptive habits (Takahashi et al. 2008). However, recent studies suggest a diverse and multilateral role for dopamine that is more consistent with encoding the precision of generalised prediction errors in the predictive coding sense (as opposed to reward prediction errors in particular). For example, punishment prediction error signals (Matsumoto and Hikosaka 2009) and mismatches between expected and experienced *information* (Bromberg-Martin and Hikosaka 2009) may be encoded in distinct anatomical populations of midbrain dopamine neurons. Furthermore, the timing of reward-related signals in VTA precludes the calculation of a reward prediction error per se (Redgrave and Gurney 2006) and may report a change in the certainty about sensory events, via cholinergic input from the pedunculopontine tegmentum (Dommett et al. 2005). Similarly, violations of perceptual expectations engage hippocampal projections to the VTA, which modulate a broad population of dopamine neurons (Lodge and Grace 2006). Human studies with functional neuroimaging suggest that the ventral striatum responds to non-rewarding, unexpected stimuli in proportion to the salience of the stimulus (Zink et al. 2006), as well as to novel stimuli (Wittmann et al. 2007). One of the proposed functions of these striatal responses is to reallocate resources to unexpected stimuli in both reward and non-reward contexts (Zink et al. 2006). Hsu et al. (2005) show that "the level of ambiguity in choices correlates positively with activation in the amygdala and orbitofrontal cortex, and negatively with a striatal system" and interpret their findings in terms of a "neural circuit responding to degrees of uncertainty, contrary to decision theory". These results suggest that rather than just coding reward prediction errors, the striatum may have a more general role in processing salient and unexpected events, under varying degrees of ambiguity or uncertainty (precision). In summary, the mesocorticolimbic dopamine system may encode numerous types of expectation violations associated with a change in the precision of top-down predictions and ensuing prediction errors (see also Schultz and Dickinson 2000; Fiorillo 2008).

Perhaps one thing to take from these considerations is the complex but intuitive interplay between the many variables that need to be encoded by the brain for optimal behaviour. This means that it may not be easy, given the present state of knowledge, to associate the algorithmic components of optimal schemes with specific neurotransmitter systems or their kinetics. Having said this, there are obvious commonalities between the dynamical simulations presented above and the more abstract formulations that rest on things like the Rescorla-Wagner model (Rescorla and Wagner 1972) and dynamic programming. All these formulations highlight the importance of prediction error on physiological states normally associated with reward. This has been nuanced in the current formulation by a focus on the precision of this prediction error as opposed to the prediction error per se. As we have noted previously, it may be that dopamine does not encode the prediction error on value but the value (precision) of prediction error. The motivation for this perspective rests on the empirical observations discussed above and, more theoretically, on symmetry arguments that place precision centre-stage in terms of amplifying expected actions and percepts. This bilateral role of neuromodulation to select actions and precepts maps nicely to a role for post-synaptic gain in intention and attention. In short, we may be looking at the same mechanism but implemented in different parts of the brain.

9.6 Conclusion

In this chapter, we have tried to cover the fundaments of adaptive behaviour starting from basic principles. We have used the imperative for biological systems to resist an increase in their entropy to motivate a free-energy principle that explains both action and perception. When this principle is unpacked, in the context of generative models the brain might use, we arrive at a fairly simple message-passing scheme based upon prediction errors and the optimisation of their precision by synaptic gain. We then considered generic forms that these models might possess, where the form itself entails prior expectations about the motion of hidden states in the world and, through active inference, behaviour. We considered fixed-point policies of the sort found in psychology and optimal control theory. We then proceeded to itinerant policies that have a more dynamic and ethologically valid flavour. The notion of itinerant policies, when combined with active inference, provides a rich framework in which to understand many aspects of behaviour. We have focused on changes in behaviour following a down-regulation or up-regulation of the precision, under which perceptual inference and learning proceeds. This was motivated by the psychopharmacology of addiction, which almost invariably involves some change in dopaminergic neurotransmission and, from an algorithmic perspective, the optimisation of precision in the brain. The results of these simulations suggest plausible explanations for bradykinetic and addictive behaviour that rest upon impaired inference and learning respectively. Both the functionalist perspective afforded by this analysis and the putative neurobiological mechanisms fit comfortably with many known facts in addiction research. However, a specific mapping between functional architectures of the sort considered here and the neurobiology of addiction clearly requires more work. Although an awful condition from a clinical point of view, addiction may be nature's most unique and pervasive psychopharmacological experiment, in which complex behaviour confounds the elemental (synaptic) mechanisms upon which it rests.

Acknowledgements The Wellcome Trust funded this work and greatest thanks to Marcia Bennett for helping prepare this manuscript.

Appendix A: Parameter Optimisation and Newton's Method

There is a close connection between the updates implied by Eq. (9.9) and Newton's method for optimisation. Consider the update under a local linearisation, assuming $\mathcal{L}_\varphi \approx \mathcal{F}_\varphi$

$$\Delta \tilde{\mu}^{(\varphi)} = \left(\exp(t\mathfrak{I}^{(\varphi)}) - I\right)\mathfrak{I}^{(\varphi)-1}\dot{\tilde{\mu}}^{(\varphi)}$$

$$\dot{\tilde{\mu}}^{(\varphi)} = \begin{bmatrix} \mu'^{(\varphi)} \\ -\mathcal{L}_\varphi - \kappa\mu'^{(\varphi)} \end{bmatrix} \quad (A.1)$$

$$\mathfrak{I}^{(\varphi)} = \frac{\partial \dot{\tilde{\mu}}^{(\varphi)}}{\partial \tilde{\mu}^{(\varphi)}} = \begin{bmatrix} 0 & I \\ -\mathcal{L}_{\varphi\varphi} & -\kappa \end{bmatrix}$$

As time proceeds, the change in generalised mean becomes

$$\lim_{t\to\infty} \Delta \tilde{\mu}^{(\varphi)} = -\mathfrak{I}^{(\varphi)-1}\dot{\tilde{\mu}}^{(\varphi)} = \begin{bmatrix} \Delta\mu^{(\varphi)} \\ \Delta\mu'^{(\varphi)} \end{bmatrix} = -\begin{bmatrix} \mathcal{L}_{\varphi\varphi}^{-1}\mathcal{L}_\varphi \\ \mu'^{(\varphi)} \end{bmatrix}$$

$$\mathfrak{I}^{(\varphi)-1} = \begin{bmatrix} -\kappa\mathcal{L}_{\varphi\varphi}^{-1} & -\mathcal{L}_{\varphi\varphi}^{-1} \\ I & 0 \end{bmatrix} \quad (A.2)$$

The first line means the motion cancels itself and becomes zero, while the change in the conditional mean $\Delta\mu^{(\varphi)} = -\mathcal{L}_{\varphi\varphi}^{-1}\mathcal{L}_\varphi$ becomes a classical Newton update. The conditional expectations of the parameters were updated after every simulated exposure using this scheme, as described in Friston (2008).

Appendix B: Simulating Action and Perception

The simulations in this paper involve integrating time-varying states in the environment and the agent. This is the solution to the following ordinary differential equation

$$\dot{u} = \begin{bmatrix} \dot{\tilde{s}} \\ \dot{\tilde{x}} \\ \dot{\tilde{v}} \\ \dot{\tilde{\omega}}^{(x)} \\ \dot{\tilde{\omega}}^{(v)} \\ \dot{\tilde{\mu}}^{(x)} \\ \dot{\tilde{\mu}}^{(v)} \\ \dot{a} \end{bmatrix} = \begin{bmatrix} \mathcal{D}\mathbf{g} + \mathcal{D}\tilde{\omega}^{(v)} \\ \mathbf{f} + \tilde{\omega}^{(x)} \\ \mathcal{D}\tilde{\mathbf{v}} \\ \mathcal{D}\tilde{\omega}^{(x)} \\ \mathcal{D}\tilde{\omega}^{(v)} \\ \mathcal{D}\tilde{\mu}^x - \mathcal{F}_{\tilde{x}} \\ \mathcal{D}\tilde{\mu}^v - \mathcal{F}_{\tilde{v}} \\ -\mathcal{F}_a \end{bmatrix}$$

$$\Im = \begin{bmatrix} 0 & \mathcal{D}\mathbf{g}_{\tilde{x}} & \mathcal{D}\mathbf{g}_{\tilde{v}} & \mathcal{D} & 0 & 0 & \cdots & 0 \\ & \mathbf{f}_{\tilde{x}} & \mathbf{f}_{\tilde{v}} & & I & & & \mathbf{f}_a \\ \vdots & & & \mathcal{D} & & \vdots & \vdots & 0 \\ & & & & \mathcal{D} & 0 & & \\ 0 & & \cdots & 0 & \mathcal{D} & 0 & \cdots & \\ -\mathcal{F}_{\tilde{x}\tilde{y}} & & \cdots & & 0 & \mathcal{D} - \mathcal{F}_{\tilde{x}\tilde{x}} & -F_{\tilde{x}\tilde{v}} & -F_{\tilde{x}a} \\ -\mathcal{F}_{\tilde{v}\tilde{y}} & & & & & -\mathcal{F}_{\tilde{v}\tilde{x}} & \mathcal{D} - \mathcal{F}_{\tilde{v}\tilde{v}} & -\mathcal{F}_{\tilde{v}a} \\ -\mathcal{J}_{a\tilde{y}} & & & & & \mathcal{F}_{a\tilde{x}} & \mathcal{F}_{a\tilde{v}} & -\mathcal{F}_{aa} \end{bmatrix}$$

(B.1)

To update these states we use a local linearisation; $\Delta u = (\exp(\Delta t \Im) - I)\Im(t)^{-1}\dot{u}$ over time steps of Δt, where $\Im = \partial \dot{u}/\partial u$ is evaluated at the current conditional expectation (Friston et al. 2010).

References

Abeles M, Hayon G, Lehmann D (2004) Modeling compositionality by dynamic binding of synfire chains. J Comput Neurosci 17(2):179–201

Ahmed SH, Graupner M, Gutkin B (2009) Computational approaches to the neurobiology of drug addiction. Pharmacopsychiatry 42(1):S144–S152 Suppl

Alcaro A, Huber R, Panksepp J (2007) Behavioral functions of the mesolimbic dopaminergic system: an affective neuroethological perspective. Brains Res Rev 56(2):283–321

Ballard DH, Hinton GE, Sejnowski TJ (1983) Parallel visual computation. Nature 306:21–26

Bellman R (1952) On the theory of dynamic programming. Proc Natl Acad Sci USA 38:716–719

Berke JD, Hyman SE (2000) Addiction, dopamine, and the molecular mechanisms of memory. Neuron 25(3):515–532

Birkhoff GD (1931) Proof of the ergodic theorem. Proc Natl Acad Sci USA 17:656–660

Breakspear M, Stam CJ (2005) Dynamics of a neural system with a multiscale architecture. Philos Trans R Soc Lond B, Biol Sci 360(1457):1051–1074

Bressler SL, Tognoli E (2006) Operational principles of neurocognitive networks. Int J Psychophysiol 60(2):139–148

Bromberg-Martin ES, Hikosaka O (2009) Midbrain dopamine neurons signal preference for advance information about upcoming rewards. Neuron 63:119–126

Coricelli G, Dolan RJ, Sirigu A (2007) Brain, emotion and decision making: the paradigmatic example of regret. Trends Cogn Sci 11(6):258–265

Camerer CF (2003) Behavioural studies of strategic thinking in games. Trends Cogn Sci 7(5):225–231

Chen X, Zelinsky GJ (2006) Real-world visual search is dominated by top-down guidance. Vis Res 46(24):4118–4133

Colliaux D, Molter C, Yamaguchi Y (2009) Working memory dynamics and spontaneous activity in a flip-flop oscillations network model with a Milnor attractor. Cogn Neurodyn 3(2):141–151

Crauel H (1999) Global random attractors are uniquely determined by attracting deterministic compact sets. Ann Mat Pura Appl 176(4):57–72

Crauel H, Flandoli F (1994) Attractors for random dynamical systems. Probab Theory Relat Fields 100:365–393

Davidson TL (1993) The nature and function of interoceptive signals to feed: toward integration of physiological and learning perspectives. Psychol Rev 100(4):640–657

Daw ND, Doya K (2006) The computational neurobiology of learning and reward. Curr Opin Neurobiol 16(2):199–204

Daw ND, O'Doherty JP, Dayan P, Seymour B, Dolan RJ (2006) Cortical substrates for exploratory decisions in humans. Nature 441(7095):876–879

Dayan P, Daw ND (2008) Decision theory, reinforcement learning, and the brain. Cogn Affect Behav Neurosci 8(4):429–453

Dayan P, Hinton GE, Neal RM (1995) The Helmholtz machine. Neural Comput 7:889–904

Dommett E, Coizet V, Blaha CD, Martindale J, Lefebvre V, Walton N, Mayhew JE, Overton PG, Redgrave P (2005) How visual stimuli activate dopaminergic neurons at short latency. Science 307:1476–1479

Eldredge N, Gould SJ (1972) Punctuated equilibria: an alternative to phyletic gradualism. In: Schopf TJM (ed) Models in paleobiology. Freeman, San Francisco, pp 82–115

Evans DJ (2003) A non-equilibrium free energy theorem for deterministic systems. Mol Phys 101:15551–15554

Feynman RP (1972) Statistical mechanics. Benjamin, Reading

Freeman WJ (1994) Characterization of state transitions in spatially distributed, chaotic, nonlinear, dynamical systems in cerebral cortex. Integr Physiol Behav Sci 29(3):294–306

Friston KJ, Tononi G, Reeke GN Jr, Sporns O, Edelman GM (1994) Value-dependent selection in the brain: simulation in a synthetic neural model. Neuroscience 59(2):229–243

Friston KJ (2000) The labile brain. II. Transients, complexity and selection. Phil Trans Biol Sci 355(1394):237–252

Friston K (2005) A theory of cortical responses. Philos Trans R Soc Lond B, Biol Sci 360(1456):815–836

Friston K (2008) Hierarchical models in the brain. PLoS Comput Biol 4(11):e1000211

Friston K, Kilner J, Harrison L (2006) A free energy principle for the brain. J Physiol Paris 100(1–3):70–87

Friston KJ, Daunizeau J, Kiebel SJ (2009) Reinforcement learning or active inference? PLoS ONE 29;4(7):e6421

Friston KJ, Daunizeau J, Kilner J, Kiebel SJ (2010) Action and behavior: a free-energy formulation. Biol Cybern [Epub ahead of print]

Fiorillo CD, Tobler PN, Schultz W (2003) Discrete coding of reward probability and uncertainty by dopamine neurons. Science 299(5614):1898–1902

Fiorillo CD (2008) Towards a general theory of neural computation based on prediction by single neurons. PLoS ONE 3:e3298

Goto Y, Yang CR, Otani S (2010) Functional and dysfunctional synaptic plasticity in prefrontal cortex: roles in psychiatric disorders. Biol Psychiatry 67(3):199–207

Gregory RL (1968) Perceptual illusions and brain models. Proc R Soc Lond B 171:179–196

Gregory RL (1980) Perceptions as hypotheses. Phil Trans R Soc Lond B 290:181–197

Gros C (2009) Cognitive computation with autonomously active neural networks: an emerging field. Cogn Comput 1:77–99
Haile PA, Hortaçsu A, Kosenok G (2008) On the empirical content of quantal response equilibrium. Am Econ Rev 98:180–200
Haken H (1983) Synergistics: an introduction. Non-equilibrium phase transition and self-organisation in physics, chemistry and biology, 3rd edn. Springer, Berlin
Herrmann JM, Pawelzik K, Geisel T (1999) Self-localization of autonomous robots by hidden representations. Auton Robots 7:31–40
Hinton GE, van Camp D (1993) Keeping neural networks simple by minimising the description length of weights. In: Proceedings of COLT-93, pp 5–13
von Helmholtz H (1866) Concerning the perceptions in general. In: Treatise on physiological optics, vol III, 3rd edn (translated by J.P.C. Southall 1925 Opt Soc Am Section 26, reprinted New York, Dover, 1962)
Henry DJ, White FJ (1995) The persistence of behavioral sensitization to cocaine parallels enhanced inhibition of nucleus accumbens neurons. J Neurosci 15(9):6287–6299
Hull C (1943) Principles of behavior. Appleton/Century-Crofts, New York
Hsu M, Bhatt M, Adolphs R, Tranel D, Camerer CF (2005) Neural systems responding to degrees of uncertainty in human decision-making. Science 310(5754):1680–1683
Jirsa VK, Friedrich R, Haken H, Kelso JA (1994) A theoretical model of phase transitions in the human brain. Biol Cybern 71(1):27–35
Johnson A, van der Meer MA, Redish AD (2007) Integrating hippocampus and striatum in decision-making. Curr Opin Neurobiol 17(6):692–697
Kelley AE, Berridge KC (2002) The neuroscience of natural rewards: relevance to addictive drugs. J Neurosci 22(9):3306–3311
Kersten D, Mamassian P, Yuille A (2004) Object perception as Bayesian inference. Annu Rev Psychol 55:271–304
Khoshbouei H, Wang H, Lechleiter JD, Javitch JA, Galli A (2003) Amphetamine-induced dopamine efflux. A voltage-sensitive and intracellular Na^+-dependent mechanism. J Biol Chem 278(14):12070–12077
Knill DC, Pouget A (2004) The Bayesian brain: the role of uncertainty in neural coding and computation. Trends Neurosci 27(12):712–719
Lapish CC, Seamans JK, Chandler LJ (2006) Glutamate-dopamine cotransmission and reward processing in addiction. Alcohol Clin Exp Res 30:451–1465
Lee TS, Mumford D (2003) Hierarchical Bayesian inference in the visual cortex. J Opt Soc Am A, Opt Image Sci Vis 20:1434–1448
Lee HJ, Youn JM, MJ O, Gallagher M, Holland PC (2006) Role of substantia nigra-amygdala connections in surprise-induced enhancement of attention. J Neurosci 26(22):6077–6081
Liss B, Roeper J (2008) Individual dopamine midbrain neurons: functional diversity and flexibility in health and disease. Brains Res Rev 58(2):314–321
Lodge DJ, Grace AA (2006) The hippocampus modulates dopamine neuron responsivity by regulating the intensity of phasic neuron activation. Neuropsychopharmacology 31:1356–1361
MacKay DM (1956) The epistemological problem for automata. In: Shannon CE, McCarthy J (eds) Automata studies. Princeton University Press, Princeton, pp 235–251
MacKay DJC (1995) Free-energy minimisation algorithm for decoding and cryptoanalysis. Electron Lett 31:445–447
Matheron G (1975) Random sets and integral geometry. Wiley, New York
Matsumoto M, Hikosaka O (2009) Two types of dopamine neuron distinctly convey positive and negative motivational signals. Nature 459:837–841
Maturana HR, Varela F (1980) De máquinas y seres vivos. Editorial Universitaria, Santiago. English version: *Autopoiesis: the organization of the living*, in Maturana, HR, and Varela, FG, Autopoiesis and Cognition. Dordrecht, Netherlands: Reidel
Maynard Smith J (1992) Byte-sized evolution. Nature 355:772–773
McDonald RJ, Ko CH, Hong NS (2002) Attenuation of context-specific inhibition on reversal learning of a stimulus-response task in rats with neurotoxic hippocampal damage. Behav Brain Res 136(1):113–126

McKelvey R, Palfrey T (1995) Quantal response equilibria for normal form games. Games Econ Behav 10:6–38

Montague PR, Dayan P, Person C, Sejnowski TJ (1995) Bee foraging in uncertain environments using predictive Hebbian learning. Nature 377(6551):725–728

Montague PR, Dayan P, Sejnowski TJ (1996) A framework for mesencephalic dopamine systems based on predictive Hebbian learning. J Neurosci 16:1936–1947

Moore CC (1966) Ergodicity of flows on homogeneous spaces. Am J Math 88:154–178

Morris R (1984) Developments of a water-maze procedure for studying spatial learning in the rat. J Neurosci Methods 11(1):47–60

Mumford D (1992) On the computational architecture of the neocortex. II. The role of corticocortical loops. Biol Cybern 66:241–251

Nara S (2003) Can potentially useful dynamics to solve complex problems emerge from constrained chaos and/or chaotic itinerancy? Chaos 13(3):1110–1121

Neisser U (1967) Cognitive psychology. Appleton/Century-Crofts, New York

Nestler EJ (2005) Is there a common molecular pathway for addiction? Nat Neurosci 8(11):1445–1449

Niesink RJ, Van Ree JM (1989) Involvement of opioid and dopaminergic systems in isolation-induced pinning and social grooming of young rats. Neuropharmacology 28(4):411–418

Niv Y, Schoenbaum G (2008) Dialogues on prediction errors. Trends Cogn Sci 12(7):265–272

Nowak M, Sigmund K (1993) A strategy of win-stay, lose-shift that outperforms tit-for-tat in the prisoner's Dilemma game. Nature 364:56–58

O'Keefe J, Dostrovsky J (1971) The hippocampus as a spatial map. Preliminary evidence from unit activity in the freely-moving rat. Brain Res 34(1):171–175

Panksepp J, Siviy S, Normansell L (1984) The psychobiology of play: theoretical and methodological perspectives. Neurosci Biobehav Rev 8(4):465–492

Panksepp J, Knutson B, Burgdorf J (2002) The role of brain emotional systems in addictions: a neuro-evolutionary perspective and new 'self-report' animal model. Addiction 97(4):459–469

Pasquale V, Massobrio P, Bologna LL, Chiappalone M, Martinoia S (2008) Self-organization and neuronal avalanches in networks of dissociated cortical neurons. Neuroscience 153(4):1354–1369

Pierce RC, Kalivas PW (1997) A circuitry model of the expression of behavioural sensitization to amphetamine-like psychostimulants. Brain Res Brain Res Rev 25(2):192–216

Porr B, Wörgötter F (2003) Isotropic sequence order learning. Neural Comput 15(4):831–864

Rabinovich M, Huerta R, Laurent G (2008) Neuroscience. Transient dynamics for neural processing. Science 321(5885):48–50

Rao RP, Ballard DH (1999) Predictive coding in the visual cortex: a functional interpretation of some extra-classical receptive-field effects. Nat Neurosci 2(1):79–87

Redgrave P, Gurney K (2006) The short-latency dopamine signal: a role in discovering novel actions? Nat Rev, Neurosci 7(12):967–975

Redish AD (2004) Addiction as a computational process gone awry. Science 306:1944–1947

Rescorla RA, Wagner AR (1972) A theory of Pavlovian conditioning: variations in the effectiveness of reinforcement and nonreinforcement. In: Black AH, Prokasy WF (eds) Classical conditioning II: current research and theory. Appleton/Century Crofts, New York, pp 64–99

Robbe D, Buzsáki G (2009) Alteration of theta timescale dynamics of hippocampal place cells by a cannabinoid is associated with memory impairment. J Neurosci 29(40):12597–12605

Salzman CD, Belova MA, Paton JJ (2005) Beetles, boxes and brain cells: neural mechanisms underlying valuation and learning. Curr Opin Neurobiol 15(6):721–729

Schultz W (1998) Predictive reward signal of dopamine neurons. J Neurophysiol 80(1):1–27

Schultz W, Dickinson A (2000) Neuronal coding of prediction errors. Annu Rev Neurosci 23:473–500

Schultz W, Dayan P, Montague PR (1997) A neural substrate of prediction and reward. Science 275:1593–1599

Seip KM, Pereira M, Wansaw MP, Reiss JI, Dziopa EI, Morrell JI (2008) Incentive salience of cocaine across the postpartum period of the female rat. Psychopharmacology 199(1):119–130

Sheynikhovich D, Chavarriaga R, Strösslin T, Arleo A, Gerstner W (2009) Is there a geometric module for spatial orientation? Insights from a rodent navigation model. Psychol Rev 116(3):540–566

Shreve S, Soner HM (1994) Optimal investment and consumption with transaction costs. Ann Appl Probab 4:609–692

Sutton RS, Barto AG (1981) Toward a modern theory of adaptive networks: expectation and prediction. Psychol Rev 88(2):135–170

Takahashi Y, Schoenbaum G, Niv Y (2008) Silencing the critics: understanding the effects of cocaine sensitization on dorsolateral and ventral striatum in the context of an actor/critic model. Front Neurosci 2:86–99

Tani J, Ito M, Sugita Y (2004) Self-organization of distributedly represented multiple behavior schemata in a mirror system: reviews of robot experiments using RNNPB. Neural Netw 17:1273–1289

Thiagarajan TC, Lebedev MA, Nicolelis MA, Plenz D (2010) Coherence potentials: loss-less all-or-none network events in the cortex. PLoS Biol 8(1):e1000278

Todorov E (2006) Linearly-solvable Markov decision problems. In: Scholkopf et al (ed) Advances in neural information processing systems, vol 19, pp 1369–1376. MIT Press, Cambridge

Traulsen A, Claussen JC, Hauert C (2006) Coevolutionary dynamics in large, but finite populations. Phys Rev E, Stat Nonlinear Soft Matter Phys 74(1 Pt 1):011901

Tschacher W, Haken H (2007) Intentionality in non-equilibrium systems? The functional aspects of self-organised pattern formation. New Ideas Psychol 25:1–15

Tsuda I (2001) Toward an interpretation of dynamic neural activity in terms of chaotic dynamical systems. Behav Brain Sci 24(5):793–810

Tyukin I, van Leeuwen C, Prokhorov D (2003) Parameter estimation of sigmoid superpositions: dynamical system approach. Neural Comput 15(10):2419–2455

Tyukin I, Tyukina T, van Leeuwen C (2009) Invariant template matching in systems with spatiotemporal coding: a matter of instability. Neural Netw 22(4):425–449

van Leeuwen C (2008) Chaos breeds autonomy: connectionist design between bias and babysitting. Cogn Process 9(2):83–92

Verschure PF, Voegtlin T, Douglas RJ (2003) Environmentally mediated synergy between perception and behavior in mobile robots. Nature 425:620–624

Watkins CJCH, Dayan P (1992) Q-learning. Mach Learn 8:279–292

Wittmann BC, Bunzeck N, Dolan RJ, Duzel E (2007) Anticipation of novelty recruits reward system and hippocampus while promoting recollection. Neuroimage 38:194–202

Zack M, Poulos CX (2009) Parallel roles for dopamine in pathological gambling and psychostimulant addiction. Curr Drug Abus Rev 2(1):11–25

Zhao Y, Kerscher N, Eysel U, Funke K (2001) Changes of contrast gain in cat dorsal lateral geniculate nucleus by dopamine receptor agonists. Neuroreport 12(13):2939–2945

Zink CF, Pagnoni G, Chappelow J, Martin-Skurski M, Berns GS (2006) Human striatal activation reflects degree of stimulus saliency. Neuroimage 29:977–983

Chapter 10
Toward a Computationally Unified Behavioral-Economic Model of Addiction

E. Terry Mueller, Laurence P. Carter, and Warren K. Bickel

Abstract This chapter describes an instance of computational constructionism applied to the understanding of drug addiction. Rather than devising models of increasingly smaller anatomical, physiological or chemical units of analysis, the practice exposited here was to expand the integrative scope of behavioral-economic concepts that have been used to describe addiction phenomena. We discussed (a) excessive and persistent consumption of substances, as studied in analysis of demand; and (b) concurrent-choice preference for immediate small and unhealthy reinforcers over delayed but large and healthy reinforcers, and reversals of preference between this two types of alternatives, as studied in the science of delay discounting. While all of these phenomena are characteristic of addiction, it is remarkable how seldom concepts for explaining (a) appear in scientific reports on (b). The notion of expanding the scope of an explanatory concept is introduced via consideration of the concept of unit price. This concept integrates numerous variables traditionally studied in isolation in addiction research, and provides the basis of more general and parsimonious explanations of addiction phenomena. The scope of the unit price concept is further expanded, as it plays a role in a computational formulation describing choice among, and reversals of preference between, concurrently available reinforcers, which are very complex aspects of behavior that are fundamental to addiction phenomena. Lastly, we discuss computational implications and cautions, and scientific and practical prospects that derive from the exercise of expanding the integrative scope of the unit price concept to a broader and more complex range of addiction phenomena. Future developments along these lines are expected to pro-

E.T. Mueller
Advanced Recovery Research Center, Virginia Tech Carilion Research Institute, Virginia Tech, 2 Riverside Circle, Roanoke, VA 24016, USA

L.P. Carter
Center for Addiction Research, Psychiatric Research Institute, University of Arkansas for Medical Sciences, Little Rock, AR 72205, USA

W.K. Bickel (✉)
Advanced Recovery Research Center, Virginia Tech Carilion Research Institute and Department of Psychology, Virginia Tech, 2 Riverside Circle, Roanoke, VA 24016, USA
e-mail: wkbickel@vtc.vt.edu

duce constructs with which preferences exhibited by drug-dependent individuals may be predicted more accurately, and may be therapeutically modified.

10.1 Introduction

> "Another and less obvious way of unifying the chaos is to seek common elements in the diverse mental facts rather than a common agent behind them, and to explain them constructively by the various forms or arrangements of these elements as one explains houses by stones and bricks."
>
> William James (James 1918, p. 1)

The scientific study of addiction presents an interesting and complex set of challenges. One important source of those challenges is the nature of the paradigms employed by the scientific disciplines addressing this disease. As with most other sciences, the science of addiction largely follows the reductionist paradigm and program. The central notion underlying reductionism is that complex whole results may be understood by studying smaller and smaller components or elements of the larger phenomena (Skurvydas 2005; Soto and Sonnenschein 2005; Strange 2005). Given the remarkable productivity of this approach, it is likely to remain the dominant paradigm for some time.

One result of the reductionist approach is a continual increase in the number of research reports published. However, this productivity might have an unintended and troubling consequence; namely, to sustain their viability in a discipline, scientists are compelled to learn more about increasingly smaller components of their primary phenomena. Said another way, they need to know more and more about less and less. A potential consequence of this specialization is the formation of intellectual silos where new knowledge fails to be communicated and disseminated outside of the relatively small community investigating that particular level of the phenomenon. These intellectual silos are not a problem if the "cause" of the disease the investigators wish to abate will become evident at some progressively finer level of analysis (Evans 2008). However, if understanding the causes of a disease will benefit from or even require the integration of multidisciplinary areas of research, then the reductionist program is at risk of missing the important commonalities and relationships between different aspects of the disease process.

In this chapter, we take an approach that is antithetical to the dominant reductionist paradigm; our paradigm could perhaps be called *computational constructionism*. Our specific goal, which is consistent with many other computational approaches, is to build a more comprehensive model by integrating seemingly distinct aspects of the addiction phenomenon, and with the resulting model, generate testable hypotheses regarding the underlying mechanisms of, and potential treatments for, addiction.

Although consistent with the goals of other computational approaches to models of addiction, the focus of this chapter can be distinguished by the level of observation at which the computational formulae are descriptive. Rather than defining addiction by modeling, for example, a neurobiological component based on dopamine to explain the observed behavioral components of addiction, the alternative approach used here will address and model the behavior of the individual intact organism. We also try to avoid defining addiction via diagnostic criteria and/or statistically-codified symptoms of addiction. Instead, we portray addiction as a result of multiple behavioral processes that have been well-described and empirically supported by behavioral-economic studies of addicted individuals. Behavioral economics is an empirical analytical discipline in which much research on the variables controlling individuals' consumption of drug and non-drug commodities has been explored and computationally described.

Behavioral economics is the study of consumption, conceptualized as decision-making within an economic framework (Bickel and Christensen 2010) which applies constraint to consumption. Behavioral economics has shown that rationality in decision making, often assumed in classical economics, is bounded; that is, consumers' decisions are constrained by several limitations, and these limitations change the definition of what is "rational" behavior as compared to classical economic definitions. The concepts from behavioral economics that are the foci of this chapter are: (a) unit price and associated concepts of demand curve analysis; and (b) delay discounting and related concepts involving inter-temporal choice. These two groups of concepts (discussed below in detail) are well described computationally. They comprise two sets of principles that have been developed in research domains largely distinct from each other, although they both have been used to compare and contrast the behavior of drug-dependent individuals with that of people who are not drug-dependent (see Bickel et al. in press, for a review). Despite commonalities of subject matter and similarities in ostensible description (both approaches are called "behavioral economics"), the procedures, analyses, and findings of the two approaches have yet to be integrated into a single computational framework, despite calls for such integration (Epstein et al. 2010). The aim of this chapter is to move toward a more unified approach to studying the phenomena of addiction by starting with the behavior of the drug-dependent individual. Such an approach is complementary to neuroscience research and computational modeling and is likely to inform our understanding of addiction at the neuroscientific level.

10.2 Behavioral Economics of Addiction I: Basic Demand Curve Analysis

Drugs are consumable commodities: there are differences in the vigor with which different drugs are consumed and with which different individuals consume them, and their consumption is sensitive to differential levels of constraint in the form of prices. In this section, these features of drugs as commodities are discussed as they are revealed in *demand curve analysis*. In economic terms, the relationship between consumption of a commodity and its price is referred to as demand, and this relationship is portrayed in a demand curve. In other words, the demand curve is the result of plotting the amount of consumption as a function of commodity price. The

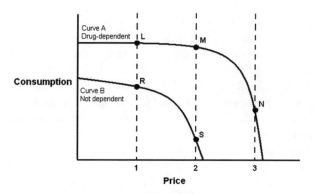

Fig. 10.1 Two hypothetical demand curves depicting drug consumption. Consumption is a continuous variable generalized from discrete measurements, and it is plotted as a function of price values generalized from discrete price points that may be implemented in a behavioral economics experiment. The scales of both axes are logarithmic, as this allows changes in the curvature of the function to be observed as straight-line slopes. *Curves A* and *B* both exhibit the negatively accelerated form that is characteristic of demand curves. Differences between *Curves A* and *B* illustrate important general characteristics of demand curves. The greater height of *Curve A* compared to *Curve B* illustrates greater demand intensity. *Curve B* is more elastic than *Curve A*, as reflected in a greater rate of decline in consumption. Point slopes (slopes of curve tangents) and curve segment slopes illustrate localized degrees of elasticity. Curve segment L–M is inelastic, as␣␣it slope is between zero and -1; curve segments M–N and R–S are elastic, as their slopes are less than -1

"law of demand" describes the reliable, generalized observation that as the price of a commodity increases the consumption of it tends to decrease. Generalized characteristics of demand curves have been determined, allowing useful analyses that compare and contrast demand curves representing the consumption of different drugs, as well as drug consumption by dependent individuals and populations as compared to consumption by those who are not drug dependent.

Behavioral economics operationalizes economists' observations of "prices" by measuring work expended to acquire a unit of a commodity. The behavioral-economic procedures for experimentally producing a demand curve for an individual's consumption of a particular commodity (Hursh 1991; Raslear et al. 1988) involve the implementation of a series of conditions differentiated by the price the individual must pay to consume the commodity. Within each condition (price point) the amount of the commodity consumed is measured. For example, Fig. 10.1 shows two hypothetical demand curves depicting drug consumption. Both curves exhibit the characteristic form of demand curves, and differences between Curves A and B illustrate important concepts used in the analysis of demand curves to describe consumption of drugs and other commodities. Two problematic behaviors associated with addiction are illustrated in Fig. 10.1: the consumption of larger and larger amounts of drug (i.e., escalation of use) and the persistence of drug-taking behavior despite increasing costs.

The characteristics of demand curves, that is, the *intensity of demand* (the height of the curve) and the *elasticity of demand* (the point slope), represent empirical generalizations that are keenly relevant to clinical care and public policy. For example, if the curves in Fig. 10.1 represent consumption by two different individuals, an intervention or policy that effectively increases the price of drug use from 1 to 2 would be expected to change consumption by a drug-dependent individual from point L to point M (Curve A) and from point R to point S in an individual who is not drug-dependent (Curve B). The predicted effects of this policy would be a decrease in consumption by individuals who are not drug-dependent because the change in price from point R to point S lies along an elastic part of that demand curve. However, very little change in consumption would be predicted for drug-dependent individuals because the change in price from point L to point M lies along an inelastic part of that demand curve. Thus, detailed knowledge of the characteristics of demand curves showing drug consumption and the location of current and projected prices along those curves for affected individuals and populations can be useful in designing and predicting the effectiveness of new therapeutic and regulatory strategies (e.g., drug immunization or taxes on drugs) to treat and prevent addiction (Bickel and DeGrandpre 1995, 1996; Hursh 1991).

The curves in Fig. 10.1 may also be used to illustrate differences between drugs with regard to problems of excessive consumption and persistence of use, also known as the abuse potential or *abuse liability* of a substance (Griffiths et al. 1979; Schuster and Thompson 1969). Curve A would represent the behavioral-economic features of a drug with a high potential for abuse (e.g., crack cocaine, which tends to be used frequently despite increasing costs), and Curve B would depict consumption patterns of a drug with comparatively less abuse potential (e.g., marijuana). Demand curve analysis has been used frequently to assess the abuse liability of drugs (e.g., Ko et al. 2002; Mattox and Carroll 1996; Winger et al. 2006). Similarly, demand curve analysis has been used to assess human subjects' *susceptibility to abusing drugs* such as opiates (Greenwald and Hursh 2006) and nicotine (MacKillop et al. 2008) and alcohol (MacKillop and Murphy 2007; Murphy and MacKillop 2006). In conformity with the law of demand, drug consumption has been shown to decrease as price is increased at both the individual (Johnson et al. 2007b) and population levels (Bach and Lantos 1999; Caulkins 2001; for a review, see Chaloupka et al. 1999; Corman et al. 2005; Darke et al. 2002a, 2002b; Longo et al. 2004; Schifano and Corkery 2008; Schifano et al. 2006).

In this section, we have discussed demand intensity and elasticity, demand curve analysis concepts relevant to study of drug consumption behaviors. These concepts describe consumption that is constrained via price manipulation. In the next section, we discuss a different approach to constraint on commodities.

10.3 Behavioral Economics of Addiction II: Analysis of Inter-temporal Choice

Addictive drugs are positive reinforcers. A drug is considered to be a positive reinforcer if an effect of consuming that drug is the increased or continued likelihood that the consumption of that drug will recur in the future. Drugs are studied as reinforcers by many researchers, and the study of drug-taking as an operant behavior has led to many advances in the understanding of addiction (Bigelow et al. 1998, 1983; Higgins et al. 1994). Addiction is often defined as persistent drug consumption in the face of undesirable consequences; however, it is typically the case that the negative consequences of drug use or addiction are located at some distance in the future, whereas the reinforcing effects of drug use (e.g., the drug high) are almost immediate. Another common feature of drug dependence or addiction is the problem of *relapse*, characterized by a period of abstinence from drug consumption and an expressed intent or preference to not use drugs, which is followed by a return to drug use. The resolve to avoid drug use that is reversed in a relapse episode is originally grounded in the individual's recognition of the larger magnitude of non-drug reinforcers; however, the influence of these large-magnitude reinforcers is significantly affected by distance in time until receipt of those reinforcers as compared to the immediate reinforcing effects of drug self-administration. In this section, we discuss the behavioral-economic approach to understanding the effects of both *reinforcement amount* and *reinforcement delay*.

Most if not all people will claim to prefer receiving $100 today as opposed to $100 one year from today. In other words, it is a generalized phenomenon that the present subjective value of a delayed amount of a reinforcer is less than the nominal amount expressed in present-day terms—a process that is more succinctly called *delay discounting* or *temporal discounting*. While it is intuitive that the value of a reinforcer would be discounted with delay, the quantitative rate of such discounting is not obvious and has been shown to vary across persons and across situations. For example, for person A, $1000 to be received in one year might be subjectively worth the same as $950 now. If this discounting rate for person A (each year of postponement equals a $50 decline in value) remains valid for a hypothetical delay of ten years, then person A would agree that $1000 to be received in ten years would be worth $500 now, since $1000 - (10 \times \$50) = \$500$. And if the rate were generally true for person A then he or she would agree that $1000 to be received after 20 years is worth nothing, since $1000 - (20 \times \$50) = \$0$. Consistency of this sort is not generally observed within an individual's choices because rates of discounting across time are not typically linear. Another factor that complicates the quantification of discounting is that rates of discounting can vary markedly from person to person. For example, addicted persons exhibit greater rates of temporal discounting as compared to individuals who are not addicted to drugs (see below for examples). Another factor that can affect the assessment of discounting is that rates of discounting for different commodities may vary even within a single person. For example, if it is determined that an individual with heroin addiction assigns a present subjective value for $1000 in currency to be received in one year at $500, it could not be inferred that person's present subjective value for a package of heroin with a street value of $1000 to be received in one year is also $500. In general, drug addicts have been shown to discount their drug of choice to a greater extent than equivalent amounts of money. In summary, the quantification of individuals' discounting rates for delayed reinforcement is constantly being refined by a subgroup of researchers within behavioral economics. This research is the *science of temporal discounting* (Bickel and Marsch 2001; Mazur 1987).

A delay discounting rate is typically determined via assessment procedures used to study an *inter-temporal choice*. Such procedures use a series of binary choice trials in which the choices are between a smaller-sooner (SS) amount and a larger-later (LL) amount. The amount of the immediate SS reinforcer is varied across trials, while the amount of the LL option remains constant. Choices in the series of trials determine the present subjective value of the delayed LL amount. Across trials, the procedure is programmed to offer "now amount" choices that become increasingly close to the point of indifference by adjusting the amount of the "now" alternatives in coordination with the participant's selection on the previous trial. Ultimately, an *indifference point* is determined, which is defined as the smallest immediate SS amount used in the procedure at which the SS amount is preferred and at which smaller SS amounts would shift preference to the larger, later amount.

Discounting assessment procedures have been devised for measuring the discounting rates of humans (Kowal et al. 2007) and non-human subjects (Mazur 1987). With human participants, the use of hypothetical reinforcers has been shown to yield the same results as studies using actual reinforcers (Bickel et al. 2009; Johnson and Bickel 2002). The end result of one application of a procedure will be the determination of a single indifference point, or one value that represents the present subjective value of the larger reinforcer if its receipt were delayed until one specific later time. A discounting rate for that amount of the reinforcer is determined by assessing indifference points for several different delays. The delay discounting procedure is repeated with the same larger reinforcer amount, but with a different delay so that several indifference points at different delays are identified. After three to seven indifference points have been determined in this way, they are plotted as a function of the delay until the LL reinforcer. The points are fitted by non-linear regression to a function, and the value of a free parameter, called k, is the subject's delay discounting rate for that particular amount of that particular reinforcing commodity.

Determining the rates for the discounting due to delay is important for an understanding of addiction because it quantifies how quickly the present subjective value of a commodity diminishes over time. This is thought to be analogous to the extent to which an individual might be influenced by the immediate reinforcing effects of drug administration (e.g., the drug high) but not by the temporally distant negative or harmful consequences of that behavior or by the delayed positive and healthy consequences of avoiding drug self-administration behavior. Research has shown that individuals who are dependent on alcohol (Bjork et al. 2004; Petry 2001; Vuchinich and Simpson 1998), nicotine (Baker et al. 2003; Bickel et al. 1999; Heyman and Gibb 2006; Johnson et al. 2007a; Mitchell 1999), heroin (Kirby et al. 1999; Madden et al. 1997; Odum et al. 2000), cocaine (Coffey et al. 2003; Heil et al. 2006; Kirby and Petry 2004), and methamphetamine (Hoffman et al. 2006, 2008; Monterosso et al. 2007) exhibit greater rates of temporal discounting as compared to individuals who are not dependent on those drugs. Moreover, higher rates of delay discounting among cigarette smokers have been shown to be associated with greater dependence (Johnson et al. 2007b; Ohmura et al. 2005) and poorer treatment outcomes (Audrain-McGovern et al. 2009; Krishnan-Sarin et al. 2007; MacKillop and Kahler 2009; Yoon et al. 2007). Also, children with mothers who smoke are at greater risk of themselves being high discounters (Reynolds et al. 2009). Together these observations suggest that the rate of delay discounting is predictive of the likelihood, severity, and prognosis of addiction as it pertains to cigarette smoking, and likely other addictions as well.

Defining the form of the regression function that most accurately represents the rate of delay discounting is an ongoing concern for the science of temporal discounting. Candidate functions fall into two major groups, based on whether the basic form is exponential or hyperbolic. An important virtue of the hyperbolic model is that it can account for the phenomenon of preference reversals, whereas exponential models generally cannot (Green and Myerson 1996). *Reversals of preference* are a hallmark feature of drug dependence and other addictive disorders, as individuals will frequently prefer to "get on the wagon" only to fall off of it, exhibiting an unfortunate reversal of their stated preference. Moreover, preference reversals appear to be an evolutionarily conserved phenomenon, as they have been demonstrated with non-human animals (Ainslie and Herrnstein 1981; Green and Estle 2003a; Green et al. 1981; Rachlin and Green 1972b) as well as humans (Brandon et al. 1990; Green et al. 1994; Kirshenbaum et al. 2009; Marlatt et al. 1988; Nides et al. 1995; Norregaard et al. 1993; Shiffman et al. 1996; Westman et al. 1997; Yoon et al. 2007). The function that relates the subjective value of a reward to delay until that reward can be interpreted as describing the process of waiting for a reward. If such functions for a larger reward and a smaller reward are hyperbolic in form, then the interaction of the two waiting processes results in reversal of preference. However, if those functions are exponential in form, then the waiting processes that they describe do not result in preference reversal. Figure 10.2 illustrates why reversals of preference could be predicted by hyperbolic delay discounting functions and not by exponential delay discounting functions.

In this third section of the chapter, we discussed the behavioral economic concepts that describe the interactions between reinforcement amount and reinforcement delay. The phenomena of delay discounting and preference reversal are interpreted in behavioral economics as resulting from such interactions. As delay discounting and preference reversals are characteristic of addiction, a better understanding of these behavioral-economic concepts will ultimately lead us to improved understanding of the dynamics of addiction. It might be interesting to note that up to this point Sects. 10.2 and 10.3 stood alone without much reference to each other. The remaining sections of this chapter will discuss how the concepts from Sects. 10.2 and 10.3 interact, how they may be integrated, and how they provide a framework for understanding addiction.

10.4 Integration of Variables I: Extension of Price to Unit Price

Unlike the theories in classical economics, behavioral economics theories should explain the behavior of non-humans as well as humans. As a result, assumptions about the meaning of "price" in this science cannot be unquestioningly

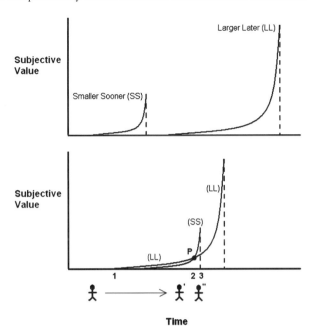

Fig. 10.2 Illustration of a preference reversal as an interaction of inverted hyperbolic delay discounting functions for a smaller-sooner (SS) reward and a larger-later (LL) reward. The functions relate subjective value of a reward to the inverse of delay till its receipt. (Delay discounting functions, by comparison, relate subjective value of a reward to delay till its receipt; they are mirror images on the *vertical axis* of functions shown in the figure.) Subjective value of a reward at a time point is represented by the height of the curve at that time. The passage of time until receipt of the reward is on the *X-axis*. A curve terminates at the nominal amount of the reward at the rightmost point on the curve; this denotes the time at which delay-to-reward has diminished to zero. The panels depict the individual processes of waiting for SS and LL rewards as inverted hyperbolic discounting functions. In the *top panel*, the increase in subjective value of a reward as delay till receipt of the reward becomes smaller is quantitatively described by hyperbolic functions; note that this quantitative character would be different for exponential functions (not shown). The *bottom panel* depicts the reversal of preference that would occur if those two processes occurred in overlapping temporal ranges and only one of the two rewards could be chosen. The relative preferences between the two rewards are depicted by the heights of the curves for each reward. The relative heights of the *two curves* are shown to be different in two time ranges: from time 1 to time 2 the individual (denoted as stick figure) prefers the LL reward; from time 2 to time 3 the same individual (denoted as stick figure double-prime) prefers the SS reward. Thus, preference reversal follows from the intersection of the *two curves* at *point P* (where the individual is depicted as stick figure prime). If the delay discounting function were exponential in form, the curves depicting the processes of waiting for SS and LL would not intersect and preference reversals would not be predicted

adopted from common knowledge about accepted mediums of exchange in human society (Hursh et al. 1988). Instead, price must be operationally defined. Initially, simple price was defined as a count of the number of responses required for reinforcer delivery, such as pecks on a key by a pigeon or presses of a lever by a rat, as a measure of the energy cost for reinforcement. The concept of unit price differs from that of simple price in that unit price is the effort or response requirement (i.e., a measure of cost) per unit of the reinforcing commodity (i.e., a measure of benefit). *Unit price* is a cost-benefit ratio that reflects the interaction of costs and benefits in a single variable (Collier et al. 1986; Hursh et al. 1988). It is a more integrative concept than simple price because it incorporates seemingly disparate variables into a single measure. In this section of the chapter, we review this conceptual integration. We discuss how unit-price demand curve analysis subsumes two traditionally distinct variables in operant research, producing results across a broad range of situations that conform to a generalized pattern; and we reveal the economy of communication, or parsimony, that the unit price notion introduces to the concepts used in the analysis of demand.

Variables that are integrated in the unit price concept are illustrated in two varieties of operant drug self-administration studies, where measures of costs and benefits are readily quantified and manipulated. In one type of traditional drug self-administration experiment, simple prices are manipulated by varying the number of responses (behavioral costs) required to produce a drug self-administration, allowing consumption to be portrayed as a function of response requirement in a standard demand curve. In a second type of traditional drug self-administration experiment, the dose of the drug (i.e., the amount of the reinforcer, a benefit variable) is manipulated, permitting rate of response to be examined as a function of dose in a traditional dose-response curve. These two variables—response requirement as cost and dose as benefit—have typically been studied in operant research paradigms as two distinct variables that affect drug self-administration.

Several lines of evidence show that unit price may be used to integrate response-requirement and drug-dose variables. For example, a review of ten self-administration studies using rats, squirrel monkeys, or rhesus monkeys showed that manipulations of response requirement and dose of drug have functionally equivalent effects on drug consumption (Bickel et al. 1990). Specifically, it was shown that doubling the drug dose had the same effect on drug consumption as requiring half as many responses per reinforcement, since both operations had the same effect on unit price, where *unit price* = (*response requirement / dose*). This review also showed that consumption plotted as a function of unit price on double logarithmic axes resulted in a negatively accelerated demand curve.

These two features of unit price analysis, *functional-equivalence* and *negatively accelerated demand curve*, were also observed in a human study in which participants self-administered nicotine in the form of cigarette puffs. In a laboratory setting where the response requirement was either 200, 400, or 1,600 responses and the reinforcer was either 1, 2, or 4 puffs, experimental parameters that resulted in the same unit prices (e.g., 200 responses/2 puffs and 400 responses/4 puffs both yield 100 responses per puff) resulted in similar levels of consumption, and the consumption data for each of these conditions converged on negatively accelerated demand curves when plotted as a function of unit price (Bickel et al. 1991).

In the drug self-administration studies described above, the costs and benefits determining unit price were operationalized as response requirement and drug dose, respectively. However, the concept of unit price has been shown to encompass a broader range of operations that function as costs or benefits in a unit price equation.

For example, in a study in which lever-presses by rats were reinforced by the delivery of food pellets, the potential costs were varied by manipulating the number of lever-presses required for reinforcer delivery (i.e., the fixed-ratio requirement) and also by manipulating the amount of force required to press the lever. The potential benefits were varied by manipulating the number of food pellets delivered per completion of a fixed ratio requirement (analogous to dose in drug studies) and also by manipulating the probability of reinforcer delivery after completion of a fixed-ratio requirement (Hursh et al. 1988). This study showed that each of these four variables comprising unit price had functionally equivalent effects on consumption to the extent that they resulted in equivalent changes in unit price. When consumption in this study was plotted as a function of unit price on double-logarithmic coordinates, the data points with similar unit prices converged on the same negatively accelerated function regardless of which variables and values for those variables were used to calculate the unit price. This report proposed the following equation to quantitatively describe the negatively accelerated function:

$$\text{Log } Q = \text{Log } L + b(\text{Log } P) - aP \tag{10.1}$$

where Q is the total consumption; P is unit price; and L, b, and a are fitted parameters for the level of consumption at a price of 1, the initial downward slope of the curve, and the acceleration by which the slope increases with changes in unit price, respectively (Hursh et al. 1988).

Another study (DeGrandpre et al. 1993) reviewed the results of 32 experiments with rats, pigeons, dogs, or monkeys as subjects. This review showed that across a variety of reinforcer types (sucrose, food, cocaine, d-amphetamine, procaine, codeine, morphine, methohexital, pentobarbital) and a variety of types of reinforcer magnitude (concentration and volume of liquid food, food pellet size, drug concentration, duration of access to food), the different types of varying reinforcer magnitude were functionally interchangeable as benefit factors in the unit price equation. This review also showed that the two most commonly used reinforcement schedules (ratio and interval schedules) result from operations that have the same effect as known costs on unit price. Furthermore, the review demonstrated the generality of the negatively accelerated demand curve for consumption where unit price rather than simple price is on the X-axis and log-log coordinates are used, suggesting that this form of the demand function occurs ubiquitously. In summary, across a diverse group of studies, the combination of cost and benefit factors into a single unit price variable has been demonstrated to parsimoniously describe the effects of a broad range of experimental conditions; and the form of the function exhibited when consumption is plotted as a function of unit price on double logarithmic coordinates is ubiquitous, appearing to be generalizable to a broad range of types of consumption (Bickel et al. 1993).

Because the negatively accelerated form for the unit-price demand curve appears to be a ubiquitous phenomenon, it is considered to be a standard form by which consumption data are recognized as exhibiting order, in a manner analogous to how data that conform to a linear function have traditionally been construed as orderly.

As a standard of orderliness, the negatively accelerated demand curve form is relevant to the treatment of addiction. Controversy surrounding the nicotine regulation hypothesis provides an example. Scientists have hypothesized that smokers regulate nicotine levels—in other words, smoking behavior occurs at a rate that will maintain a characteristic level of nicotine in the smoker's body. Many studies have attempted to settle this controversy. Each of the studies examined nicotine intake or consumption of nicotine by smokers as a dependent variable; however, the studies differed in the way that the nicotine dose could be manipulated as an independent variable. The general conclusion from this body of research was that blood levels of nicotine are regulated by smokers; however, the conclusion could not be stated forcefully because between-study discrepancies and contradictions were evident when it was assumed that consistent results should converge on a linear function relating nicotine consumption to nicotine yield (dose) in cigarettes (Gritz 1980; Henningfield 1984; McMorrow and Foxx 1983; Moss and Prue 1982).

It is also possible that the relationship between nicotine consumption and nicotine yield (dose) in cigarettes was not linear. An alternative approach to analyzing the data from the nicotine regulation studies (DeGrandpre et al. 1992) did not presume that results consistent with each other would converge on a linear function. In that review, the independent variable (nicotine yield) was taken to be a benefit variable in the unit price equation (unit price = 1/nicotine yield) and the nicotine yield (consumption) data were reanalyzed as a function of unit price. That review found that the data from the studies were much more consistently described by a negatively accelerated unit-price demand curves as specified by Eq. (10.1). Curves in Figs. 10.3a and 10.3b, reprinted from DeGrandpre et al. (1992), illustrate this point. The curves reflect 25 reinterpreted data sets from the 17 reviewed studies (Ashton et al. 1979; Benowitz et al. 1986; Creighton and Lewis 1978; Frith 1971; Goldfarb et al. 1976; Griffiths et al. 1982; Gust and Pickens 1982; Haley et al. 1985; Henningfield and Griffiths 1980; Hill and Marquardt 1980; Jarvik et al. 1978; Russell et al. 1980, 1973, 1975; Stepney 1981; Turner et al. 1974; Zacny and Stitzer 1988). Nicotine consumption is plotted as a function of unit price. The variance accounted for (VAC) in the individual data sets (R^2) by Eq. (10.1) was very high (average VAC: 99.7%; range: 96.4–100%; $P < 0.01$ in all cases). However, high R^2 values might have simply been a result of fitting individual curves to a relatively small number of data points (three data points were used in the majority of these cases). As a more stringent test of Eq. (10.1) as a standard of orderliness, DeGrandpre et al. (1992) also computed a version of Eq. (10.1) in which the parameters that were assumed in the equation were the means of the parameters that resulted from fitting the individual data sets. The R^2 values reflecting the variance in the individual data sets accounted for by this "mean equation analysis" were also high (average VAC: 96.2%; range: 89.0–100%; $P < 0.05$ in 18 of 25 cases). DeGrandpre et al. (1992) suggested that some VAC values may have been even higher but for the likelihood that some studies did not include a broad enough range of unit prices to produce a complete demand curve.

Further support for Eq. (10.1) as a standard of orderliness comes from five reviewed studies that reported measurements of blood nicotine in combination with a

Fig. 10.3 Graphic illustration of convergence on Eq. (10.1) of 25 nicotine-regulation-experiment data sets from 17 studies. Nicotine consumption (intake) is plotted as a function of unit price in log-log coordinates, where unit price = (1/nicotine yield). The curve drawn through each data set is an approximate line of best fit. Nicotine regulation studies were included in the reanalysis unless (a) the study assessed an insufficient number of nicotine yields (unit prices)—at least three unit prices are required to establish a demand curve; (b) nicotine consumption in the experiment could not be determined; (c) the nicotine yield manipulation of the study was difficult to quantify as a unit price (e.g., there were nicotine preloads); or (d) nicotine consumption in the study was confounded by a variable other than unit price (e.g., nicotine yield values were grouped and thus precluded determination of an exact unit price or subjects were trying to quit smoking). The 17 studies are grouped according to the method for manipulation of nicotine intake (brandswitching versus shortened cigarettes). For each panel, the publication is shown. The method of estimation of nicotine intake for each data set is indicated in the *lower left area in each panel*: NI-C = cigarette consumption; NI-P = smoke consumption taken from puff number, volume and/or duration; NI-BN = blood nicotine levels. The illustrated convergence to Eq. (10.1) reflects variances accounted for by Eq. (10.1). Percent of variance is based on comparison, using linear regression analysis, between observed nicotine consumption and those predicted by Eq. (10.1) when values are assumed for the fitted parameters a and b and an intercept at unit price = 1. When a and b were fitted from the individual data set the mean percentage VAC was 99.7% (range: 96.4–100%; $P < 0.01$ in all cases). When a and b parameters used in Eq. (10.1) were the means of those values from the individual data sets, the mean percentage variance accounted for was 96.2% (range: 89.0–100%; $P < 0.05$ in 18 of 25 cases)

behavioral measure (smoke consumption or cigarette consumption) of nicotine consumption. For these studies, the mean VAC for the behavioral measure was 95.6%, and for the blood nicotine measure it was 93.9%. This consistency of VAC values across these measures also supports Eq. (10.1) as a standard of orderliness for nicotine intake data.

In summary, when the standard of orderliness for data was assumed to be a negatively accelerated function described by Eq. (10.1), the data from various studies of nicotine regulation show greater consistency (i.e., conform to a generalized pattern) than when the standard of orderliness is a linear function. This 1992 review by DeGrandpre and colleagues helped to substantiate the generality that smokers regulate their consumption of nicotine based on the nicotine content of the cigarettes smoked. It showed that smokers' nicotine consumption appears to obey a unit-price-based law of demand, as does consumption of other reinforcing commodities, generally. The review demonstrated that the manipulation of the unit price of nicotine (by manufacturers or via public policy) has predictable effects on drug consumption, and therefore such manipulations potentially have important implications for public health.

In this section of the chapter, we described the expansion of the economic notion of price to the behavioral-economic concept of unit price, which is a concept that integrates cost and benefit variables that interact in determining rate of consumption. Thus, we find that the independent variable of the law of demand, "price," is being expanded in two directions that are important to the understanding of addiction: (a) Toward incorporating a broader scope of what defines the costs of drug consumption; these expanded "prices" may include fewer and poorer quality social relations, diminished income production, antagonistic relations with legal authorities, psychological problems caused by drug consumption, tolerance to positive sub-

298 E.T. Mueller et al.

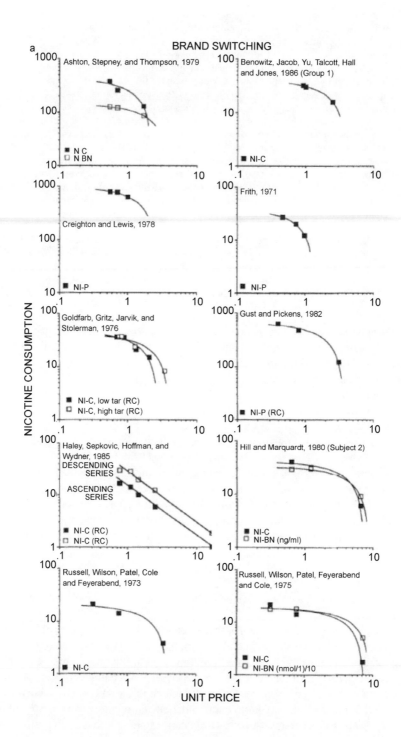

10 Toward a Computationally Unified Behavioral-Economic Model of Addiction

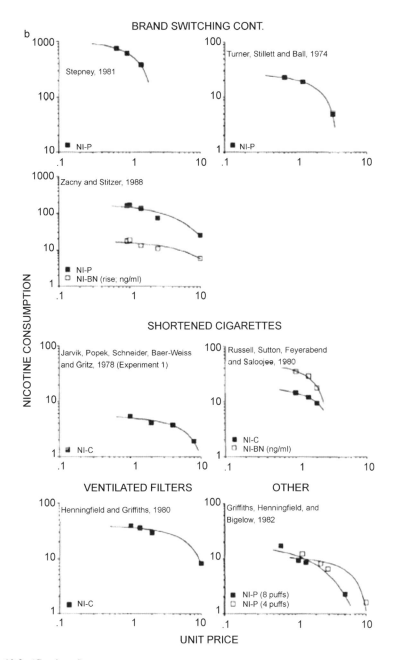

Fig. 10.3 (Continued)

jective effects, and a predominance of negative reinforcing effects; and (b) toward incorporating as drug-related "benefits" in the life of a drug dependent individual not only the reinforcing effects of drug consumption (e.g., the drug "high"), but also other sources of reinforcement (e.g., social relationships and status, escape from demands on one's time and effort, high monetary income in an illicit marketplace). This conceptual expansion supports the proposition that the analysis of demand for drug-related reinforcers may be a fundamental scientific activity in a comprehensive understanding of drug dependency. In the next section, we carry our conceptual expansion further to incorporate reinforcers in general, be they drug-related or non-drug-related.

10.5 Integration of Variables II: "Costs" and "Benefits" in Concurrent Choice

Although laboratory models of drug self-administration have been shown to have excellent internal, external, and predictive validity, there are a number of social, cultural, and environmental factors that can affect drug consumption outside of the controlled laboratory setting (Carter and Griffiths 2009). One of these factors is the availability of alternative non-drug reinforcers. For example, presenting drug-dependent individuals with a choice to either use drugs or receive money has repeatedly been shown to delay or decrease the use of drugs (Mueller et al. 2009; Silverman 2004). Outside of the laboratory, in the "real world," individuals are faced with a wide variety of choices among concurrently available drug and non-drug reinforcers. Thus, any computational approach to understanding drug use at the behavioral level must be able to account for situations of choice among and between different reinforcers (Bickel et al. 1993).

In Sect. 10.3 of this paper, we pointed out that the behavioral-economic study of delay discounting is essentially the quantitative study of choices whose outcomes are different from each other because they incorporate different combinations of delay and amount of a reinforcer. In Sect. 10.4, we described the generalization of the "cost" and "benefit" notions in demand-curve analysis and showed how a broader range of variables have been integrated into the unit price concept. In this section, we expand the integrative scope of the unit price concept to include situations of choice.

The science of delay discounting has been derived from learning theory that, in turn, was influenced by the matching law. This approach to learning partitioned the notion of reinforcement into several distinct dimensions (e.g., rate, probability, amount, and delay of reinforcement) and undertook the analysis of the effects of those dimensions as variables that would affect choice among reinforcers. The science of delay discounting shows that preference reversals occur from the interaction of the amounts and delays of a larger-later (LL) reinforcer and a smaller-sooner (SS) reinforcer. A formulation of the matching law that predicts preference reversals as resulting from such interactions is

$$\frac{D_{LL}}{A_{LL}} \leq \frac{D_{SS}}{A_{SS}}, \tag{10.2}$$

where D_{LL} and A_{LL} are the delay and amount values of the LL reinforcer and D_{SS} and A_{SS} are the delay and amount values of the SS reinforcer. Equation (10.2) predicts that under conditions in which there is the option between a LL reinforcer and a SS reinforcer, the proportion of selections of the LL option will be greater than or equal to 0.5 (qualitative preference for LL) if the term on the left of the inequality is smaller than the term on the right of the inequality. Conversely, the proportion of selections of the LL option would be predicted to be less than 0.5 (qualitative preference for SS) if the term on the right of the inequality is smaller.

Notice that if delay to reinforcement is construed as a cost of consumption and reinforcer amount is construed as a benefit of consumption, the left- and right-hand terms in Eq. (10.2) are cost-benefit ratios. That is, the left- and right-hand terms in Eq. (10.2) are unit prices of LL and SS reinforcers, respectively. Thus, Eq. (10.2) is effectively equivalent to Eq. (10.3)

$$UP_{LL} \leq UP_{SS}, \tag{10.3}$$

where UP_{LL} and UP_{SS} are the unit prices of larger-later and smaller-sooner reinforcers, respectively. Thus, Eq. (10.3) is at once a formulation expressed in terms integral to demand-curve analysis (unit prices) and also a formulation that is predictive of phenomena studied in the science of delay discounting (preference reversals).

A good testing ground for the predictive success of Eq. (10.3) would entail experimental observations of preference reversals in situations of choice. Such studies have been conducted and have experimentally demonstrated that subjects' preference between SS and LL reinforcers in different conditions may reverse across the conditions if those conditions entail effectively different combinations of delay and amount for SS and LL reinforcers. Thus, the ability of Eq. (10.3) to predict the conditions under which preference will reverse can be assessed in such experiments.

In preference reversal experiments subjects are characteristically offered choices between a LL reinforcer and a SS reinforcer (see Fig. 10.2 for a depiction of a hypothetical preference reversal). Across conditions of the experiment the amount and delay characteristics of the LL and SS reinforcers are controlled in this way: the reinforcer amounts of the LL reinforcer and the SS reinforcer remain constant across conditions; the difference between the delay until the SS reinforcer and the delay until the LL reinforcer also remains constant; whereas, the delay until the delivery of the SS reinforcer is varied across conditions. Thus, the primary independent variable in a preference reversal experiment is the delay until the SS. Within each of the conditions in which the delay until the SS reinforcer is delivered varies, the proportion of choices in which the LL reinforcer was selected is recorded (LL proportion). These LL proportion values can then be plotted as a function of the delay until SS. Figure 10.4 portrays data from a hypothetical preference reversal experiment in a manner typical of such experiments.

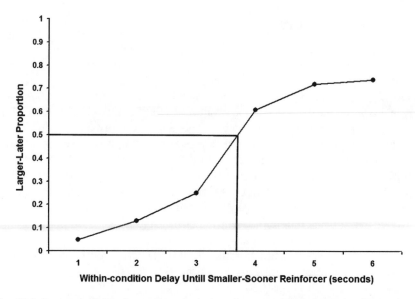

Fig. 10.4 Portrayal of data from a hypothetical preference reversal experiment. Proportion of choices in which larger-later reinforcer is selected (larger-later [LL] proportion) is plotted as a function of within-condition delay until the smaller-sooner (SS) reinforcer. Location of data points in relation to the horizontal line at the LL proportion of 0.5 indicates preference; *points above the line* (>0.5) indicate preference for the LL option; *points below the line* (<0.5) indicate preference for SS reinforcers; *points at larger-later proportion* = 0.5 indicate indifference (illustrated in Fig. 10.2 as point P). Preference reversals occur across conditions in which the curve crosses the line at larger-later proportion = 0.5. The *vertical line* in the figure that extends downward from the function to the *X-axis* indicates an estimate of delay-until-SS value at which the preference reversal would be expected to occur on the basis of the experimentally obtained data

A recent study (Mueller and Bickel 2010) reviewed the results from 17 experiments reported in nine published studies (Ainslie and Herrnstein 1981; Green and Estle 2003b; Green et al. 1981, 1994; Green and Snyderman 1980; Logue and Pena-Correal 1985; Navarick and Fantino 1976; Rachlin and Green 1972a; Snyderman 1983) of preference reversal to examine whether Eq. (10.3) could predict the points at which preference reversals were observed to occur. In this reanalysis, experimentally implemented delays until LL and SS reinforcer delivery were interpreted as behavioral-economic costs, and reinforcer amounts were interpreted as benefits, to create unit prices for the LL and SS reinforcers in each condition of the experiments. These were substituted into Eq. (10.3) as the left- and right-hand terms, respectively. Equation (10.3) was used to make predictions regarding the hypothesized preference for the LL versus the SS reinforcer within the conditions of the experiments. Data from this reanalysis showed that comparing UP_{LL} and UP_{SS} values resulted in correct predictions of observed preference (proportion of choices < or ≥ 0.5) in 619 of the 786 experimental conditions (78.8%) across all experiments. Figure 10.5 displays the percentages of correct predictions for each individual study examined in this analysis. In the subset of cases in which both a preference reversal was predicted and occurred, the correlation between the delays until the SS

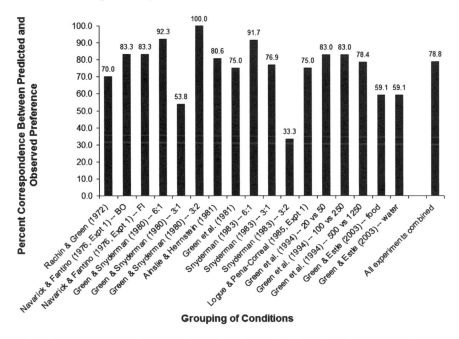

Fig. 10.5 Summary of results in a review of experimentally induced preference reversal experiments. On the *X-axis* are groupings of conditions across which an experiment may have demonstrated a reversal of preference between smaller-sooner reinforcers and larger-later reinforcers. The height of each *bar* represents the percentage of conditions in the group for which Eq. (10.3) correctly predicted empirically observed preferences

at which a preference reversal was predicted to occur and the observed delays until the SS at which a preference reversal occurred was 0.95 [Pearson $r(50) = 0.95$]. These findings provide strong empirical support for the value of comparing unit prices of LL and SS reinforcers as a means for predicting cross-condition reversals of preference between them.

In this section, we extended the scope of the unit price concept to include measures from operations involved in situations of choice. We demonstrated the effectiveness of comparing unit prices of larger-later versus smaller-sooner reinforcers to predict laboratory-induced preference reversals. Since choice among numerous concurrent alternatives is a prominent feature of the lives of individuals dependent on drugs, this extension of the unit price construct is another step toward a comprehensive understanding of drug addiction through the analysis of demand for reinforcers.

10.6 Conclusions and Future Directions

In Sects. 10.2 and 10.3 of this chapter, we discussed two broad classes of behavioral-economic research that appear to be remarkably disjointed from each other. One class has unit-price demand curve analysis at its core; this class is directly influenced by consumer demand theory of traditional economics and it conceptualizes the consequences of behavior as the receipt of commodities (Bickel et al. 1993; Hursh 1980; Hursh and Silberberg 2008). The other class can be described as the study of inter-temporal choice (Bickel and Marsch 2001). Its underlying conceptual framework construes the consequences of behavior as reinforcement. Inter-temporal choice studies explore the effects of the interaction between reinforcer amount and delay on choices among alternatives that vary on those dimensions. The experimental study of preference reversals is clearly in the latter class of research, since that phenomenon is characterized as the result of the interaction of reinforcer amount and delay. As described above, the reinterpretation of the results of experimentally induced preference reversal experiments in terms of unit prices of reinforcers represents a computational integration of two major classes of behavioral-economic research and theory that have heretofore remained largely distinct.

The study of inter-temporal choice encompasses the quantitative study of how delayed reinforcers are discounted—that is, the assessment of delay discounting by individuals. As discussed earlier, the assessment of a subject's rate of delay discounting is accomplished by offering several series of choices between receiving a large reinforcer later versus a small reinforcer sooner (where "sooner" is most often operationalized as "now") and detecting in each series the point of indifference between the two alternatives. Note that an indifference point is commonly indicated to the researcher by the point at which the subject reverses preference between larger-later reinforcement and smaller-sooner reinforcement. The discounting of delayed reinforcers can thus be recognized as a phenomenon involving preferences and their potential reversal. This suggests that the science of delay discounting may be advanced by applying the unit price concept to the design and interpretation of delay discounting studies.

This suggestion is supported by a relationship discovered in discounting assessment data from 26 cigarette smokers whose discounting of hypothetical rewards of $50 and $1000 was assessed when they were nicotine-deprived and when they were satiated (Mueller et al. 2010). The assessment procedures were as described in Sect. 10.3 of this chapter and indifference points were indicated by participants' cross-trial reversals of preference between hypothetically receiving LL reinforcers or SS reinforcers "now." The experimental procedures entailing the delays and amounts for the SS reinforcers and the LL reinforcers were interpreted as behavioral-economic costs and benefits as described above to calculate unit prices for the SS and LL options in each trial of the procedure. Such computations were used to predict points of preference reversal, or indifference points, in this group of delay discounting assessments. Rank correlations of the theoretical and empirically determined indifference points provided strong support for the accuracy of the unit price construct in predicting participants' expressions of preference that are implicit in delay discounting assessments.

Our use of the unit price construct to predict reversals of preference has implications for the future of that construct. In the above reinterpretation of delay discounting assessment data, there was no evidence that the correlation between predicted and observed indifference points depended upon either the amount of the reinforcer (e.g., $50 or $1000) for which the discounting rate of a participant was being assessed, or the state of the participant (satiated or deprived). This discovery suggests a limitation of the unit price construct as it has been formulated to date. Unit price consists of environmental cost and benefit variables, but does not incorporate variables that reflect the characteristics of the organism. In the future, refined versions of the unit price construct (or related constructs) might incorporate terms that allow an accounting of subject states (e.g., drug withdrawal) or traits (e.g., "addictive personality").

While the current formulation of the unit price construct incorporates environmental variables, it may account for them in ways that do not reflect scientific conclusions about how the environmental variables affect behavior. For example, in applications of unit price to date, delay until reinforcement has been incorporated as a cost variable. This formulation of unit price treats all delays to reinforcement as having effects on behavior that simply differ by a constant proportion. The science of delay discounting strongly suggests that this is not an accurate quantification of the effects of reinforcement delay. Future versions of unit price and related concepts may be improved, for example, by the inclusion of terms and/or parameters whose values reflect the hyperbolic nature of delay discounting.

The unit price construct has thus far been applied to the analysis of preference among qualitatively identical commodities. Behavioral economics refers to such commodities as "substitutes." The scientific meaning of substitutes and related concepts is readily suggested through example: Coca-Cola and Pepsi could be classified as "substitutes" because one is consumed in place of the other; hot dog franks and hot dog buns could be classified as "complements" because the consumption of hot dog buns closely co-varies with the consumption of hot dog franks; and paper clips and cheese could be classified as "independent commodities" because there is typically no relationship between the consumption of paper clips and the consumption of cheese. Experiments involving decisions between substitutes represent only one type of interaction (and perhaps the most simplified type), whereas reinforcers are often consumed as complements or as independent commodities outside of the laboratory. In computationally accounting for substitutes, units of measure are the same for many factors in the equation and thus cancel out, minimizing a potential source of appreciable complexity. Accounting for the complexities of complementary or independent commodities in a computation model of addiction might prove to be quite challenging. However, it is behavioral economics that makes us aware of such complexities (Green and Freed 1993; Hursh 1980; Rachlin et al. 1981), and behavioral economics has methods for approaching them with quantitative sophistication (Bickel et al. 1995, 1992; Johnson et al. 2004). Behavioral economics may thus be well positioned to further extend its analytical and computational toolbox so that preference among qualitatively different reinforcers is explained.

Science has tended to characterize addiction as a static phenomenon rather than as a dynamic process. The behavioral-economic notion of substitution among reinforcing commodities may be central to elaborating the *process* of becoming and remaining addicted. This chapter has pointed out how from the perspective of demand curve analysis addiction entails relatively less elasticity of demand for the addictive substance, and how from the perspective of the science of delay discounting addiction entails a greater valuation of immediate versus delayed reinforcers. The processes assessed with measures of elasticity and discounting may be interdependent via the process of substitution. It has been shown that elasticity of demand is positively correlated with the availability of substitutes (Bickel et al. 1995; Hursh 1978; Johnson and Bickel 2003; Johnson et al. 2004; Shahan et al. 2001, 2000). It may be important to recognize, furthermore, that one way substitution can occur is through *inter-temporal* exchange. For example, in the process of homeostatic regulation, the immediate consumption of a reinforcing commodity may be forestalled or diminished by the availability of the commodity at a later time; that is, the later consumption *substitutes* for immediate consumption. Some aspects of homeostatic regulation may thus be an effect of inter-temporal substitution that can be described in behavioral-economic terms as greater elasticity *of immediate consumption* compared to the elasticity *of delayed consumption*. The extent of this effect, it should be emphasized, has been shown to be influenced by the discounting rate characteristics of the consumer. If the degree of delay discounting changes, either as a motivational or developmental process within the consumer, or across different consumers, the magnitude of this substitution effect will change also. In this chapter we discussed that delay discounting rates for individuals who are substance-dependent are higher compared to individuals who are not dependent. We also promoted the merits of unit price that are based on conceptual commonalities for understanding demand elasticity of abused substances and delay discounting. Because of the conceptual commonalities, future versions of the unit price construct have the potential to incorporate variables that may appear in quantitative descriptions of regulated processes such as the motivation to consume an abused substance, or developmental aspects of addiction, or individual differences in discounting rates of the substance-dependent. Such a future construct would permit a more accurate computational analysis of the dynamic interaction of the processes that are assessed as the demand elasticity and discounting rates for abuses substances and those who abuse them. To the extent that such analyses explore dynamic reinforcement variables that result in the *removal* of a phenomenon (e.g., the removal of nicotine withdrawal symptoms), they may be addressing the concept of *negative* reinforcement—a concept heretofore unaddressed by demand curve analysis because "removal" seems antithetical to the acquisitive nature of "consumption." Regardless of the variables that are interpreted as differential elasticities for immediate versus delayed consumption, the analyses would clarify the *process* of addiction.

A better understanding of the process of addiction will certainly entail appreciation for the different time scales over which choices are made. At one point in his life, for example, an alcoholic may find himself challenged by the choice between having a first drink or not. Years later and while in recovery he may in hindsight come to believe that decision was tantamount to deciding between a decade of gainful employment versus living hand to mouth on the streets. While it may be self-evident that the larger time scale is constituted of decisions on a much smaller time scale, the task before science is to devise concepts that integrate the smaller-scale events over time in such a way that (a) the larger pattern is predictable from the character of small-scale events, and (b) operations to modify the character of the small-scale events (i.e., substance abuse therapies) can be taken so as to prevent the large-scale phenomenon. We suggest that the present version of the unit price concept is good first step in these directions.

This chapter has described and clarified an instance of computational constructionism applied to the understanding of drug addiction. The exercise was shown to involve (a) the examination of a handful of behavioral phenomena associated with drug addiction (excessive consumption of a substance, persistent consumption of a substance in the face of increasing costs, high rates of delay discounting, and frequent reversals of preference), any one of which may traditionally be scientifically studied in an "intellectual silo"; (b) the extension of the scope of the behavioral-economic unit price concept so that it integrates a larger number of experimental variables and incorporates features that are common to drug-addiction phenomena; and (c) the use of the unit price construct to begin to explain the extremely complex phenomena of choice among concurrently available reinforcers, and the process by which one becomes addicted. Future developments along these lines are expected to produce constructs with which preferences exhibited by drug-dependent individuals may be predicted more accurately and may be therapeutically modified.

Acknowledgements The writing of this chapter was supported by National Institute on Drug Abuse Grants R37 DA 006526-18, R01 DA 11692-10, R01 DA022386-02, R01 DA024080-01A1, Wilbur Mills Chair Endowment, and in part by the Arkansas Biosciences Institute, a partnership of scientists from Arkansas Children's Hospital, Arkansas State University, the University of Arkansas-Division of Agriculture, the University of Arkansas, Fayetteville, and the University of Arkansas for Medical Sciences. The Arkansas Biosciences Institute is the major research component of the Tobacco Settlement Proceeds Act of 2000.

References

Ainslie G, Herrnstein RJ (1981) Preference reversal and delayed reinforcement. Anim Learn Behav 9(4):476–482

Ashton H, Stepney R, Thompson JW (1979) Self-titration by cigarette smokers. Br Med J 2(6186):357–360

Audrain-McGovern J, Rodriguez D, Epstein LH, Cuevas J, Rodgers K, Wileyto EP (2009) Does delay discounting play an etiological role in smoking or is it a consequence of smoking? Drug Alcohol Depend 103(3):99–106

Bach PB, Lantos J (1999) Methadone dosing, heroin affordability, and the severity of addiction. Am J Publ Health 89(5):662–665

Baker F, Johnson MW, Bickel WK (2003) Delay discounting in current and never-before cigarette smokers: Similarities and differences across commodity, sign, and magnitude. J Abnorm Psychol 112(3):382–392

Benowitz NL, Jacob P, Yu L, Talcott R, Hall S, Jones RT (1986) Reduced tar, nicotine, and carbon monoxide exposure while smoking ultralow- but not low-yield cigarettes. JAMA J Am Med Assoc 256(2):241–246

Bickel WK, Christensen DR (2010) Behavioral economics of drug addiction. In: Stolerman IP (ed) Encyclopedia of psychopharmacology. Springer, Heidelberg

Bickel WK, DeGrandpre RJ (1995) Price and alternatives: Suggestions for drug policy from psychology. Int J Drug Policy 2:93–105

Bickel WK, DeGrandpre RJ (1996) Basic psychological science speaks to drug policy: Drug cost and competing reinforcement. In: Bickel WK, DeGrandpre RJ (eds) Drug policy and human nature: Psychological perspectives on the control, prevention and treatment of illicit drug use. Plenum, New York, pp 31–52

Bickel WK, Marsch LA (2001) Toward a behavioral economic understanding of drug dependence: delay discounting processes. Addiction 96(1):73–86

Bickel WK, DeGrandpre RJ, Higgins ST, Hughes JR (1990) Behavioral economics of drug self-administration. I. Functional equivalence of response requirement and drug dose. Life Sci 47(17):1501–1510

Bickel WK, DeGrandpre RJ, Hughes JR, Higgins ST (1991) Behavioral economics of drug self-administration. II. A unit-price analysis of cigarette smoking. J Exp Anal Behav 55(2):145–154

Bickel WK, Hughes JR, DeGrandpre RJ, Higgins ST, Rizzuto P (1992) Behavioral economics of drug self-administration. IV. The effects of response requirement on the consumption of and interaction between concurrently available coffee and cigarettes. Psychopharmacology (Berl) 107(2–3):211–216

Bickel WK, DeGrandpre RJ, Higgins ST (1993) Behavioral economics: a novel experimental approach to the study of drug dependence. Drug Alcohol Depend 33(2):173–192

Bickel WK, DeGrandpre RJ, Higgins ST (1995) The behavioral economics of concurrent drug reinforcers: a review and reanalysis of drug self-administration research. Psychopharmacology (Berl) 118(3):250–259

Bickel WK, Odum AL, Madden GJ (1999) Impulsivity and cigarette smoking: delay discounting in current, never, and ex-smokers. Psychopharmacology (Berl) 146(4):447–454

Bickel WK, Pitcock JA, Yi R, Angtuaco EJC (2009) Congruence of BOLD response across intertemporal choice conditions: fictive and real money gains and losses. J Neurosci 29(27):8839–8846

Bickel WK, Mueller ET, MacKillop J, Yi R (in press). Behavioral economic and neuro-economic perspectives on addiction. In: Sher K (ed) Oxford handbook of substance use disorders. Oxford University Press, New York

Bigelow GE, Griffiths RR, Stitzer ML, Liebson IA (1983) Development of clinical procedures for abuse liability assessment: progress report from the Behavioral Pharmacology Research Unit of the Johns Hopkins University School of Medicine and Baltimore City Hospitals. NIDA Res Monogr 43:125–131

Bigelow GE, Brooner RK, Silverman K (1998) Competing motivations: drug reinforcement vs non-drug reinforcement. J Psychopharmacol 12(1):8–14

Bjork JM, Hommer DW, Grant SJ, Danube C (2004) Impulsivity in abstinent alcohol-dependent patients: relation to control subjects and type 1-/type 2-like traits. Alcohol 34(2–3):133–150

Brandon TH, Tiffany ST, Obremski KM, Baker TB (1990) Postcessation cigarette use: the process of relapse. Addict Behav 15(2):105–114

Carter LP, Griffiths RR (2009) Principles of laboratory assessment of drug abuse liability and implications for clinical development. Drug Alcohol Depend 105:S14–S25

Caulkins JP (2001) Drug prices and emergency department mentions for cocaine and heroin. Am J Publ Health 91(9):1446–1448

Chaloupka FJ, Grossman M, Bickel WK, Saffor M (eds) (1999) The economic analysis of substance use and abuse: An integration of econometric and behavioral economic research. University of Chicago Press, Chicago

Coffey SF, Gudleski GD, Saladin ME, Brady KT (2003) Impulsivity and rapid discounting of delayed hypothetical rewards in cocaine-dependent individuals. Exp Clin Psychopharmacol 11(1):18–25

Collier GH, Johnson DF, Hill WL, Kaufman LW (1986) The economics of the law of effect. J Exp Anal Behav 46(2):113–136

Corman H, Noonan K, Reichman NE, Dave D (2005) Demand for illicit drugs among pregnant women. Adv Health Econ Health Serv Res 16:41–60

Creighton DE, Lewis PH (1978) The effect of different cigarettes on human smoking patters. In: Thornton RE (ed) Smoking behavior: physiological and psychological influences. Churchill, Edinburgh

Darke S, Kaye S, Topp L (2002a) Cocaine use in New South Wales, Australia, 1996–2000: 5 year monitoring of trends in price, purity, availability and use from the illicit drug reporting system. Drug Alcohol Depend 67(1):81–88

Darke S, Topp I, Kaye H, Hall W (2002b) Heroin use in New South Wales, Australia, 1996–2000: 5 year monitoring of trends in price, purity, availability and use from the Illicit Drug Reporting System (IDRS). Addiction 97(2):179–186

DeGrandpre RJ, Bickel WK, Hughes JR, Higgins ST (1992) Behavioral economics of drug self-administration. III. A reanalysis of the nicotine regulation hypothesis. Psychopharmacology (Berl) 108(1–2):1–10

DeGrandpre RJ, Bickel WK, Hughes JR, Layng MP, Badger G (1993) Unit price as a useful metric in analyzing effects of reinforcer magnitude. J Exp Anal Behav 60(3):641–666

Epstein LH, Salvy SJ, Carr KA, Dearing KK, Bickel WK (2010). Food reinforcement, delay discounting and obesity. Phys Behav 100:438–445

Evans JA (2008) Electronic publication and the narrowing of science and scholarship. Science 321(5887):395–399

Frith CD (1971) The effect of varying the nicotine content of cigarettes on human smoking behaviour. Psychopharmacologia 19(2):188–192

Goldfarb T, Gritz ER, Jarvik ME, Stolerman IP (1976) Reactions to cigarettes as a function of nicotine and "tar". Clin Pharmacol Ther 19(6):767–772

Green L, Estle SJ (2003a) Preference reversals with food and water reinforcers in rats. J Exp Anal Behav 79(2):233–242

Green L, Estle SJ (2003b) Preference reversals with food and water reinforcers in rats. J Exp Anal Behav 79(2):233–242

Green L, Freed DE (1993) The substitutability of reinforcers. J Exp Anal Behav 60(1):141–158

Green L, Myerson J (1996) Exponential versus hyperbolic discounting of delayed outcomes: Risk and waiting time

Green L, Snyderman M (1980) Choice between rewards differing in amount and delay—toward a choice model of self-control. J Exp Anal Behav 34(2):135–147

Green L, Fisher EB, Perlow S, Sherman L (1981) Preference reversal and self-control—choice as a function of reward amount and delay. Behav Anal Lett 1(1):43–51

Green L, Fristoe N, Myerson J (1994) Temporal discounting and preference reversals in choice between delayed outcomes. Psychon Bull Rev 1(3):383–389

Greenwald MK, Hursh SR (2006) Behavioral economic analysis of opioid consumption in heroin-dependent individuals: Effects of unit price and pre-session drug supply. Drug Alcohol Depend 85(1):35–48

Griffiths RR, Brady JV, Bradford LD (1979) Predicting abuse liability of drugs with animal drug self-administration procedures: Psychomotor stimulants and hallucinogens. In: Thompson T, Dews PB (eds) Advances in behavioral pharmacology, vol 2. Academic Press, New York

Griffiths RR, Henningfield JE, Bigelow GE (1982) Human cigarette smoking: manipulation of number of puffs per bout, interbout interval and nicotine dose. J Pharmacol Exp Ther 220(2):256–265

Gritz ER (1980) Smoking behavior and tobacco abuse. In: Mello NK (ed) Advances in substance abuse. JAI Press, Greenwich, pp 91–158

Gust SW, Pickens RW (1982) Does cigarette nicotine yield affect puff volume? Clin Pharmacol Ther 32(4):418–422

Haley NJ, Sepkovic DW, Hoffmann D, Wynder EL (1985) Cigarette smoking as a risk for cardiovascular disease. Part VI. Compensation with nicotine availability as a single variable. Clin Pharmacol Ther 38(2):164–170

Heil SH, Johnson MW, Higgins ST, Bickel WK (2006) Delay discounting in currently using and currently abstinent cocaine-dependent outpatients and non-drug-using matched controls. Addict Behav 31(7):1290–1294

Henningfield JE (1984) Behavioral pharmacology of cigarette smoking. In: Thompson T, Dews PB, Barrett JE (eds) Advances in behavioral pharmacology, vol 4. Academic Press, Orlando

Henningfield JE, Griffiths RR (1980) Effects of ventilated cigarette holders on cigarette smoking by humans. Psychopharmacology 68(2):115–119

Heyman GM, Gibb SP (2006) Delay discounting in college cigarette chippers. Behav Pharmacol 17(8):669–679

Higgins ST, Budney AJ, Bickel WK (1994) Applying behavioral concepts and principles to the treatment of cocaine dependence. Drug Alcohol Depend 34(2):87–97

Hill P, Marquardt H (1980) Plasma and urine changes after smoking different brands of cigarettes. Clin Pharmacol Ther 27(5):652–658

Hoffman WF, Moore M, Templin R, McFarland B, Hitzemann RJ, Mitchell SH (2006) Neuropsychological function and delay discounting in methamphetamine-dependent individuals. Psychopharmacology (Berl) 188(2):162–170

Hoffman WF, Schwartz DL, Huckans MS, McFarland BH, Meiri G, Stevens AA et al (2008) Cortical activation during delay discounting in abstinent methamphetamine dependent individuals. Psychopharmacology (Berl) 201(2):183–193

Hursh SR (1978) The economics of daily consumption controlling food- and water-reinforced responding. J Exp Anal Behav 29(3):475–491

Hursh SR (1980) Economic concepts for the analysis of behavior. J Exp Anal Behav 34(2):219–238

Hursh SR (1991) Behavioral economics of drug self-administration and drug abuse policy. J Exp Anal Behav 56(2):377–393

Hursh SR, Silberberg A (2008) Economic demand and essential value. Psychol Rev 115(1):186–198

Hursh SR, Raslear TG, Shurtleff D, Bauman R, Simmons L (1988) A cost-benefit analysis of demand for food. J Exp Anal Behav 50(3):419–440

James W (1918) The principles of psychology, vol 1. Holt, New York

Jarvik ME, Popek P, Schneider NG, Baer-Weiss V, Gritz ER (1978) Can cigarette size and nicotine content influence smoking and puffing rates? Psychopharmacology 58(3):303–306

Johnson MW, Bickel WK (2002) Within-subject comparison of real and hypothetical money rewards in delay discounting. J Exp Anal Behav 77(2):129–146

Johnson MW, Bickel WK (2003) The behavioral economics of cigarette smoking: The concurrent presence of a substitute and an independent reinforcer. Behav Pharmacol 14(2):137–144

Johnson MW, Bickel WK, Kirshenbaum AP (2004) Substitutes for tobacco smoking: a behavioral economic analysis of nicotine gum, denicotinized cigarettes, and nicotine-containing cigarettes. Drug Alcohol Depend 74(3):253–264

Johnson MW, Bickel WK, Baker F (2007a) Moderate drug use and delay discounting: a comparison of heavy, light, and never smokers. Exp Clin Psychopharmacol 15(2):187–194

Johnson MW, Bickel WK, Baker F (2007b) Moderate drug use and delay discounting: A comparison of heavy, light, and never smokers. Exp Clin Psychopharmacol 15(2):187–194

Kirby KN, Petry NM (2004) Heroin and cocaine abusers have higher discount rates for delayed rewards than alcoholics or non-drug-using controls. Addiction 99(4):461–471

Kirby KN, Petry NM, Bickel WK (1999) Heroin addicts have higher discount rates for delayed rewards than non-drug-using controls. J Exp Psychol Gen 128(1):78–87

Kirshenbaum AP, Olsen DM, Bickel WK (2009) A quantitative review of the ubiquitous relapse curve. J Subst Abuse Treat 36(1):8–17

Ko MC, Terner J, Hursh S, Woods JH, Winger G (2002) Relative reinforcing effects of three opioids with different durations of action. J Pharmacol Exp Ther 301(2):698–704

Kowal BP, Yi R, Erisman AC, Bickel WK (2007) A comparison of two algorithms in computerized temporal discounting procedures. Behav Process 75(2):231–236

Krishnan-Sarin S, Reynolds B, Duhig AM, Smith A, Liss T, McFetridge A et al (2007) Behavioral impulsivity predicts treatment outcome in a smoking cessation program for adolescent smokers. Drug Alcohol Depend 88(1):79–82

Logue AW, Pena-Correal TE (1985) The effect of food-deprivation on self-control. Behav Process 10(4):355–368

Longo MC, Henry-Edwards SM, Humeniuk RE, Christie P, Ali RL (2004) Impact of the heroin 'drought' on patterns of drug use and drug-related harms. Drug Alcohol Rev 23(2):143–150

MacKillop J, Kahler CW (2009) Delayed reward discounting predicts treatment response for heavy drinkers receiving smoking cessation treatment. Drug Alcohol Depend 104(3):197–203

MacKillop J, Murphy JG (2007) A behavioral economic measure of demand for alcohol predicts brief intervention outcomes. Drug Alcohol Depend 89(2–3):227–233

MacKillop J, Murphy JG, Ray LA, Eisenberg DT, Lisman SA, Lum JK et al (2008) Further validation of a cigarette purchase task for assessing the relative reinforcing efficacy of nicotine in college smokers. Exp Clin Psychopharmacol 16(1):57–65

Madden GJ, Petry NM, Badger GJ, Bickel WK (1997) Impulsive and self-control choices in opioid-dependent patients and non-drug-using control participants: drug and monetary rewards. Exp Clin Psychopharmacol 5(3):256–262

Marlatt GA, Curry S, Gordon JR (1988) A longitudinal analysis of unaided smoking cessation. J Consult Clin Psychol 56(5):715–720

Mattox AJ, Carroll ME (1996) Smoked heroin self-administration in rhesus monkeys. Psychopharmacology 125(3):195–201

Mazur JE (1987) An adjusting procedure for studying delayed reinforcement. In: Commons ML, Mazur JE, Nevin JA, Rachlin H (eds) Quantitative analysis of behavior: Vol 5: The effect of delay and of intervening events on reinforcement value. Erlbaum, Hillsdale, pp 55–73

McMorrow MJ, Foxx RM (1983) Nicotine role in smoking—an analysis of nicotine regulation. Psychol Bull 93(2):302–327

Mitchell SH (1999) Measures of impulsivity in cigarette smokers and non-smokers. Psychopharmacology (Berl) 146(4):455–464

Monterosso JR, Ainslie G, Xu JS, Cordova X, Domier CP, London ED (2007) Frontoparietal cortical activity of methamphetamine-dependent and comparison subjects performing a delay discounting task. Hum Brain Mapp 28(5):383–393

Moss RA, Prue DM (1982) Research on nicotine regulation. Behav Ther 13:31–46

Mueller ET, Bickel WK (2010). An opportunity cost reinterpretation of preference reversal experiments. Paper present at the 33rd annual meeting of the Society for Quantitative Analysis of Behavior, San Antonio, Texas

Mueller ET, Landes RD, Kowal BP, Yi R, Stitzer ML, Burnett CA et al (2009) Delay of smoking gratification as a laboratory model of relapse: effects of incentives for not smoking, and relationship with measures of executive function. Behav Pharmacol 20(5–6):461–473

Mueller ET, Bickel WK, Landes RD (2010) Smoker's delay discounting indifference points are associated with changes in opportunity-cost-informed price. Paper presented at the 72nd annual meeting of the college on problems of drug dependence, Scottsdale, Arizona

Murphy JG, MacKillop J (2006) Relative reinforcing efficacy of alcohol among college student drinkers. Exp Clin Psychopharmacol 14(2):219–227

Navarick DJ, Fantino E (1976) Self-control and general models of choice. J Exp Psychol, Anim Behav Processes 2(1):75–87

Nides MA, Rakos RF, Gonzales D, Murray RP, Tashkin DP, Bjornson-Benson WM et al (1995) Predictors of initial smoking cessation and relapse through the first 2 years of the lung health study. J Consult Clin Psychol 63(1):60–69

Norregaard J, Tonnesen P, Petersen L (1993) Predictors and reasons for relapse in smoking cessation with nicotine and placebo patches. Prev Med 22(2):261–271

Odum AL, Madden GJ, Badger GJ, Bickel WK (2000) Needle sharing in opioid-dependent outpatients: psychological processes underlying risk. Drug Alcohol Depend 60(3):259–266

Ohmura Y, Takahashi T, Kitamura N (2005) Discounting delayed and probabilistic monetary gains and losses by smokers of cigarettes. Psychopharmacology (Berl) 182(4):508–515

Petry NM (2001) Delay discounting of money and alcohol in actively using alcoholics, currently abstinent alcoholics, and controls. Psychopharmacology (Berl) 154(3):243–250

Rachlin H, Green L (1972a) Commitement, choice, and self-control. J Exp Anal Behav 17:15–72

Rachlin H, Green L (1972b) Commitment, choice and self-control. J Exp Anal Behav 17(1):15

Rachlin H, Battalio R, Kagel J, Green L (1981) Maximization theory in behavioral psychology. Behav Brain Sci 4(3):371–388

Raslear TG, Bauman RA, Hursh SR, Shurtleff D, Simmons L (1988) Rapid demand curves for behavioral economics. Anim Learn Behav 16(3):330–339

Reynolds B, Leraas K, Collins C, Melanko S (2009) Delay discounting by the children of smokers and nonsmokers. Drug Alcohol Depend 99(1–3):350–353

Russell MA, Wilson C, Patel UA, Cole PV, Feyerabend C (1973) Comparison of effect on tobacco consumption and carbon monoxide absorption of changing to high and low nicotine cigarettes. Br Med J 4(5891):512–516

Russell MA, Wilson C, Patel UA, Feyerabend C, Cole PV (1975) Plasma nicotine levels after smoking cigarettes with high, medium, and low nicotine yields. Br Med J 2(5968):414–416

Russell MA, Sutton SR, Feyerabend C, Saloojee Y (1980) Smokers' response to shortened cigarettes: dose reduction without dilution of tobacco smoke. Clin Pharmacol Ther 27(2):210–218

Schifano F, Corkery J (2008) Cocaine/crack cocaine consumption, treatment demand, seizures, related offences, prices, average purity levels and deaths in the UK (1990–2004). J Psychopharmacol 22(1):71–79

Schifano F, Corkery J, Deluca P, Oyefeso A, Ghodse AH (2006) Ecstasy (MDMA, MDA, MDEA, MBDB) consumption, seizures, related offences, prices, dosage levels and deaths in the UK (1994–2003). J Psychopharmacol 20(3):456–463

Schuster CR, Thompson T (1969) Self administration and behavioral dependence on drugs. Annu Rev Pharmacol 9(1):483–502

Shahan TA, Bickel WK, Badger GJ, Giordano LA (2001) Sensitivity of nicotine-containing and de-nicotinized cigarette consumption to alternative non-drug reinforcement: a behavioral economic analysis. Behav Pharmacol 12(4):277–284

Shahan TA, Odum AL, Bickel WK (2000) Nicotine gum as a substitute for cigarettes: a behavioral economic analysis. Behav Pharmacol 11(1):71–79

Shiffman S, Paty JA, Gnys M, Kassel JA, Hickcox M (1996) First lapses to smoking: within-subjects analysis of real-time reports. J Consult Clin Psychol 64(2):366–379

Silverman K (2004) Exploring the limits and utility of operant conditioning in the treatment of drug addiction. Behav Anal 27(2):209–230

Skurvydas A (2005) New methodology in biomedical science: methodological errors in classical science. Medicina (Kaunas) 41(1):7–16

Snyderman M (1983) Delay and amount of reward in a concurrent chain. J Exp Anal Behav 39(3):437–447

Soto AM, Sonnenschein C (2005) Emergentism as a default: cancer as a problem of tissue organization. J Biosci 30(1):103–118

Stepney R (1981) Would a medium-nicotine, low-tar cigarette be less hazardous to health? Br Med J (Clin Res Ed) 283(6302):1292–1296

Strange K (2005) The end of "naive reductionism": rise of systems biology or renaissance of physiology?. Am J Physiol, Cell Physiol 288(5):C968–C974

Turner JA, Sillett RW, Ball KP (1974) Some effects of changing to low-tar and low-nicotine cigarettes. Lancet 2(7883):737–739

Vuchinich RE, Simpson CA (1998) Hyperbolic temporal discounting in social drinkers and problem drinkers. Exp Clin Psychopharmacol 6(3):292–305

Westman EC, Behm FM, Simel DL, Rose JE (1997) Smoking behavior on the first day of a quit attempt predicts long-term abstinence. Arch Intern Med 157(3):335–340

Winger G, Galuska CM, Hursh SR, Woods JH (2006) Relative reinforcing effects of cocaine, remifentanil, and their combination in rhesus monkeys. J Pharmacol Exp Ther 318(1):223–229

Yoon JH, Higgins ST, Heil SH, Sugarbaker RJ, Thomas CS, Badger GJ (2007) Delay discounting predicts postpartum relapse to cigarette smoking among pregnant women. Exp Clin Psychopharmacol 15(2):176–186

Zacny JP, Stitzer ML (1988) Cigarette brand-switching: effects on smoke exposure and smoking behavior. J Pharmacol Exp Ther 246(2):619–627

Chapter 11
Simulating Patterns of Heroin Addiction Within the Social Context of a Local Heroin Market

Lee Hoffer, Georgiy Bobashev, and Robert J. Morris

Abstract This study illustrates how the social structure of the heroin market can impact the physiology of heroin addiction and how heterogeneity of addiction patterns can be shaped by market dynamics. We use a novel agent-based modeling (ABM) approach to simulate possible neurophysiologic functions based on the collective self-organizing behavior of market agents. The conceptual model is based on three components: biological, behavioral, and social. Biological components are informed by mechanistic animal studies, behavioral component relies on studies of real-life human experiences with addiction, and social aspects are based on market research that describes the transactional and decision-making processes associated with the distribution of drugs within local drug markets. Using ABM, this paper unifies these three components to simulate how heroin addiction patterns are generated and shaped through heroin markets. The market model is based on data from an ethnographic study of a local heroin market and includes customers (users), street and private dealers, street brokers, police, and other potential market actors. Behavioral data is based on converting narrative descriptions and fieldwork observations into formal states and transitions, and a simple model of addiction process for the drug users is based on published peer-reviewed literature. Analysis of model-based simulations reveals "binge/crash," "stepped," and "stable" patterns in customer addiction levels.

11.1 Introduction

Substance use has both social and individual aspects. Internal pharmacological and neurobiological factors act as strong self-administration drivers (Koob and Le Moal 2005; Ahmed et al. 2007; Gutkin et al. 2006). At the same time, human substance

L. Hoffer
Department of Anthropology, Case Western Reserve University, 11220 Bellflower Rd., Cleveland, OH, 44106, USA

G. Bobashev (✉) · R.J. Morris
RTI International, Research Triangle Park, NC, 27709, USA
e-mail: bobashev@rti.org

use depends on drug availability (i.e., drug markets) and social environment (Hoffer 2006). Drug availability, in turn, is intertwined with drug demand and law enforcement activities. In this paper, we address a challenge of linking individual consumption patterns with both internal physiological mechanisms of addiction and external market activities. In particular, we attempt to answer the question: how do heroin markets shape users' addictions? While there is a broad body of science that tries to predict behavior based on neurobiology and physiology, our approach reverses this order by deducing physiological patterns from social behavior.

The choice of heroin for our research is based on two factors: the importance of heroin research to both basic science and public policy, and the availability of ethnographic data. Heroin, 6-monoacetyl morphine (6-MAM), is a psychoactive substance that both directly and indirectly influences the daily lives of millions of people each year. Complicating efforts to reduce the public health harms associated with its use is that addiction to heroin, like that to all psychoactive substances, involves complex biological and social/environmental factors. On one hand, the importance of the biology and addiction cannot be ignored. Clearly, addiction is a "brain disease." On the other hand, people use drugs in very specific social, cultural, and political contexts, and thereby experience addiction differently. These social environments affect drug use behaviors, as well as how harm associated with drug use is experienced by the user.

In the United States, as in most countries, an unavoidable component of heroin use is the illegal markets through which the drug is acquired. Since the Harrison Narcotic Act in 1914 making the drug illegal (Hanson et al. 2006) and Richard Nixon's inauguration of the war on drugs in 1969 (Musto 1987), more and more resources are spent each year in the United States unsuccessfully trying to undermine illegal drug distribution activities. The economic cost of this policy and its burden on our judicial system are well established (Musto 1987; Singer 2006). Research also has demonstrated how drug market activities, often framed and manipulated by this war, have negatively influenced the health behaviors of drug users (Kerr et al. 2005; Koester 1994; Zule et al. 2002).

As the hallmark method in cultural anthropology, ethnographic research entails gaining perspective on a social group's beliefs and behaviors by the researcher becoming an ad hoc member of the group under study (Bernard 1988; Patton 1980; Hammersley and Atkinson 1993; Fetterman 1998). This is an inductive and interpretive methodology. Since the 1960s, ethnography has been the primary methodology used for in-depth research studies of illegal drug distribution and dealing operations (Preble and Casey 1969; Agar 1973; Adler 1985; Bourgois 1997). Researchers take considerable time and effort with participants to overcome suspicion and develop the rapport necessary to collect credible accounts of illegal drug dealing. The method requires observing interactions and behaviors to validate findings. Because of the significant rapport necessary to collect this type of data, ethnography has been suggested as the only valid research approach in this context (Bourgois 1997).

Hoffer's ethnographic studies (Hoffer 2006) suggest that acquiring heroin through a market is highly variable and inherently promotes intense short-term fluctuations in individual addiction levels. Heroin users, however, attempt to maintain their addictions, which engage them in a dynamic interaction between their biology and the market. Agent-based modeling techniques allow one to describe and simulate individual behavior as well as the interactions between individuals. At the same time, agent-based models (ABM) allow incorporation of the dynamic of drug effects *within* an individual. Thus, ABMs provide a link across temporal and physical scales as well as combine research findings from tangential disciplines such as the social and basic sciences. In a previous publication (Hoffer et al. 2009), we have concentrated on the market aspects of social dynamics. In this paper, we focus on the inverse relationship, that is, the individual neurophysiologic factors that are shaped by market activities.

11.1.1 Individual and Social Patterns Impacting Heroin Addiction

Despite the considerable harm associated with heroin use, some people addicted to the drug seem able to maintain stable addictions over considerable time periods. However, for most people this stability is unachievable. Many factors often conspire against heroin addicts: money runs out, the drug habit gets too unmanageable, or both. Eventually, some users willingly or unwillingly change their habits or even quit. Even dealers and others with apparently limitless access to the drug can get "tired" of using, attempt to cut back, or otherwise reduce consumption. Under societal and health pressures, users can go into drug-treatment programs, or they may be incarcerated, both of which can effect drug habits.

Because of the variety of patterns in heroin use over time and the challenges of measuring them accurately, characterizing these patterns using conventional methodologies has proven extremely difficult. Typically for behavioral studies, data are extrapolated by asking heroin users to self-report how many days in the last 30 days they used a drug[1] and how much they typically used each day/occasion. These methods are clearly inexact and do not adequately address variations in use. For example, a typical user's response to the questions above might be, "I use heroin twice a day, but there were several days last month when I only used once and a few where I used three times." Such responses do not fit formal survey question formats.

A large body of biological research focuses on the animal models of drug self-administration under controlled experimental conditions. However, human heroin users do not simply receive heroin whenever they want it. Rather, they purchase it within the context of a market and through relationships with other market participants, that is, dealers. This aspect of maintaining an addiction introduces considerable variation. In human studies, researchers are only able to capture a rough estimate of how much heroin a user consumes over a specific timeframe (i.e., 30 days),

---

[1]This paraphrases a question from the RBA (Risk Behavior Assessment) questionnaire (National Institute on Drug Abuse 1993), commonly administered by professional interviewers and considered the gold standard in community-based research studies funded by the National Institute on Drug Abuse. Other instruments use different questions but similarly remain oriented to a specific timeframe.

but little information concerning their overall pattern of use. Questions remain, such as: how consistently do heroin users ingest the drug? How often are such patterns disrupted? To generate answers to these questions, ABMs can simulate these behaviors by designing agents who are addicted to heroin and then allowing them to acquire and use the drug in association with other agents. In addition to building agents with heroin addictions, our model includes resources (cash and heroin) that could be exchanged by the agents, the relationships necessary to acquire the drug, and the places where users must go to facilitate transactions. This more *contextualized* version of heroin addiction is what our simulation attempts to characterize.

The heroin market modeled in the Illicit Drug Market Simulation (IDMS) project detailed below is not intended to answer all the questions about the interactions between the individual and the market; however, this is the first attempt to apply a multiscale systems approach to addiction modeling. The approach provided is experimental and could be used to generate hypotheses, as well as provide useful insights for designing prevention and treatment interventions. For example, among the agents in our model, there were critical times when their addiction became out of control. In the real world, these situations might cause users to (1) enter treatment in an effort to reduce their addiction, (2) combine their resources with other users, potentially increasing their health risks, and/or (3) commit crimes to support their increasing drug habit.

As a result, policymakers can gain a more detailed understanding of what is required to construct effective policy by looking at simulations of the patterns in heroin addiction and how these patterns are shaped and vary over time. Instead of being restricted to individual narratives of addiction, this simulation demonstrates that addictions are dynamic and patterned through the market context in which they exist.

11.1.2 Internal Components of Heroin Addiction

A large body of biological, clinical, neuropharmacologic, and ethnographic research describes heroin addiction from individual perspectives. In our study, we simulate experienced users and do not focus on heroin initiation and neurophysiologic processes associated with initial euphoria in a naïve subject. We only consider the three main features associated with repeated heroin use, such as withdrawal, tolerance, and a user's drug habit, that is, addiction.

Withdrawal is a key feature of most addiction, but is particularly relevant for those addicted to heroin. After a certain length of time of repeatedly using the substance, a person addicted to heroin must continue to use it or else become sick. This withdrawal syndrome involves intense symptoms such as chills, fever, runny nose, intense muscle pains, headaches, constipation and/or diarrhea, insomnia, anxiety, depression, and crawling flesh. Withdrawal from heroin is incapacitating and persists for approximately 2 weeks after an addict discontinues use. However, symptoms are alleviated seconds after an addict in withdrawal uses the drug. Thus, heroin addicts commonly use the terms "fixing" and "getting well" to refer to injecting heroin, implying correcting and/or returning to a state of well-being.

Tolerance is another important feature of heroin addiction, and refers to needing ever-increasing amounts of the drug to feel the euphoria it produces. After consistently using a certain amount, a user must use more (or a higher level of potency) to experience a high. Tolerance is a neuropsychological consequence of consistent use. Eventually, because achieving a high may require using amounts of heroin that have become unaffordable to the user, "staying well" or maintaining their addiction to avoid withdrawal, is often a primary motive for addicts. As a result of tolerance, people addicted to heroin often use the drug to prevent withdrawal rather than to induce euphoria. While, on balance, avoiding withdrawal might be more motivating to heroin addicts who are maintaining their drug habits, this does not mean users are not still motivated to get high, that is, overcome their tolerance. After all, the euphoria keeps heroin users wanting more.

Both tolerance and withdrawal are diagnostic terms characterizing substance dependence disorders, classified by both DSM-IV (American Psychiatric Association 2000) and ICD-10 (Isaac et al. 1994) diagnostic manuals. People who are dependent on heroin frequently report withdrawal and tolerance to the drug, as well as cross-tolerance to other opiates, and these criteria are often the two most common indicators of heroin addiction.

Another term relevant to this paper is "heroin habit." Unlike terms defined by the medical community, the term heroin habit is part of the lexicon of heroin users. In short, a heroin habit is how heroin users describe their addiction and typically refers to an amount of drug that addicts believe they need to consistently use to avoid withdrawal. Another way to understand a heroin habit is the typical dose a user reports using in one day.

Because heroin is a commodity, heroin habits are often framed monetarily. A "$20-per-day-habit" straightforwardly means the user is using approximately $20 worth of heroin per day. In the author's research (Hoffer 2006), this $20 dose corresponded to a "pill" of black tar heroin, the smallest unit of heroin sold in the market. If a heroin addict does not use equal to or more than their habit, they will eventually get sick and go into withdrawal. Maintaining a habit refers to a heroin user's ability to consistently use this baseline level of heroin. Simply, a user's habit is what heroin addicts seek to maintain, and is the subject of this paper.

Finally, we introduce a working technical term, "addiction level," which describes the amount of drug (in standard units: 1/120 of a gram) a user consumes per day. It is closely related to habit and, in the case of stable price-to-dose ratio, there is an exact one-to-one relationship. However, when price varies for the same dose depending on the market condition, addiction level becomes a better measure of use than a habit expressed in dollar amounts. Addiction level is thus a convenient measure (similar to standard drinks in alcohol research) to use in modeling and analysis.

11.1.3 Social and Market Components of Heroin Addiction

The social environment surrounding the user can be quite complex and can involve drug-use "buddies," drug dealers, friends, relatives, neighbors, co-workers, treatment professionals, and law enforcement professionals. While it is challenging to identify the main social groups and components associated with the individual's drug use, in our research we consider the closed social circle that controls the supply of the drug. A real-life heroin market differs from the controlled supply environment in animal experiments and is a product of social behavior.

The global economy of heroin involves production, sales/exchanges, and use. Local social markets are strongly impacted by both regional drug trends (Curtis and Wendel 2000; Agar et al. 1998; Agar and Reisinger 2002; Hamid 1992) and local use patterns. Because this study focuses on individual addiction patterns, we make certain assumptions about the stability of drug supply and consider a micro-level *social space* in which transactions occur in the local distribution of a drug. Unlike buying a product in a store, buying and selling heroin is not a depersonalized activity. In this context, prices, logistics, and credit are all unregulated and potentially negotiable (Johnson et al. 1985).

For this modeling project, we made decisions about the level of detail describing the market. We considered market complexity, relevance of the details to the research question, and availability of the data. For example, although people who use heroin also often use other drugs, for the purpose of this model, we only consider their heroin addiction. We also don't consider the intricacies of deal negotiation, rather, we focus on the fact that the deal was eventually made and heroin was obtained by the user.

11.2 The Data

Because heroin use and heroin dealing are illegal and stigmatizing behaviors, especially in the United States, collecting data on these activates is no simple matter. Conventional methodologies such as surveys do not accurately capture these complexities; one cannot realistically survey the participants of a heroin market.

Information on the heroin market presented in this manuscript was collected during eighteen months of ethnographic research conducted with heroin users and dealers in Denver, CO (Hoffer 2006). Because of the ethnographic nature of the study, information was obtained through narratives of and fieldwork observations with participants. One of the innovations of our study was to convert the narratives into events, transitions, and transition probabilities that allowed us to build an agent-based model. In-depth, extensive, and historical accounts of heroin users, dealers, and middlemen (i.e., brokers), and their interactions in the market over time were used for programming these groups as "agents" in the simulated market. The advantage of ethnographic research over surveys is that the data are collected on social actions linked in sequence, as well as to participant belief systems. What heroin users are describing and how they are acting are easier to interpret if the researcher is "there" when the action takes place. This perspective was essential for determining and setting simulation parameters, evaluating agent logics, debugging programs, and most important, interpreting results.

A description of the heroin user's daily activities can be transformed into model parameters in two ways. One way is the logical sequence of events related to market activities. For example, when users want to buy drugs, they can go to a private dealer, if they know one, or go to the open marketplace. At the market, users can search for a street dealer and engage in a deal if they find a dealer with heroin. Alternatively, they can go to a street broker. A broker is a drug user who connects customers and dealers.

The second way is providing distributions of numeric parameters, such as frequencies of events and conditional probabilities. Although these parameters are not measured exactly, ethnographic studies can provide ideas about the ranges and distributions of parameter values. For example, several drivers can make a user seek heroin. One driver is related to a user's withdrawal. This driver is guided by habit and heroin pharmacodynamics from the last use. Another driver could be related to the external cueing or other reasons not associated with the last dose. In the latter case, the frequency of use or the distribution of such events is difficult to assess precisely, but long-term ethnographic observations can provide reasonable insight about the range and shape of the distribution. In the case of withdrawal as the driver, we have used published pharmacokinetic data calibrated to humans. Because each individual has different metabolism and response to a drug, we used a distribution with reasonable parameter ranges.

11.3 The Model

While ethnographic description provides considerable information regarding various aspects of heroin addiction, we have selected only those elements that are relevant to market functioning and individual behaviors associated with selling, buying, and using heroin. Other components such as treatment, criminal justice processes, and family relations were not included in the model. Thus, the agent-based model developed for the Illicit Drug Market Simulation (IDMS) contains six different agent types. In particular, customers, brokers, sellers, and private dealers are the most behaviorally complex agents. These agents can learn about market, change their level of addiction based on heroin use, choose transaction partners, and modify their activities based on past experiences. Police and homeless agents are less complex. Hoffer et al. (2009) describes more specific details. In the simulation study presented here, we used 360 agents: $n = 200$ customers, $n = 25$ private dealers, $n = 20$ street dealers, 15 street brokers, and $n = 100$ homeless agents.

Agents in the IDMS function according to agent rules. These rules specify the states in which the agent can be (e.g., seeking money, seeking heroin, buying heroin, using heroin) and transitions between these states (e.g., *if* found a dealer *and* the dealer has enough heroin *then* purchase the desired amount of heroin or *if* purchased heroin can *either* use right away *or* go home, with certain probabilities of making the choice). These states, and conditional decisions and the rates of transitions between the states for an agent, are represented in a form of a state diagram (Fig. 11.1). For example, a drug user could be in a satiated state but, after a while, the withdrawal symptoms become more pronounced according to a within-agent pharmacological model. When the level of withdrawal reaches a certain point, the user changes the state to "seeking the drug."

Because of the complexity of the full market model and the space limitation of this manuscript, the authors are unable to provide complete documentation for the IDMS model here. To find a narrative description, table of parameters, and the programming code for this model, the authors invite researchers to visit http://www.case.edu/artsci/anth/Hoffer.html. Copies of the simulation also are available upon request.

In this paper, we focus only on the 200 customer agents, or the main users, of heroin. Their actions are primarily focused on (1) maintaining a drug habit (set randomly for each agent as an initial condition); (2) using more heroin than their habit when possible to get high; and (3) attempting to avoid running out of heroin and entering withdrawal. If users consistently use more heroin to feel euphoria, the increased use can reset their heroin habit to the greater amount. If users return to their previous amount, they will get sick until their body readjusts. Such increments in use are not random. Rather, they tend to be organized by the heroin units sold in the market. Using the "$20-per-day-habit" as an example, if heroin users want to feel high, they cannot simply purchase a little more heroin, that is, less than a pill. Instead, they would have to purchase two pills. Increments (and decrements) in use correspond to the heroin available.

Furthermore, the heroin-using agents in our simulation were dependent on the capacity of the market; while some would use all their resources to purchase the maximum available drug, others could be more prudent and purchase the amount they expect to need in the near future. Potency of the drug was standardized and linearly associated with heroin unit size.

Market organization shapes a user's ability to acquire the drug (Hoffer 2006; Curtis and Wendel 2000). For example, in local communities that have an open-air market in which anyone can buy heroin at any time, the features of the market (i.e., open access, constant availability) will directly influence a user's ability to maintain his or her addiction. Similarly, if no open-air markets exist, such as in many smaller, rural communities, a heroin user's personal relationship with a dealer(s) directly determines access. We consider contingencies that organize a user's ability to acquire the drug structural components of heroin markets.

A no less relevant issue concerns how heroin users navigate social relationships within these market structures. In other words, while heroin users might have biological motivations to use and knowledge about where to acquire the drug, these prerequisites are put into action only within a set of social norms and/or belief systems. In other words, there are rules customers (and dealers) must know and follow (Hoffer 2006).

While the complete model and market simulation outcomes are presented elsewhere (Hoffer et al. 2009), the findings below only concentrate on how customer

11 Simulating Patterns of Heroin Addiction Within the Social Context 321

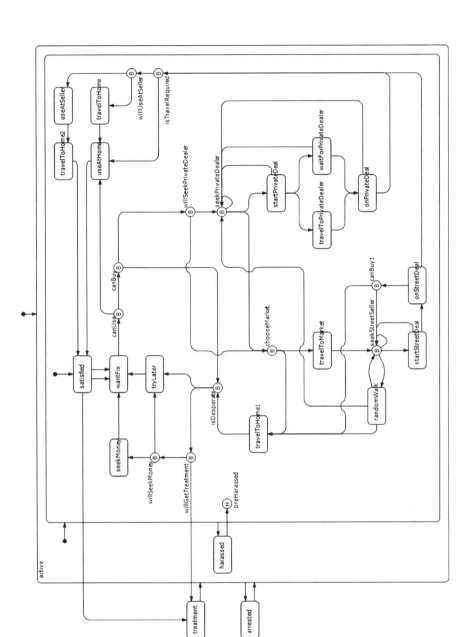

Fig. 11.1 A state diagram representing the states and potential transitions between the states for a heroin user (customer)

agents' addictions were operationalized. Each customer agent in the simulation was assigned four initial conditions, including (1) an addiction level, (2) a drug concentration level, (3) an inventory of drug, and (4) money. A customer agent's addiction level is measured in drug units the customer sought to maintain the addiction. In the model each agent was initially assigned a random number drawn from a normal distribution with the mean of 120 and a standard deviation of 55. This addiction level roughly reflects an addict's "heroin habit", as described above. The concentration level represented the current amount of drug in the agent's body. This number also was drawn from a normal distribution with a mean of 15 and a standard deviation of 45.

The processes of satiation, withdrawal, tolerance, and habit are the key features that distinguish our model from usual behavioral models. By modeling these processes, we introduce neurobiological and physiological drivers that create sustainable support for drug use. Because this is an introductory "pilot" model, we used a simple functional relation based on drug concentration in the body to describe satiation and withdrawal. Assuming that drug injection increases drug concentration very quickly, we ignore the processes of rapid increase of concentration in blood and assume that it happened instantly after the injection. The concentration C stays constant for some short period of time τ, and then gradually decreases according to the first order pharmacokinetic equation:

$$dC/dt = -\lambda Ct \quad \text{for } (t > \tau),$$

where λ is the rate at which drug concentration decreases and depends on the individual metabolism. If drug concentration is below a certain threshold, the message for seeking a fix is generated and the agent starts looking for the drug. Although this process does not represent the actual heroin concentration, it mimics the dynamics and the timing of the need for the new fix. Tolerance is introduced in an implicit way through the change in the addiction (or habit) level. The habit depends on the accumulated dose over a prolonged period and represents the change in the habit, that is, it is linearly increased as the user consumes more heroin than targeted by his or her addiction level, and linearly decreases in the event that the customer did not use heroin.

The heroin used and sold by agents in the IDMS corresponds to the units of heroin available in the market: 10-, 30-, 40-, 60-, 120-, and 360-unit bundles. These bundles parallel actual units sold in the market in both cost and proportionality, namely, $20 for a pill (10 units); $40 for 2 pills (30 units); $50 for 3 pills (40 units); $70 for a half-gram (60 units); $130 for a gram (120 units); and $330 for 3 grams (360 units). Considerable data from the ethnographic research (Isaac et al. 1994) contributed to determining this pricing scheme.

In the event that a customer agent had an addiction level occurring between these amounts, they had to select what amount to purchase and use. If they had the resources, customer agents always selected the larger amount. For instance, an agent with a 37-unit addiction level could only buy 30 or 40 units worth of heroin, and if he or she had $50 or more the customer agent purchased 40 units (3 pills). Customer agents used heroin units that best aligned with their addiction level in this manner until they ran out of inventory. If they possessed fewer units of drug than what they wanted to use, they used the remainder of their inventory. Customers also used heroin randomly, that is, when they were not otherwise motivated to do so.

After running out inventory, the customer evaluated if they could purchase heroin. Customers with money either purchased as much drug as they could or just enough to cover their addiction level. This decision was randomly determined. Customers could only purchase heroin at the full retail prices, and most customers received an income that they used for this purpose. Incomes were distributed based on one of three pay schedules: weekly, every 2 weeks, or monthly. To reflect the potential for less-organized income sources, IDMS also included times when customers randomly received money.

Customer agents could engage in the market and purchase heroin in one of three ways: going into the public (open-air) market and purchasing from a street seller, using a broker (middleman) in this market, or using a private dealer not in an open-air market. Purchases from a street seller were straightforward and direct; customers could see street sellers. If a customer agent transacted with a broker, the sales process was indirect. The customer located a broker who was visible to him or her, but the broker made the sale with a street seller or private dealer. Customer sales through private dealers required a customer knowing the location of the dealer. These locations were identified after customers used a broker for a certain number of sales with that dealer. Private dealers and customer agents transacted in two ways: either the private dealer delivered drugs to the customer or, more typically, the customer traveled to the dealer's location.

Agents selling heroin processed each transaction separately and often had competing transactions that delayed completion of sales. Street sellers and brokers also were subject to arrest from police agents roaming the open-air market and thus would change location. Customers maintained an up-to-date list of sellers and locations in which they had successful transactions, returning to locations in descending order of success the next time they needed to purchase the drug.

The agent rules directing seller and customer behaviors clearly do not include all decision-making processes or the complexity associated of heroin-market operations. Nonetheless, they do include a fairly robust and dynamic set of criteria and realistically incorporate variation in these behaviors. It was not always easy for customer agents to purchase the heroin they wanted to maintain their heroin addiction levels.

11.4 Results

The initial and final distributions of addiction levels for 200 customers in the 12-month simulation are compared in Fig. 11.2. The mean addiction level for the sample at baseline was 122 units and 156 units at 12 months. With one exception, these distributions are relatively similar. The exception was that at 12 months, considerably more customer agents had addiction levels of 250 or above ($n = 1$ vs. 25).

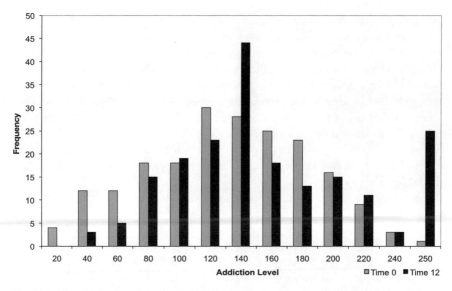

Fig. 11.2 The distribution of customer-addiction levels, in units, for the sample of $N = 200$ customers at baseline (Time 0) and 12 months postbaseline (Time 12)

However, this finding is not out of line after considering individual variations observed in the patterns of addiction over time.

Although the overall distribution of addiction levels was important and within the boundaries of what would be expected in a real sample of heroin users, the major focus of this manuscript is to identify patterns in addiction levels among agents over time. Using the initial mean addiction level as a guide, we randomly selected 15 customers around the mean, as well as two standard deviations above and below it, for more detailed investigation. Although the complete analysis of all 200 agents incorporating multiple simulation runs is ongoing, several trends in this initial random sample are reported below. Figures 11.3, 11.4, 11.5 and 11.6 show individual customer agent addiction levels over the entire 12 months of one random simulation run.

First, some customer agents exhibited an intense and brief time period in which their addiction level increased and then decreased. These long-term binges were defined when an agent's addiction doubled or more over a timeframe of 1 month. These increases were then immediately followed by a sharp decline in addiction level, usually of the same magnitude. Figures 11.3 and 11.4 show agents representing this pattern.

This pattern likely contributed significantly to the addiction level discrepancy between the initial and final distributions. Ending the simulation at 12 months was arbitrary, and addiction levels of 250 and above are likely for customers who are on the upward slope of a binge (Fig. 11.5).

Overall, this binge/crash pattern can be imagined behaviorally as customers who are trying to overcome their tolerance rapidly, but then run out of money to consistently maintain such a high addiction level. Heroin users often report these sorts

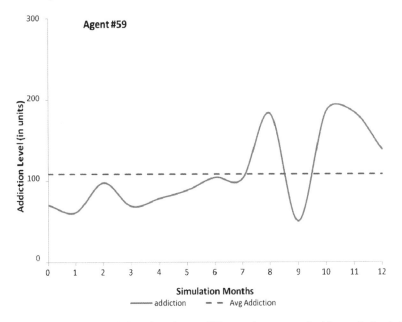

Fig. 11.3 The addiction level, in units, of agent #59 over the course of a 12-month simulation. This agent shows a binge/crash pattern in the agent's addiction level

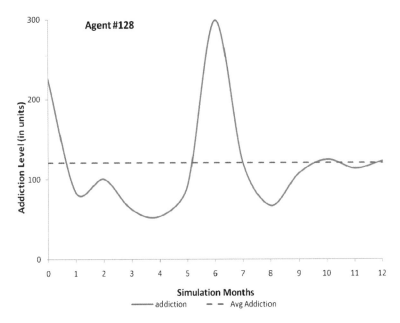

Fig. 11.4 The addiction level, in units, of agent #128 over the course of a 12-month simulation. This agent shows a binge/crash pattern in the agent's addiction level

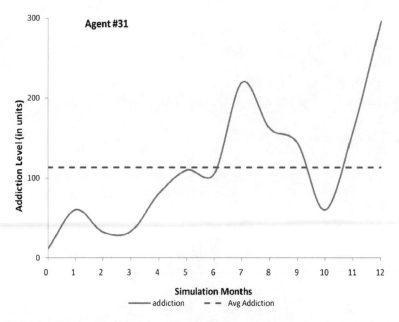

Fig. 11.5 The addiction level, in units, of agent #31 over the course of a 12-month simulation. This agent shows a stepped pattern to increasing the agent's addiction level. This agent also may be exhibiting the beginning of a binge/crash pattern at month 10

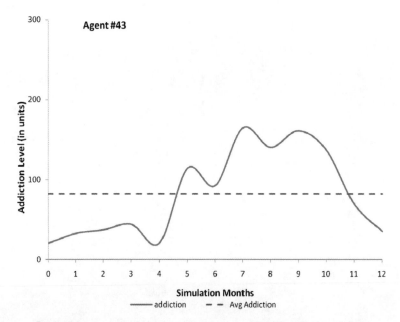

Fig. 11.6 The addiction level, in units, of agent #43 over the course of a 12-month simulation. This agent shows a stepped pattern increasing the agent's addiction level

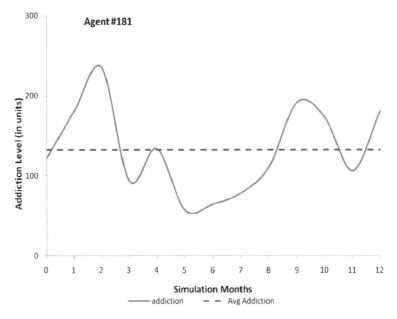

Fig. 11.7 The addiction level, in units, of agent #181 over the course of a 12-month simulation. This agent shows a stepped pattern reducing the agent's addiction level

of binge/crashes, especially when they buy heroin using money that they acquire haphazardly or unexpectedly from crime.

Contrasting this binge/crash pattern, another pattern observed was a gradual or stepped pattern of addiction levels. In this pattern, addiction levels increase for 1 month, are followed by a plateau period, and then followed by another increase. The pattern is then repeated as the addiction level decreases. As in the binge/crash pattern, the addiction level is also increasing and decreasing; however, the timeframe here is more extended. Figures 11.5 and 11.6 show agents with this pattern.

Most stepped patterns produced overall increases in addiction levels. However, in a few instances, addiction levels decreased in this manner. Overall, increases and reductions seemed to correspond to the initial addiction level of the agent. If the addiction level started low, it could increase gradually, while if it started high, it might diminish over time, as shown in Fig. 11.7.

The gradual increases in addiction levels presented in the stepped pattern correspond to a common narrative in becoming addicted. Users often report gradual increases in heroin use over several months: they might start out "chipping" (i.e., using only on the weekends) and then proceed over time to everyday use. Many heroin addicts report that often they slowly increase the amount of heroin that they use to overcome tolerance.

The final pattern of customer agent addictions is shown in Fig. 11.8. This pattern represents stability over time, meaning that the agent's addiction level remains relatively constant throughout the entire 12 months of the simulation. Our initial analysis suggests that this pattern appears less frequently than the other patterns.

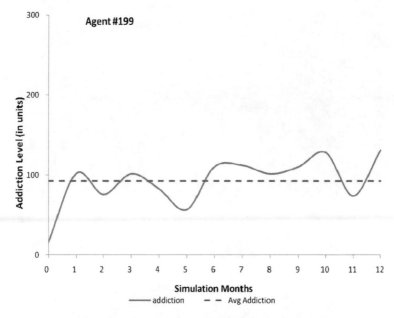

Fig. 11.8 The addiction level, in units, of agent #199 over the course of a 12-month simulation. This agent shows a stable pattern in the agent's addiction level

It also seems most dependent on an agent's initial addiction level, occurring most often at relatively lower initial addiction levels. This result has certain face validity because large addictions are inherently more expensive to maintain.

Stable patterns in addiction levels also are noted in the behavior of heroin users. Heroin users who have considerable experience with their addiction show remarkable resilience in maintaining affordable and manageable levels of use. Although rare in our initial analysis, it is important to recognize agents who have developed behaviors to meet the needs of their addiction and who could do so consistently through the contingencies of the market.

As noted, a complete analysis incorporating multiple simulation runs and investigating all 200 customer agents is in progress. Nonetheless, initial patterns identified in this preliminary analysis are promising. The authors hope to better characterize the binge/crash, stepped, and stable patterns, as well as recognize additional patterns using ABM.

11.5 Limitations

It is important to recognize that the model presented here addresses *only* a limited dimension of the market's overall influence on heroin users' addictions. One important missing component is the influence of other heroin users on a heroin user's behavior. People addicted to heroin often rely on their associates to share the cost of drugs. This is particularly relevant when users run out of their own funds (see below). The current version of the model is focusing on the incorporation of the full range of processes associated with purchasing the drug. It also is currently unclear how factors associated with the cost or the processes of sales contribute to the patterns observed. We are currently working on the incorporation of this component into the next version of the model.

Other issues not included in the model involve how heroin users actively seek to reduce their level of addiction. Heroin users who believe their addiction is becoming out of control often enter treatment to reduce their use and/or substitute other opiate class substances to reduce their dependence on heroin (Koester et al. 1999). We model entering and leaving treatment as a random process associated with the level of addiction and availability of resources to buy the drug, however, we don't track the detailed process of relapse and reentry as was considered, for example, by Zarkin et al. (2005). Polydrug use is not currently included in the model.

In this model, the variety of alternatives for obtaining money and heroin is represented in two random processes: receiving windfalls and seeking money, when the agent is actively seeking money and can find it with some probability. In the real world, heroin addicts are much more industrious and find ways to acquire heroin in these situations. Some of the ways users acquire heroin without money include partnering with fellow heroin users (as noted above), committing crimes to get money, or getting heroin on credit from dealers (Hoffer 2006). Currently these behaviors are not included in the model, and the windfall and seeking money processes are calibrated to represent the frequency with which a user is able to acquire drugs.

Despite the limitations of IDMS, this model takes an important step toward providing a comprehensive understanding of heroin addiction by specifically incorporating individual features of a heroin addiction expressed within the context of a dynamic social environment (the market), in which heroin and cash are exchanged.

As this project remains ongoing, the authors hope to address some of these important limitations and to verify findings from this model through independent data collection. In particular, combining ethnographic research with Ecological Momentary Assessment, where the users record their daily heroin use and purchasing patterns, the authors will obtain the data to validate decisions and patterns assumed in the model.

11.6 Conclusion

This paper illustrates how the social fabric of a heroin market can impact the physiology of heroin addiction and how heterogeneity of addiction patterns can be shaped by the events in the market. Two innovations are specifically notable: first is a new conceptual use of ABMs, where the focus is on the within-agent dynamics under the influence of the surrounding social landscape. This approach complements the more traditional use of ABM, which studies self-organization of the complex system due to the interaction of its members. The second innovation is an attempt to uncover physiological trends and patterns that are generally not measurable in real life. Future studies might provide ways to measure partial components of the model such as drug concentration or craving to validate patterns. However, this research will never be able to provide an uninterrupted history of these dynamics available via ABM. This study shows the utility of the systems approach, where simple but realistic rules and equations provide insight into a real-life problem.

Using the results of this study, we would argue that heroin addiction and perhaps some other addictions, such as addiction to crack or methamphetamine, cannot be fully understood without incorporating ways in which drug users acquire the drug within these environments. Thus, as the next stage in our project, we are collecting daily drug use reports from active heroin users via interactive voice response (IVR) or smartphones. These data will greatly facilitate both setting market simulation parameters, as well as collecting the data necessary to validate simulated addiction patterns and outcomes.

Finally, the real-world application of the findings reported here are significant in a number of ways. For addicts attempting to manage their addiction, uncontrolled increases and decreases in their level of addiction are nearly unavoidable. For frontline health professionals working with addicts, being able to identify and work with specific patterns of addiction in devising strategies to minimize harm might be extremely beneficial. These patterns also may prove useful in anticipating opportunities when addicts might be more receptive to treatment alternatives. Developing ABMs that included biologically motivated and socially meaningful behavior represents an exciting and novel approach for researching the complexity of drug addiction.

References

Adler PA (1985) Wheeling and dealing: an ethnography of an upper-level drug dealing and smuggling community. Columbia University Press, New York

Agar M (1973) Ripping and running. Academic Press, New York

Agar M, Reisinger HS (2002) A heroin epidemic at the intersection of histories: the 1960s epidemic among African Americans in Baltimore. Med Anthropol 21:189–230

Agar M, Bourgois P, French J, Murdoch O (1998) Heroin addict habit size in three cities: context and variation. J Drug Issues 28:921–940

Ahmed SH, Bobashev GV, Gutkin BS (2007) The simulation of addiction: Pharmacological and neurocomputational models of drug self-administration. Drug Alcohol Depend 90:304–311

American Psychiatric Association (2000) Diagnostic and statistical manual of mental disorders: DSM-IV-TR. Author, Washington

Bernard HR (1988) Research methods in anthropology. Sage, Newbury Park

Bourgois P (1997) In search of respect: selling crack in El Barrio. Cambridge University Press, New York

Curtis R, Wendel T (2000) Toward the development of a typology of illegal drug markets. In: Natarajan M (ed) Illegal drug markets: from research to prevention policy. Criminal Justice Press, Monsey, pp 121–152

Fetterman D (1998) Ethnography: step by step. Sage, London

Gutkin BS, Dehaene S, Changeux JP (2006) A neurocomputational hypothesis for nicotine addiction. Proc Natl Acad Sci 103:1106–1111

Hamid A (1992) The development cycle of a drug epidemic: the cocaine smoking epidemic of 1981–1991. J Psychoact Drugs 24:337–348

Hammersley M, Atkinson P (1993) Ethnography: principals in practice, 2nd edn. Routledge, New York

Hanson GR, Venturelli PJ, Fleckenstein AE (2006) Drugs and society, 9th edn. Jones and Bartlett, London

Hoffer L (2006) Junkie business: The evolution and operation of a heroin dealing network. Wadsworth, Belmont

Hoffer LD, Bobashev G, Morris RJ (2009) Researching a local heroin market as a complex adaptive system. Am J Community Psychol 44:273–286. doi:10.1007/s10464-009-9268-2

Isaac M, Janca A, Sartorius N (1994) ICD-10 symptom glossary for mental disorders. Division of Mental Health, World Health Organization, Geneva

Johnson BD, Goldstein PJ, Preble E, Schmeidler J, Lipton DS, Spunt B, Miller T (1985) Taking care of business: the economics of crime by heroin abusers. D.C. Heath and Company, Lexington

Kerr T, Small W, Wood E (2005) The public health and social impacts of drug market enforcement: A review of the evidence. Int J Drug Policy 16:210–220

Koester SK (1994) Copping, running and paraphernalia laws: contextual variables and needle risk behavior among injection drug users in Denver. Human Organ 53:287–295

Koester SK, Anderson KT, Hoffer L (1999) Active heroin injectors' perceptions and use of methadone maintenance treatment: cynical performance or self-prescribed risk reduction. Subst Use Misuse 34:2135–2153

Koob GF, Le Moal M (2005) Neurobiology of addiction. Elsevier, New York

Musto D (1987) The American disease: origins of narcotic control. Oxford University Press, New York

National Institute on Drug Abuse (1993) Risk behavior assessment, 3rd edn. National Institutes of Health, Rockville

Patton MQ (1980) Qualitative evaluation and research methods, 2nd edn. Sage, London

Preble E, Casey J (1969) Taking care of business: The heroin user's life on the street. Int J Addict 4:1–24

Singer M (2006) Something dangerous: emergent and changing illicit drug use and community health. Waveland Press, Long Grove

Zarkin GA, Dunlap LJ, Hicks KA, Mamo D (2005) Benefits and costs of methadone treatment: results from a lifetime simulation model. Health Econ 14(11):1133–1150

Zule WA, Desmond DP, Neff JA (2002) Syringe type and drug injector risk for HIV infection: a case study in Texas. Soc Sci Med 55:1103–1113

Index

A

ABMs, 313, 315, 316, 318, 319, 329, 330
Abuse liability, 99, 289
Acetylcholine, *see* ACh
ACh, 112, 118–126, 128, 129, 131–134, 136
actions
    nAChR, 117, 120, 121, 131
    nAChR-mediated, 129, 133
    nicotinic, 126, 137
Activation, 4, 45, 113, 116, 118–122, 126, 128, 129, 131, 132, 134, 136, 190, 196, 277
    fast, 120, 122, 126, 133, 134, 138
    mesolimbic dopaminergic, 199
Activation time constant, 119, 136
Active inference, 237, 240, 243, 247, 249, 253, 254, 262, 270, 274, 277, 278
Actor, 169, 191, 219, 220, 223
Actor components, 218–220
Actor-critic models, 206–208, 218–220, 222, 224
Actor's preference, 220, 222, 223
Adaptation, vii, 19, 21, 23–25, 32, 35, 38, 46, 49, 50, 78, 258
    fast, 32
    slow, 32, 33, 37, 53
Adaptive process, 23–25
    regulated, 21, 31, 34, 49
Adaptive regulator, 30–34, 37, 45, 50–52
    fast, 36, 37
Addiction, v, vii, viii, 19, 20, 59, 61, 86, 87, 163, 164, 172–174, 181, 189–192, 194, 205–209, 220–225, 286–290, 313–317, 328–330
    animal models of, 210, 227
    autoshaping model of, 85, 86
    behavioral, 164, 172, 173
    characteristic of, 63, 285, 292
    computational models of, 163, 170, 182, 207, 209
    models of, 57, 58, 63, 66, 68, 87, 89, 96, 99, 100, 209, 286
    theories of, 57, 59, 80, 82, 84, 96, 102–105, 226
Addiction levels, 317, 319, 322–330
    agent's, 325–328
Addiction phenomena, 105, 285–287
Addictive behavior, 17, 57–59, 67, 80, 82, 84, 96, 97, 99, 102, 104, 138, 182, 208, 209, 215, 237, 238, 276
Addictive drugs, 32, 98, 145, 165, 169, 181, 205, 206, 221, 226, 289
Addicts, vi, 39, 71, 163, 165, 168, 173, 174, 178, 179, 181, 182, 201, 205, 210–212, 218, 226, 227, 316, 317, 330
Administration, 15, 20, 21, 26, 27, 31, 36, 40, 42, 43, 45, 72
Affinity
    high, 112, 117, 118, 129
    low, 112, 117, 118
Agent
    biological, 239, 248, 250
    heroin-using, 320
    homeless, 319
Agent-based models, *see* ABMs
Agonist, 4, 11, 12, 15–17, 122
    indirect, 15
Agonist concentration ratio, 3, 12, 13, 15
Agonist concentrations, 10, 12, 122
    equiactive, 3, 11–13, 17
Alcohol, 173, 266, 289, 291
Allostasis, 23, 83, 84
Amphetamine, 167, 173, 197, 199, 214, 266

acute, 196, 199
Amphetamine administration, 194, 196, 199, 200
   acute, 197, 198
Antagonists, 3, 11–13, 15–17, 116, 126, 128
   competitive, 3, 11, 12, 15
   concentration, 3, 11–13, 15, 17
   dose of, 11, 14
   injection of, 12, 13
   K_{dose}, 11, 16
   potencies, 11, 12, 16
Anticipation, 19, 45, 74, 189
Apomorphine, 4, 12, 14
Ascending limb, 6, 7
Assessment, 211, 290, 304
Associations, 15, 16, 70, 150, 168, 207, 210, 213–216, 219, 224, 226, 261, 270, 273, 274, 316
   action-outcome, 150, 151
   cue-reinforcer, 171, 172
   cue-reward, 190
Associative plasticity, 246, 265, 274, 275
Attraction, 255, 259, 261
   basin of, 258
Attractors, 249, 258–260, 262–264
   appropriate, 263, 264
Autopoiesis, 258, 259
Autoshaping, 57, 58, 84, 85, 88, 97, 98, 100, 103

B

Basal reward level, 216–218
behavior
   adaptive, 238, 253–255, 267, 277
   compulsive, 151, 210–212, 220, 222, 228
   drug-associated, 114, 215
   drug-related, 115, 215
   drug-taking, 3, 145, 146, 288
   goal-directed, 146, 152, 155, 157, 226
   habitual, 137, 145, 146, 151, 227
   lever-pressing, 5, 6, 10
   normal, 238, 268
   reinforced, 6, 111
   repetitive, 57, 96
   seeking, 114, 210, 212, 228
Behavioral economics, 287, 291, 292, 305
Bellman equation, 148, 153, 155
Blocking effect, 173, 215, 218
Bradykinesia, 266–269

C

Cells, 125, 134, 155, 171, 246, 266, 274, 276
Choices
   addictive, 165, 173, 181
   behavioral, 113, 115
   larger-later, 176, 179, 180
   proportion of, 301, 302
   smaller-sooner, 176, 179, 180
Cholinergic drive, 131, 132, 135
Cholinergic input levels, 132–134, 137, 139
Cocaine, 4, 6, 7, 9–12, 14–16, 23, 59, 84, 96, 167–169, 182, 214, 216, 226, 228, 291, 295
   unit dose of, 4, 5, 7
Cocaine concentrations, 3, 8, 10, 11, 13, 15
Cocaine levels, 10, 11
Cocaine satiety thresholds, 11, 13, 14
Cocaine self-administration, 4, 8, 11, 221
Cognitive maps, 150, 151
Commodities, 287–290, 304–306, 317
   independent, 305
Compensatory response, 20, 28, 29, 35–38, 41–46, 49
Compulsion zone, 3, 10
Compulsivity, 210, 212, 224, 226
Concentrations, 3, 5, 6, 8, 10, 13, 136, 295, 322
Conditional expectations, 241, 242, 245, 246, 262–264, 272–275, 278
Conditioned response, *see* CR
Conditioned stimulus, *see* CS
confidence intervals
   conditional, 271, 272, 274
Constant cholinergic drive, 131, 132, 134
Constant cholinergic tone, 128, 129, 139
Constant drug effect, 36, 37
Control
   cognitive, 226, 228
   loop, 53, 54
   optimal, 63, 248, 249, 252, 255, 256
Control action, 64, 65, 68, 71
Control systems, 65–69, 89
   hybrid, 68
Control theoretic implementations, 58, 59, 81, 104
Control theoretic models, 62, 68, 78, 80, 94, 103, 104
Control theory, vi, 33, 57–60, 63, 67–69, 71, 76, 98, 102, 104, 105
CR, 73, 80, 83–85, 96, 97, 207
   compensatory, 76, 83
Craving, 6, 63, 67, 71, 75, 81, 83, 85, 86, 96, 146, 155, 156, 165, 171, 330
Critic, 169, 191, 218–223
Critic component, 218, 219
CS, 76, 84, 85, 166, 190, 193, 196, 198, 200, 201, 207, 208, 210, 219

Index

D
DA, 111, 113, 116, 117, 133, 134, 137, 214, 266
DA activity, 117, 120, 125, 129, 131–135, 137, 138
DA cell activity, 120, 133
DA cells, 115–118, 120, 125, 126, 129, 131–134, 137–139
DA neuron activity, 116, 129, 131, 134
DA neurons, 112, 113, 116, 117, 121, 125, 126, 131, 134, 138, 217, 266, 272
Dealers, 315, 318–320, 323, 329
  private, 313, 319, 323
  street, 319
Decision-making, v, vi, 163, 164, 169, 170, 172, 287
Decision-making model, 163, 164, 181, 182
Decision-making systems, 163, 165, 172, 182, 218
  multiple, 164, 181
  normal, 209
Deliberation, 170, 171
  benefit of, 225, 226
  cost of, 226, 227
Deliberation time, 226, 227
Delivery, 58, 66, 169, 209, 210, 295, 301
Demand
  elasticity of, 288, 306
  law of, 288, 289, 297
Demand curves, 287–289, 297
  characteristics of, 288, 289
Dependence, 19, 20, 39, 42, 45, 46, 61, 86, 98, 147, 213, 329
  conditional, 250, 251, 257
Dependent variables, 5, 6, 70, 296
Desensitization, 118–121, 124–126, 132, 134, 136, 139
Desensitized state, 118, 119, 122–124, 134
  deeper-level, 118
Devaluation, 80, 150, 156, 157, 192, 195
Digestive tract, 34, 49, 50, 53
Direct stimulation, 120, 125, 129, 131–137, 139
Discounting functions, 167, 175, 176, 292, 293
Discounting rates, 167, 175, 176, 180, 290, 291, 305, 306
Disinhibition, 116, 120, 129, 131–135, 137–139
Dissociation, 15, 16
Distribution, 5, 7, 8, 12, 15, 16, 33, 49, 133, 134, 176, 199, 239, 313, 319, 323, 324
Disturbances, 21, 22, 24, 25, 27–30, 32–36, 50, 53, 78

Divergence, 239, 251, 252, 257, 261
Divergence-constraint, 248, 252, 254, 256, 265
Dopamine, vii, 10, 15, 75, 112, 145, 151, 152, 157, 169, 189, 191, 194, 196, 199, 215, 276, 277
  extracellular, 227
  phasic, 139, 168, 226
Dopamine concentrations, 10, 11, 214
Dopamine neurons, 112, 146, 153, 166, 194, 198, 214, 219, 224, 276
  midbrain, 151, 270, 276
Dopamine receptors, 4, 11, 217, 221, 222
  availability of, 217, 222
Dopamine signaling, 75, 76, 113
Dopamine signals, 140, 166
Dopamine system, 111, 151, 169, 195
Dopamine transporters, 10, 11, 15, 165
Dopaminergic, 113, 116, 117, 237, 246, 266, 276
Dopaminergic neurons, 153, 167, 192
Dorsal striatum, 113, 114, 176, 208, 220, 221, 224, 226
Dose
  actual, 5, 38, 43
  antagonist, 11, 13, 14, 16
  first, 25, 40
  initial, 39, 91
  large, 47, 216
  larger, 39, 40
  next, 5, 9, 92–95
Dose–duration function, 5, 9, 10, 14
Dose–response curve, 39–43, 120, 122
Drug abuse, 61, 87, 145–147, 151, 152, 155–157
Drug action, 24, 38, 41, 42, 114, 157, 169
Drug addiction, 59, 63, 64, 68–70, 73, 78, 81, 85, 102, 115, 169, 195, 205, 209, 216, 217, 221, 228
Drug administrations, 20, 22, 24, 25, 30, 32, 33, 39, 44, 47, 49, 50, 52, 53, 58, 291
Drug concentration, 4, 5, 8, 15, 52, 295, 322, 330
Drug consumption, 209–211, 215, 216, 220, 224, 227, 287–290, 294, 297, 300
Drug cues, 80, 82, 84, 85, 212, 227
Drug delivery, 5, 168, 171, 210
Drug dependence, 21, 46, 290, 292
Drug dose, 19, 29, 36, 38–45, 47, 48, 90–92, 294
Drug effect, 5, 19, 20, 22, 25, 27, 28, 31, 32, 35–41, 43–45, 47, 50–53, 64, 73, 76, 79, 81, 98
  normal, 36, 41, 44

paradoxical, 36, 44
Drug intoxication, 189, 201
Drug regulator, 50, 51
Drug reinforcer, 80, 81, 209, 218
Drug self-administration, 3, 59, 60, 67, 81, 83, 84, 86, 93, 94, 96, 227, 290, 294, 300, 315
   model of, 59, 72
   rate of, 63, 102, 103
Drug sensitization, 73, 81, 82, 90, 191, 194
Drug tolerance, 19–21, 23–25, 27, 30, 46, 59, 73, 78–80, 100
   development of, 22, 23, 27, 30
   model of, 19, 20, 30, 32, 33, 53
Drug use, 57, 58, 80, 83, 85–87, 94, 97, 99, 168, 182, 212, 217, 289, 290, 314, 322
   early stages of, 84, 97, 98
Drug users, 313, 314, 319, 320, 330
Drug withdrawal, 35, 46–49, 74, 92, 190, 305
Drug withdrawal syndrome, 74, 80, 82, 84
Drug-seeking, 137, 215, 225, 226
Drug-seeking behaviors, 82, 145, 146, 152, 157, 172, 181, 210, 220, 221, 227, 228
Drugs
   administered, 19, 44, 45, 49
   compulsive, 212, 221, 224
   pharmacological effect of, 205–207, 209, 215, 220, 221, 224
   psychostimulant, 191, 196
Drugs of abuse, 61, 80, 86, 87, 98, 103, 145, 146, 151–153, 155–157, 163–165, 168, 173, 176, 190, 192, 205, 206, 226

E
Elimination, 9, 12, 15
Empirical evidence, 57, 59, 76, 104
Empirical results, 59
Ensemble density, 248, 251, 252
Entropy, 239, 240, 277
Environmental cues, 19, 20, 26, 27, 34, 42, 43, 45, 74, 139, 215
Equilibrium, 15, 24, 67, 252, 254, 255
Equilibrium density, 239, 252, 255, 256, 265
Error signal, 71, 89–92, 151, 167, 215–219
   critic's, 220, 221
   reward prediction, 167, 214
Euphoria, 6, 63, 64, 71, 77, 87, 89, 91, 92, 317, 320
Exogenous substance, 25–28, 30, 31, 34, 35, 42, 43

Expectations, 148, 169–171, 178, 193, 253, 254, 270, 272, 274, 279
Exposure, 115, 122, 131, 132, 196, 211, 227
   long-term, 216, 217

F
Fast regulator, 33, 50–52
Feedback loops, 21, 51, 60, 63, 64, 66, 75, 82, 83, 92, 104
   negative, 66, 75, 87, 88
   positive, 74, 75, 80, 83, 89, 93
Feedback models, 20, 58, 70, 79
   electronic, 20
   positive, 81, 83, 85, 87, 88
   simple negative, 72
Feedback systems, 33, 57, 60, 61, 67, 70, 104
   adaptive, 33, 61
   closed-loop, 67
   positive, 66, 81–84
Free-energy formulations, 237, 238

G
GABA cells, 115, 116, 120, 126, 131, 133, 137–139
GABAergic activity, 132, 134
GABAergic cells, 112, 117, 126, 128, 129, 131–134, 138
GABAergic input, *see* IG
GABAergic neurons, 116, 117, 121, 131
Gain, 45, 74, 91, 93–95, 113, 216, 247, 252, 254, 270, 276, 316
   open-loop, 36
   post-synaptic, 266, 276, 277
   synaptic, 246, 247, 266, 277
Gain function, 91, 93–95
Gambling, 164, 172, 173, 224
Generative models, 241, 244, 248, 253, 260, 262, 265, 270, 277
   agent's, 248, 251, 253, 262
GLU, 112, 113, 117, 118, 120, 128, 133, 139, 228
Glutamatergic, *see* GLU
Glutamatergic input, *see* IGlu
Goal-state, 67, 71, 74, 79, 83

H
Habits, 87, 113, 139, 150, 151, 154, 213, 223, 315, 317, 319, 320, 322
Habitual, 145, 146, 181, 205, 206, 208, 214, 220, 226, 228
Habitual system, 145, 181, 206, 224–228
Habituation, 20, 24, 80, 81, 93, 94, 100, 137
Heroin, 11, 70, 71, 169, 209, 266, 290, 291, 314–320, 322, 323, 327–329

Index 337

purchase, 323
Heroin addiction, 70, 71, 97, 100, 290, 313, 316–319, 329, 330
Heroin addiction model, 57, 58, 70–72, 74, 78, 88, 89, 98, 99
Heroin addicts, 315, 317, 329
Heroin habit, 317, 320, 322
Heroin markets, 313, 316, 318, 320, 329
Heroin users, 314, 315, 317–320, 324, 328, 329
Heroin withdrawal syndrome, 71
Hill equations, 7, 119, 120
Homeopathy, 43, 44
Homeostasis, 20–23, 25, 61, 70, 78
Homeostatic models, 23, 63, 72, 73, 78, 79, 84
Homeostatic theories, 67, 82, 84
Hormesis, 43

I

IDMS, 316, 319, 322, 323, 329
IG, 75, 116–118, 126, 128, 133, 134
IGlu, 117, 118, 126, 128, 138
Illicit drug market simulation, *see* IDMS
Immediate rewards, 153, 214, 218
  small, 164, 174, 211
Implications, 24, 75, 117, 138, 157, 225, 304
  behavioral, 215, 224
Impulsivity, 164, 174, 179, 211, 212, 215
Incentive-sensitization, 97, 100, 189, 191
Inhibition, 133, 134, 138
  impaired, 211, 212
Injection, 4, 13, 27, 84, 322
Input signal, 32, 49–51, 53, 60, 69
Instrumental behavior, 150, 207, 208
Instrumental conditioning, 57, 58, 61, 80, 81, 85, 89, 92–100, 102, 207, 208
Instrumental drug conditioning, 81, 95
Intellectual silos, 286, 307
Inter-temporal choice, 287, 289, 291, 304
Interactive voice response, *see* IVR
Intermittent adaptation, 19, 21
Internal model, 64, 79, 87, 238
Interval regulator, 50–52
Itinerant policies, vii, 237, 249, 251, 257, 260, 262, 263, 265, 270, 278

L

Liking, 6, 97, 98, 190
Lossy accumulator, 89–95, 99

M

Maintenance, 8, 76, 84, 111, 113, 114, 276
Manifold, 250, 258, 259, 262, 264
Manifold-states, 249, 250, 259, 261, 262, 264

Market model, 313, 320
Markov decision processes, *see* MDPs
MDPs, 147–150
Mesolimbic activation, 192, 197–199
Model
  dual-process, 206, 224, 226, 227
  incentive-sensitization, 80, 82
  linear dynamical, 62, 97, 104
  neurocomputational, 75, 76, 205
  nonhomeostatic, 59, 86, 102
  quantitative, 23, 163
  satiety threshold, 3, 7, 15, 17
  temporal difference, 170, 176, 180, 193, 256
Model estimation, 50, 51
Model of addiction, 72, 75, 212
Model simulation, 38, 40, 43, 45
Model-based system, 145, 151–157
Model-based value, 153, 155
Morphine, 4, 76, 266, 295
Motivation, 72, 79, 80, 86, 87, 97, 111, 153, 156, 189, 191, 192, 199–201, 218, 239, 272, 277, 306
Motivational values, 114, 153, 190, 194–196, 199

N

nAcc, 11, 111, 139, 192, 198, 220, 227, 266
nAChR, 75, 111–124, 126, 128, 129, 131–134, 136–139
nAChR activation, 115, 117, 118, 120, 128, 136, 139
nAChR desensitization, 112, 116, 129, 133
nAChR kinetics, 120, 137
nAChR knockout mice, 125, 129
nAChR subtypes, 112, 116, 122–124
Navigation function, 256, 257
Negative feedback, vi, 21, 22, 30, 61, 64, 66, 71, 74, 75, 81–84, 88, 89, 94, 96, 97, 99, 104
Negative feedback models, 71, 86
Negative feedback systems, 22, 53, 65, 81
Negative reactions, 33, 38, 39, 41, 42, 46–48, 53
Neurons, v, 117, 125, 138, 139, 197, 214, 217, 226, 276
Nic, vi, 75, 111–126, 128, 129, 131–134, 136–140, 266, 289, 291, 296, 297
Nicotine, *see* Nic
Nicotine action, 111, 116, 120, 122, 125, 131, 137
Nicotine addiction, 75, 111–113
Nicotine applications, 125, 126, 131, 132, 135, 137

Nicotine concentrations, 119, 120, 122, 124, 126, 131, 132, 134, 135, 138
  applied, 120, 128, 134
Nicotine consumption, 296, 297
Nicotine levels, 132, 138, 296
Nicotine regulation studies, 296, 297
nicotine self-administration, 76, 114, 115, 137
Nicotinic acetylcholinergic receptor, *see* nAChR
Nucleus accumbens, *see* NAcc

O

Observations, 28, 64–66, 166, 172, 173, 219, 286, 288, 292
Observer, 64, 65, 87, 88
Occupancy, 5, 10, 11, 273, 274
  fractional, 5, 7, 10–12, 15
Open loop gain, 37, 44–46
Open-air market, 320, 323
Opponent process theory, 57, 58, 72–78, 80, 81, 85, 89, 94, 96, 97, 99, 100
Opponent process theory model, 89, 94
Opponent processes, 75, 83, 97, 100, 103, 113–115
Optimisation, 238, 243, 246, 247, 257, 265, 270, 277, 278
Optimising, 237, 241, 243, 246, 248, 257, 276
Option
  larger-later, 178, 179
  smaller-sooner, 176, 178–180

P

Patterns, vi, 23, 93, 145, 163, 259, 315, 316, 324, 327, 329, 330
  binge/crash, 313, 324–327
  stable, 313, 328
  stepped, 313, 326, 327
Pavlovian, 49, 74, 76, 77, 85, 198, 201, 205–207, 219, 228
Pavlovian conditioning, 20, 207
Pavlovian cue, 189, 190
Pavlovian-instrumental transfer, *see* PIT
Perception, 60, 237–240, 242, 243, 265, 277, 279
Perceptual inference, 237, 238, 240, 243, 247, 270, 274, 278
Perceptual learning, 238, 246, 262, 265, 270, 271
PFC, 112, 121, 151, 192, 220
Pharmacodynamic potencies, 16, 17
Pharmacodynamics, 4, 5, 58, 71, 137
Pharmacokinetics, 13, 16, 17, 93, 97, 98
Phasic activity, 214, 217, 224
Phasic dopamine signal, 115, 139, 168

Physiological processes, 20, 25, 29, 31, 33, 36, 49, 54, 61
  regulated, 21, 29
Physiological states
  current, 192, 194, 201
  expected, 262, 264
  motion of, 274, 275
  true, 262, 264
Pigeons, 179, 293, 295
Pill, 317, 320, 322
PIT, 156, 170, 208, 228
Plasticity, 82, 113, 114
Policies
  fixed-point, 237, 248, 251, 256, 257, 265, 277
  optimal, 148, 175
  sort of, 249, 252, 258, 259
  value-based, 248, 254, 256, 259
Positive feedback, 61, 64, 66, 68, 74–76, 82, 85, 88, 89, 94, 96, 97, 99, 100
Positive reactions, 38, 39
Precommitment, 179, 180
Prediction error signal, 169, 214, 215, 218–220, 224
Prediction errors
  action-dependent, 221, 222
  amplitude of sensory, 267, 269
  appropriate, 274, 275
  associated, 262, 274, 275
  underlying sensory, 267, 269
Predictions, 6, 7, 59, 114, 132, 133, 136–138, 153, 168, 171, 191, 207, 214–216, 218, 219, 226, 227, 243–246, 267, 274, 275
preference reversal experiments
  hypothetical, 301, 302
  induced, 303, 304
Preference reversals, 174, 175, 179, 285, 292, 293, 300–304
Prefrontal cortex, *see* PFC
principles
  free-energy, 238–240, 242, 277
Prior expectations, 167, 218, 237, 238, 251, 253, 257, 267, 269, 277
Priors, 237, 244, 248, 249, 254
  formal, 248, 251, 265
  simple, 253, 254
Process output, 21, 22, 30, 31, 35, 46, 50
Processes
  addictive, 75, 77, 78, 100, 165
  decision-making, 163, 170, 313, 323
  goal-directed, 206, 208
  habitual, 206–208
  habituation, 81, 93

Index

Processes (cont.)
  sensitization/habituation, 93–95
  slow, 39, 218
Psychobiological mechanisms, 96, 98, 105
Psychological approach, 5, 7, 15, 103

R
Random fluctuations, 243–245, 248, 252, 255, 265, 276
Ratio, 12, 13, 15, 133, 271, 272, 295
Receptor population, 5, 7, 11, 12, 16
Receptor site, 26, 28, 32
Recognition density, 239–242
Recognition dynamics, 242, 243, 245
Reduction, 38, 39, 42–44, 173, 175, 217, 226, 266, 269, 272, 327
  progressive, 267, 269, 274
Regulated process, 20, 25, 30–32, 35, 306
Regulated systems, 33, 53, 57, 67, 68, 105
Regulation, 21–23, 25, 32, 34, 36–38, 46, 53, 79
  adaptive, 23, 37
  physiological, 36
Regulation loop, 33, 34, 36, 44, 45, 49, 53, 54
Regulator, 25, 26, 30, 33, 50, 54
  slow, 33, 35, 46, 50, 53
Regulator theory, 58, 63
Reinforcement, 87, 100, 115, 139, 166, 189, 214, 293, 294, 300, 301, 304, 305
  negative, 82, 83, 97, 306
Reinforcement delay, 290, 292, 300, 305
Reinforcement learning, see RL
Reinforcer amounts, 301, 302, 304
Reinforcer delivery, 293, 295
Reinforcers
  delayed, 304, 306
  larger-later, 302, 303
  natural, 168, 210, 218
  non-drug, 290, 300
  positive, 45, 289
  smaller-sooner, 301, 303
Reinforcing commodities, 294, 297, 305, 306
Reinstatement, 171, 211, 228
Relapse, 85, 97, 165, 166, 172, 201, 212, 215, 227, 290
Remission, 181, 182
Rescorla-Wagner model, 166, 168, 277
Research
  addictions, 67, 105
  behavioral-economic, 303, 304
  ethnographic, 314, 316, 318, 322, 329
Respondent conditioning, 57, 58, 73, 76, 81, 84, 85, 96, 97, 99, 100

Respondent conditioning model, 73, 76–78, 80, 85, 96
Reward expectations, 139, 167, 178, 180
Reward functions, 147, 149, 152–157, 225
Reward learning, 166, 189, 190
Reward prediction, 166, 191–193
Reward prediction errors, 151, 256, 276
Reward stimuli, 87, 192
Reward system, 87, 216, 217
Reward values, 80, 81, 153, 156, 216
Rewarding stimuli, 113, 115, 221
RL, vii, 146–149, 164, 166, 173, 177–181, 191, 192, 206, 211–214, 237, 248, 249, 255–257, 265
  computational theory of, 206, 207, 228
  model-free, 146, 149, 155
  temporal difference, 166, 213
RL framework, 206, 207, 213, 228
RL models, 113, 175, 180
RL state-space, 176, 179
RL theory, 146, 193, 205, 206
Route, 19, 27, 34, 150
  natural, 26, 27

S
sA, 68, 69, 147–150, 154–156, 166–169, 172, 173, 175, 176, 178–180, 191, 218–220, 223–225, 239, 240, 242–246, 248, 249, 251–256, 258, 259, 319, 320
Satiation, 195, 322
Satiety, 6, 61, 85, 150, 192
Satiety threshold, 3, 10–13, 17
Schild equation, 12, 15
Schild method, 11, 15
Schild plots, 14, 15
Scope, 59, 62, 99, 102, 104, 285, 297, 303, 307
  integrative, 285, 300
Self-administration, vi, 5, 8–11, 13, 68, 79, 93, 94, 99, 113–115
  maintained, 3, 9, 10, 15
Self-administration behavior, 3–5, 17, 114
Self-organisation, 239, 258
Self-perceived fitness, see SPFit
Sensitivity, 26, 31, 32, 105, 209–211, 276
Sensitization, 73, 80–82, 85, 91, 93, 94, 115, 119, 122, 156, 189, 194–201
Sensor, 21, 25, 26, 29–32, 54
Sensor signal, 32, 50, 53
Sensory evidence, 267, 272, 274
Sensory inputs, 238, 239, 253, 262, 263
Sensory prediction errors, 247, 267, 269

Simulations, 19, 22, 33, 34, 36–38, 41, 46, 47, 49, 58, 60, 91, 92, 263, 266–268, 270–272, 278, 279, 316, 320
    12-month, 323, 325–328
Slopes, 7, 14, 15, 90, 288, 295
Smaller-sooner, *see* SS
Social environment, 88, 314, 318
SPFit, 86–88, 97, 100, 102, 103
SPFit theory, 61, 86–88, 97, 100, 103
Stability, 34, 36, 49, 69, 78, 81, 84, 91, 93, 174, 255, 315, 318, 327
    loop, 53, 54
Stabilization, 58, 63, 64, 115
State, *see* sA
State transitions, 151, 175, 177, 178, 238, 248, 251
State-action values, 148, 149, 151–156, 216
    cached, 153, 155
State-space, 163, 167, 171–173, 177–181, 248, 249, 251, 252, 254, 257, 258, 265
States
    attractive, 237, 248, 259
    current, 64, 147, 148, 191, 193, 194
    current mesolimbic, 195, 201
    environmental, 193, 240
    expected, 242, 243
    hyper-dopaminergic, 272, 274, 275
    initial, 177, 178, 180
    internal, 74, 242
    mesolimbic-activated, 200, 201
    multiple, 175, 177, 180
    natural appetite, 189, 191
    neutral, 71, 78, 82
    quiescent, 63, 70, 73
    sensitized, 119, 121, 126
    single, 175, 177
    steady, 20, 24
    time-varying, 240, 279
Stimuli, 5, 22, 24, 33, 34, 37, 41, 43, 49, 85, 166, 168, 177, 189–191, 196, 197, 207, 213–215
Street brokers, 313, 319
Street sellers, 323
Stress, 82–84, 212, 213, 227
Striatum, 151, 154, 165, 176, 208, 214, 217, 219, 220, 277
Substance, v, 25–32, 34, 35, 44, 45, 50, 99, 209, 285, 289, 307, 316
    addictive, vi, viii, 306
    endogenous, 19, 29–32, 35
    messenger, 25, 26, 30
    regulated, 31
Surrogate, 66, 102, 103
Surrogate measures, 102–104

    common, 102, 103
Systems
    biological, 69, 277
    chaotic, 68, 104
    digestive, 49, 53
    goal-directed, 206, 224–228
    homeostatic, 64, 79
    model-free, 152–155
    neural, 82, 151, 156, 163, 209

T
TD, 146, 163, 164, 191, 193
TD learning, 166, 167, 170, 171, 174, 192–194
TD models, 163, 193
TDRL, 166, 177, 213, 214
    average-reward, 216
Temporal difference, *see* TD
Temporal difference reinforcement learning, *see* TDRL
Temporal distance, 175, 178, 179
Temporal dynamics, 73, 116, 117, 131, 134
Time constant, 22, 33, 36, 41, 49, 53, 120, 126, 128, 136
    long, 35, 46
    minimal, 119, 124
    recovery, 119, 132
Timeframe, 315, 324, 327
Tolerance development, 19–24, 26, 27, 30, 34, 35, 40–42, 49, 51
Tolerance mechanism, 27, 35, 41, 43
Tolerance process, 27, 41
Transfer function, 32, 54
Transition function, 147–149, 153

U
Unconditioned response, *see* UR
Understanding addiction, 205, 206, 292
Unit doses, 4, 5, 7, 8, 10, 15
Unit price concept, 285, 294, 300, 303, 304, 307
Unit price equation, 294–296
UR, 73, 76, 83, 84, 97, 207

V
Value
    critic's, 219, 221–223
    discounted, 175, 195
    high, 200, 227, 228
    inflated, 155, 156
    intermediate, 117, 131
    model-free, 151, 153, 154, 157
    of perfect information, *see* VPI
    parameter, 38, 47, 99, 104, 105, 319
    true, 214, 215, 253, 271, 272
Value function, 150, 152, 191, 193, 256, 257

state-action, 148, 149
Ventral, 219, 220, 224, 226
Ventral striatum, 99, 154, 171, 176, 191, 196, 276
Ventral tegmental area, *see* VTA
VP, 142, 189, 197, 198, 202
VP neurons, 198, 199
VPI, 225–227
VTA, 75, 111–113, 115–118, 120–122, 124, 125, 131, 132, 134, 135, 137–139, 191, 198, 208, 266, 276

VTA DA cells, 115, 126, 128, 139

W

Withdrawal, vi, 36, 42, 45–49, 71, 75, 86, 89, 92, 97, 190, 191, 212, 316, 317, 319, 320, 322
Withdrawal effects, 59, 91, 92
Withdrawal syndrome, 71, 72, 77, 79, 99, 316